T0145362

Advances in Intelligent Systems and Computing

Volume 986

The series "Advances in Intelligent Systems and Computing" contains publications on theory, applications, and design methods of Intelligent Systems and Intelligent Computing. Virtually all disciplines such as engineering, natural sciences, computer and information science, ICT, economics, business, e-commerce, environment, healthcare, life science are covered. The list of topics spans all the areas of modern intelligent systems and computing such as: computational intelligence, soft computing including neural networks, fuzzy systems, evolutionary computing and the fusion of these paradigms, social intelligence, ambient intelligence, computational neuroscience, artificial life, virtual worlds and society, cognitive science and systems, Perception and Vision, DNA and immune based systems, self-organizing and adaptive systems, e-Learning and teaching, human-centered and human-centric computing, recommender systems, intelligent control, robotics and mechatronics including human-machine teaming, knowledge-based paradigms, learning paradigms, machine ethics, intelligent data analysis, knowledge management, intelligent agents, intelligent decision making and support, intelligent network security, trust management, interactive entertainment, Web intelligence and multimedia.

The publications within "Advances in Intelligent Systems and Computing" are primarily proceedings of important conferences, symposia and congresses. They cover significant recent developments in the field, both of a foundational and applicable character. An important characteristic feature of the series is the short publication time and world-wide distribution. This permits a rapid and broad dissemination of research results.

**** Indexing: The books of this series are submitted to ISI Proceedings, EI-Compendex, DBLP, SCOPUS, Google Scholar and Springerlink ****

More information about this series at http://www.springer.com/series/11156

Radek Silhavy
Editor

Cybernetics and Automation Control Theory Methods in Intelligent Algorithms

Proceedings of 8th Computer Science On-line Conference 2019, Vol. 3

Editor
Radek Silhavy
Faculty of Applied Informatics
Tomas Bata University in Zlín
Zlín, Czech Republic

ISSN 2194-5357 ISSN 2194-5365 (electronic)
Advances in Intelligent Systems and Computing
ISBN 978-3-030-19812-1 ISBN 978-3-030-19813-8 (eBook)
https://doi.org/10.1007/978-3-030-19813-8

This Springer imprint is published by the registered company Springer Nature Switzerland AG
The registered company address is: Gewerbestrasse 11, 6330 Cham, Switzerland

Preface

This book constitutes the refereed proceedings of the Cybernetics and Automation Control Theory Methods in Intelligent Algorithms section of the 8th Computer Science On-line Conference 2019 (CSOC 2019), held on-line in April 2019.

CSOC 2019 has received (all sections) 198 submissions, 120 of them were accepted for publication. More than 59% of accepted submissions were received from Europe, 34% from Asia, 5% from America and 2% from Africa. Researchers from more than 20 countries participated in CSOC 2019.

CSOC 2019 conference intends to provide an international forum for the discussion of the latest high-quality research results in all areas related to computer science. The addressed topics are the theoretical aspects and applications of computer science, artificial intelligences, cybernetics, automation control theory and software engineering.

Computer Science On-line Conference is held on-line and modern communication technology, which are broadly used improves the traditional concept of scientific conferences. It brings equal opportunity to participate all researchers around the world.

I believe that you will find following proceedings interesting and useful for your own research work.

March 2019 Radek Silhavy

Organization

Program Committee

Program Committee Chairs

Petr Silhavy	Faculty of Applied Informatics, Tomas Bata University in Zlin
Radek Silhavy	Faculty of Applied Informatics, Tomas Bata University in Zlin
Zdenka Prokopova	Faculty of Applied Informatics, Tomas Bata University in Zlin
Roman Senkerik	Faculty of Applied Informatics, Tomas Bata University in Zlin
Roman Prokop	Faculty of Applied Informatics, Tomas Bata University in Zlin
Viacheslav Zelentsov	Doctor of Engineering Sciences, Chief Researcher of St. Petersburg Institute for Informatics and Automation of Russian Academy of Sciences (SPIIRAS)

Program Committee Members

Boguslaw Cyganek	Department of Computer Science, AGH University of Science and Technology, Krakow, Poland
Krzysztof Okarma	Faculty of Electrical Engineering, West Pomeranian University of Technology, Szczecin, Poland
Monika Bakosova	Institute of Information Engineering, Automation and Mathematics, Slovak University of Technology, Bratislava, Slovakia

Pavel Vaclavek	Faculty of Electrical Engineering and Communication, Brno University of Technology, Brno, Czech Republic
Miroslaw Ochodek	Faculty of Computing, Poznan University of Technology, Poznan, Poland
Olga Brovkina	Global Change Research Centre Academy of Science of the Czech Republic, Brno, Czech Republic; Mendel University, Brno, Czech Republic
Elarbi Badidi	College of Information Technology, United Arab Emirates University, Al Ain, United Arab Emirates
Luis Alberto Morales Rosales	Head of the Master Program in Computer Science, Superior Technological Institute of Misantla, Mexico
Mariana Lobato Baes	Superior Technological of Libres, Mexico
Abdessattar Chaâri	Laboratory of Sciences and Techniques of Automatic Control & Computer Engineering, University of Sfax, Tunisia
Gopal Sakarkar	Shri. Ramdeobaba College of Engineering and Management, India
V. V. Krishna Maddinala	GD Rungta College of Engineering & Technology, India
Anand N. Khobragade	Maharashtra Remote Sensing Applications Centre, India
Abdallah Handoura	Computer and Communication Laboratory, Telecom Bretagne, France

Technical Program Committee Members

Ivo Bukovsky
Maciej Majewski
Miroslaw Ochodek
Bronislav Chramcov
Eric Afful Dazie
Michal Bliznak
Donald Davendra
Radim Farana
Martin Kotyrba
Erik Kral
David Malanik
Michal Pluhacek
Zdenka Prokopova
Martin Sysel
Roman Senkerik
Petr Silhavy
Radek Silhavy
Jiri Vojtesek
Eva Volna
Janez Brest
Ales Zamuda
Roman Prokop
Boguslaw Cyganek
Krzysztof Okarma
Monika Bakosova
Pavel Vaclavek
Olga Brovkina
Elarbi Badidi

Organizing Committee Chair

Radek Silhavy Faculty of Applied Informatics, Tomas Bata University in Zlin

Conference Organizer (Production)

OpenPublish.eu s.r.o.
Web: http://www.openpublish.eu
Email: csoc@openpublish.eu

Conference Web site, Call for Papers

http://www.openpublish.eu

Contents

Novel Filtering-Based Approach Using Fuzzy Logic for Prevention of Adversaries in Sensory Application

N. Tejashwini[1(✉)], D. R. Shashi Kumar[2], and K. Satyanarayan Reddy[3]

[1] Visvesvaraya Technological University, Belgaum, Karnataka, India
tejashwini.n@gmail.com
[2] Department of Computer Science and Engineering, Cambridge Institute of Technology, Bengaluru, Karnataka, India
[3] Department of Information Science and Engineering, Cambridge Institute of Technology, Bengaluru, Karnataka, India

Abstract. Security problems associated with the Wireless Sensor Network (WSN) have yet not met it fail-proof solution against the most lethal threat. After reviewing existing security solution, it was felt that there is a wide scope of improvement as existing system still cannot offer certainty in its decision of adversaries, whether it is identification or prevention of any threat. Therefore, the proposed system introduces a novel mechanism where fuzzy logic has been applied to filter down the decisions on the basis of unique pattern of energy and distance factors to identify and prevent the adversary from stealing the public keys within the sensor node. The study outcome shows that proposed system offers significant security without affecting the communication performance. In fact, proposed system exhibits good residual energy and sustaining data delivery performance for a longer iteration.

Keywords: Wireless Sensor Network · Security · Public key cryptography · Security · Attack · Fuzzy logic · Attack behaviour

1 Introduction

The area of Wireless Sensor Network (WSN) has attracted higher degree of attention owing to its cost-effective sensing mechanism [1]. There have been various segments of research-based solution towards different problems associated with WSN [2–4]. However, problems associated with security issues are yet to be completely addressed by existing research community irrespective of number of research-based techniques [5–8]. The primary problems associated with the security issues in WSN are because of the reason that a sensor node is not capable enough to execute a strong cryptographic protocol [9]. Although, there are some potential encryption algorithms in other networks e.g. [10], but they cannot be executed on a sensor node that has lower degree of computational capability as well memory. Hence, it is a great deal of challenge to design such a light weight encryption mechanism that maintains a good balance with strong security feature and computational complexity. Apart from this, it has also been

R. Silhavy (Ed.): CSOC 2019, AISC 986, pp. 1–10, 2019.
https://doi.org/10.1007/978-3-030-19813-8_1

noticed that key management plays a significant role in encryption mechanism and public key encryption is one of the frequently used scheme. However, usage of public key encryption is also associated with various loopholes e.g. till date majority of the public key encryption offer strong security but doesn't emphasize on the communication-computation performance in WSN. Another significant loopholes in public key encryption is the assumptions itself about the public keys where there is nowhere description of any form of extra protection towards such public keys nor there is any assurance that such keys will never be compromised.

Therefore, this paper introduces one of the simple and novel mechanism to perform filtering of the adversaries on the basis of multiple factors e.g. distance and energy in order to protect public keys. Section 2 discusses about the existing research work followed by problem identification in Sect. 3. Section 4 discusses about proposed methodology followed by elaborated discussion of algorithm implementation in Sect. 5. Comparative analysis of accomplished result is discussed under Sect. 6 followed by conclusion in Sect. 7.

2 Related Work

There are various research works towards addressing the security problems in WSN and this section brief of most recent trends of literature. The concept of micro key management has been introduced by Aissani et al. [11] that uses *re-keying process* in WSN. The emphasis on security problems on upcoming IoT is discussed by Burg et al. [12], which gives an insight that there is a long way to go to make WSN secure enough for cyber-physical system. Study towards master key management using *hierarchical approach* was proven to offer less computational complexity as discussed in work of Gandino et al. [13]. Gopalakrishnan and Bhagyaveni [14] have used chaotic approach for secure routing without any dependencies on pre-shared keys. Gope and Hwang [15] have presented an unique key exchanging mechanism where a special form of emergency key is used for enhanced security measures. Study towards addressing wormholes using analytical modeling is carried out by Sayad Haghighi et al. [16]. Huang et al. [17] has used the concept of group key management mainly focusing on the strengthening authentication system. Kesavan and Radhakrishnan [18] have presented a dynamic keying processing using clustering technique for addressing security problems in case of mobility aspect present in WSN. Similar form of research work towards grouped-based secure communication has been also carried out by Li et al. [19]. Study towards securing physical layer has been investigated by Moara-Nkwe et al. [20] where the authors have presented a key generation system concerning both energy and data quality. Symmetric key usage has been reported to effectively control the network admission. This fact was established by Oliveira et al. [21]. Computational problems associated with identity-based encryption have been presented by Shim et al. [22] using real-time sensor motes. Apart from this, there are various other approaches to address security loopholes in WSN e.g. multifactor authentication (Shin et al. [23]), optimization-based approach [24], public-key encryption (Al-Turjman et al. [25]), hierarchical-based approach for attack identification (Wu et al. [26]), linear programming (Yildiz et al. [27]), time and energy-based approach (Zhang et al. [28]), composite key distribution (Zhao et al. [29]), randomization approach (Zhen et al. [30]). Hence, there are various literatures that has discussed about security techniques in WSN.

3 Problem Description

From the prior section, it can be seen that existing security system suffers from following limitation e.g. (i) lack of schemes that offers decisive-based approach for adversary identification as well as prevention, (ii) existing schemes doesn't focus on computational performance excellence and focus more on introducing complex encryption schemes, (iii) no studies being carried out towards safeguarding the public keys in public key encryption, (iv) absence of any such mechanism that ensures route security on the basis of progressive filtering process. Hence, the statement of problem could be "*Constructing a cost effective security scheme without using complex cryptography in order to offer top level of security to public key is quite a challenging task.*" The next section outlines the methodology implemented to solve this problem.

4 Proposed Methodology

The proposed study implementation is an extension of our prior work [31] that uses public key encryption. However, our prior work is extended with a possible scenario of stealing of public keys. Hence, proposed system adopts an analytical research approach to investigate this fact. The scheme is shown in Fig. 1.

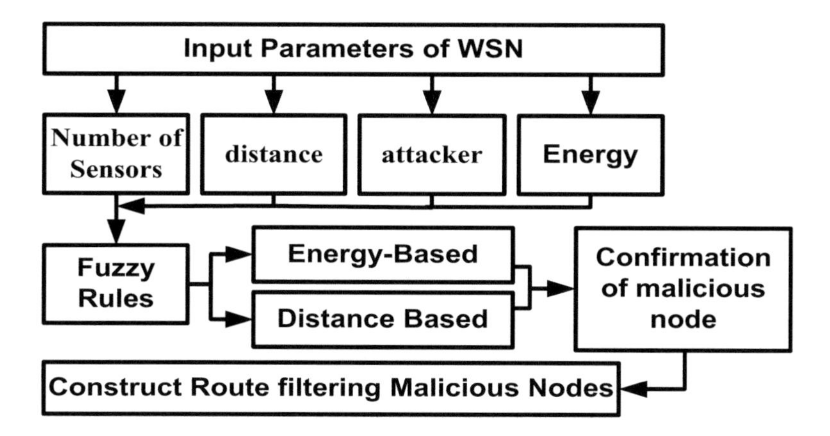

Fig. 1. Proposed analytical scheme

Basically, proposed system introduced a simple security approach where fuzzy logic has been used for crafting rules for identifying the presence of malicious node on the basis of energy as well as distance. Initially, the malicious nodes are tracked on the basis of residual energy which is usually higher than normal nodes and then the system further confirms their presence by exploring the degree of spread of attacker (or compromised nodes) on the basis of distance, which is normally higher than regular node. Distance will represent the extent of attack as normal node will only choose paths to forward data to sink node while malicious node will have objective to attack. Hence, such distance factor for malicious nodes will be always higher than normal node. The next section briefs about algorithm implementation.

5 Algorithm Implementation

The design of the proposed security feature for safeguarding the communication process in WSN is carried out using a very light weight public key cryptography process. The study assumes that the cluster head is always a prominent target of attack while performing data aggregation and hence the aim of this algorithm is to safeguard all sorts of communication leading to and from cluster head. The significant steps of the proposed system are as follows:

Algorithm for Fuzzy-based secure routing

Input:*n, d, a*

Output:*conn*

Start

1. initn,
2. **For** i=1: n
3. $\alpha_1 \rightarrow \sum d$
4. $\alpha_2 \rightarrow d_{ns}$
5. Apply θ_{rule}
6. **If** $n_i = a$
7. Compute $R_{con} \rightarrow \Delta R$
8. break conn;
9. **Else**
10. Establish conn
11. **End**

End

The proposed system takes its necessary input i.e. *n* (Number of Sensors), *d* (distance), *a* (attacker) which after processing yields a secured connection *conn*. The proposed system constructs a fuzzy rule system considering where resources are considered a primary constrains to develop the rule θ_{rule} (Line-5). The secondary constraint considered in setting the rule is basically distance related parameters. For all the sensor nodes (Line-2), the proposed system computes distance among all the sensors *d* and reposit in one matrix called as core distance i.e. α_1 (Line-3). The next part of the algorithm considers distance between all the sensors to the destination node i.e. α_2 (Line-4). The above two forms of computation of distance is carried out in order to facilitate triangulation of the possible attacker node without using any form of sophisticated cryptography techniques. Finally, the algorithm applies fuzzy-based rule that considers energy factor for the process of triangulation. According to this rule, following operations are carried out: (1) First Stage of Tracking: All the nodes are evaluated to check for their residual energy. Considering the phenomenon that all the cluster head are synchronized with each other and each cluster head bears information about their member nodes, in such case, the residual energy is in constant tracking process as behaviour of maliciousness. In case if there is a presence of certain node, whose residual energy is found to be abnormally high, it givens an indication that the identified node could be malicious node. However, such identification could be not concluded of confirmed instantly. (2) Second Stage of Tracking: In order to confirm the information captured in the first stage of tracking, the algorithm adds more number of distance-based constraints to offer better behavioral definition of the nodes. The logic of this stage is that malicious node will not instantly intrude the network nor it will stay as an inactive node for a long time. The utility of malicious node could only increase if they can steal or corrupt more number of public keys. In order to achieve this objective, the malicious node will need to spread its malicious activity as far as possible. Hence, the second rule of the fuzzy logic offers more insight on the spread of the malicious node by evaluating the intermittent links. The presence of intermittent links are direct indications of malicious/victim nodes at the end.

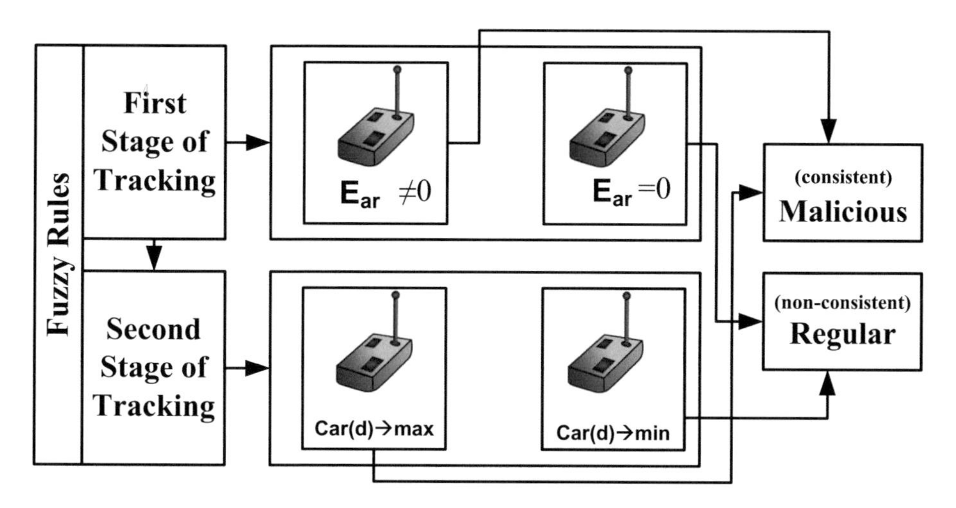

Fig. 2. Scheme of secure tracking of malicious node

According to the above scheme (Fig. 2), in first stage of tracking, if there is more consistency of retention of energy, it is a positive indication of malicious node or else it is a regular node. Consecutively, in next stage, if the cardinality of the distance is found consistently to be increasing, it represents spreading of attack by compromising number of public keys. Hence, regular nodes are characterized by lesser consistency in growing distance among different nodes, while malicious node will need to capture as more number of nodes as possible. Upon identification of malicious node, the proposed algorithm aborts any form of connection (Line-8) or else they establish the connection. Hence, the proposed system offers higher degree of security over the public keys which are highly essential to implement public key encryption in WSN.

6 Results Discussion

This section discusses about the results being accomplished after implementing the proposed algorithms in MATLAB. The study considers the presence of 50–500 sensor nodes bearing the standard MEMSIC node characteristics. The simulation also considers 0.5 J of initialized energy, 2500 bits of packet size, and 20% of presence of attacker nodes in the simulation area of 1000×1000 m^2. For effective study analysis, the outcomes are studied with respect to energy parameters and compared with most standard SecLEACH algorithm [32].

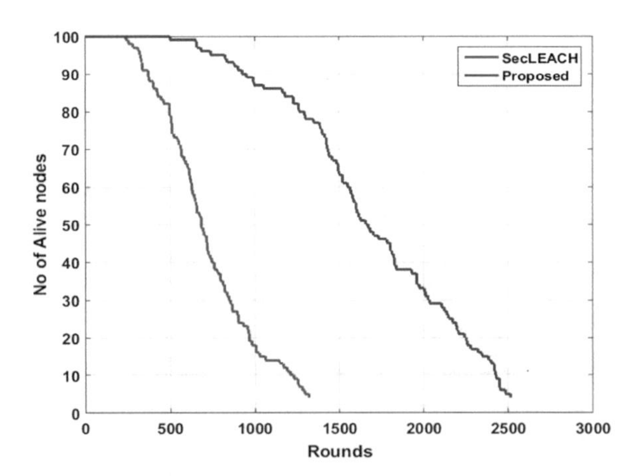

Fig. 3. Comparative analysis of alive nodes

Figures 3 and 4 corresponds to the energy-based analysis, where it can be seen that proposed system offers better rounds of sustenance of sensor node as compared to standard SecLEACH. The prime reason behind this is SecLEACH offers energy efficiency but doesn't offer capability of tracking the behaviour of malicious node unlike proposed system. Therefore, proposed system exhibits practicality in implementation of such algorithm which will not degrade the network lifetime.

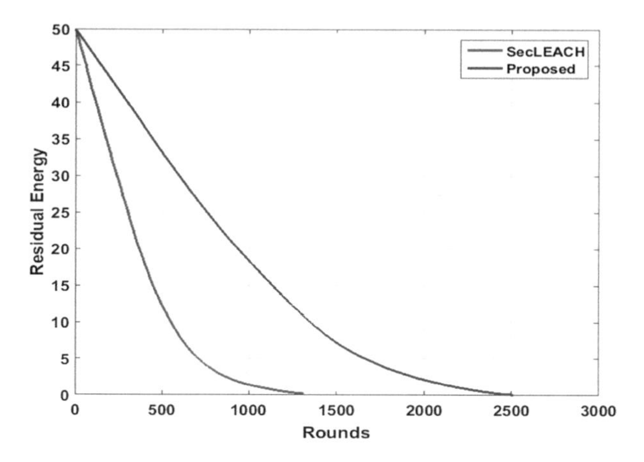

Fig. 4. Comparative analysis of residual energy

Figures 5 and 6 highlights that proposed system has good packet forwarding feature that is because proposed system is devoid of any complex cryptographic algorithm implementation. It simply uses rules to track the malicious behaviour to steal the public keys from the sensors. At the same time, proposed system also shows reduces occurrences in energy variance, which is one of the essential parameter to decide the energy efficiency among the nodes. If the variance of the energy is high that it exhibits inapplicability of SecLEACH to offer resource friendly security features towards protecting stealing of public keys. Moreover, adoptions of fuzzy rules are not found to adversely affect the security as well as communication performance of the WSN. Hence, proposed system can be claimed to offer a cost effective security protocol that is resistive against majority of the attacks related to public key stealing in WSN.

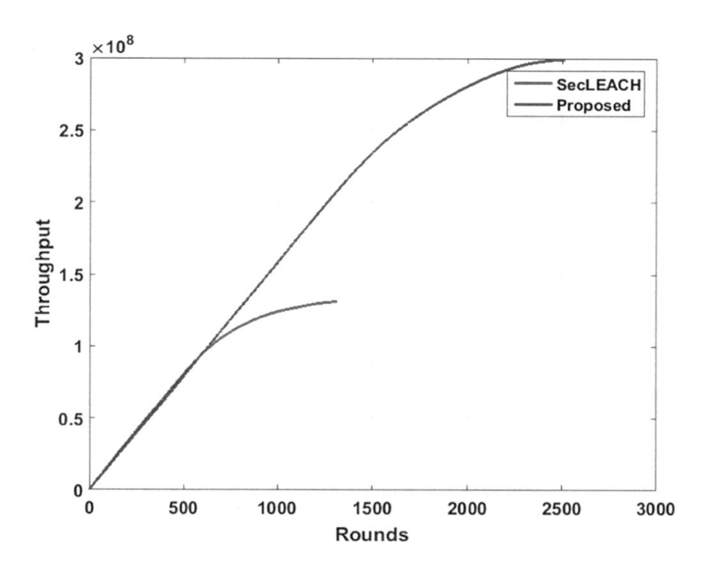

Fig. 5. Comparative analysis of throughput

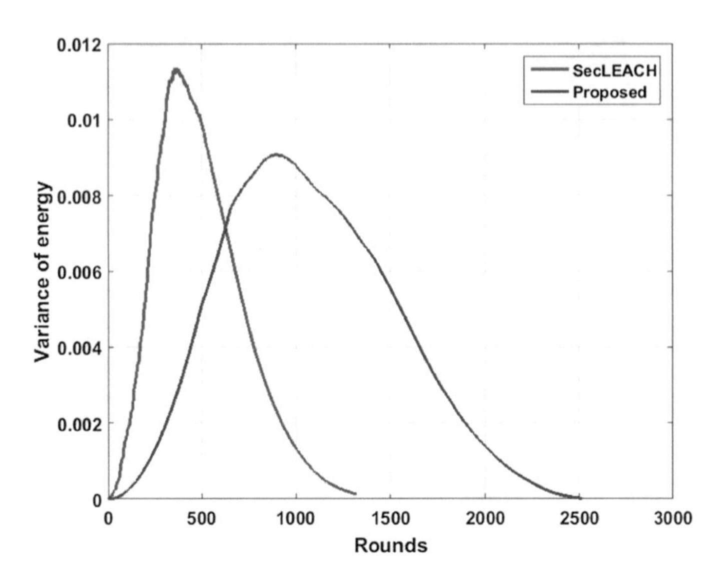

Fig. 6. Comparative analysis of variance

7 Conclusion

The complete concept of trapping the adversary in the proposed system is based on the fact that if public key is compromised than the adversary can actually understand the pattern of key management within the other nodes to some extent. Hence, without using any form of encryption, the proposed system assumes that there is a presence of attacker nodes within the simulation area and no one is aware of its presence. The primary concept of confirming the presence of adversary node is that they will attempt higher sustenance and look for massive spread. It will mean that energy consumption rate of adversary will be slower compared to regular nodes and similarly, their presence on different location (i.e. compromised nodes) will keep on increasing. Hence, a fuzzy logic based approach has been utilized where this pattern of energy and distance is used for identify the susceptible nodes. The communication is established with only regular nodes. The study outcomes also show better balance between data delivery performance and energy efficiency in contrast to existing security protocol.

References

1. Xu, G., Shen, W., Wang, X.: Applications of wireless sensor networks in marine environment monitoring: a survey. Sensors **14**(9), 16932–16954 (2014)
2. Kobo, H.I., Abu-Mahfouz, A.M., Hancke, G.P.: A survey on software-defined wireless sensor networks: challenges and design requirements. IEEE Access **5**, 1872–1899 (2017)
3. Ruiz, M., Alvarez, E., Serrano, A., Garcia, E.: The convergence between wireless sensor networks and the Internet of Things; challenges and perspectives: a survey. IEEE Lat. Am. Trans. **14**(10), 4249–4254 (2016)

4. Kumar, S.A.A., Ovsthus, K., Kristensen, L.M.: An industrial perspective on wireless sensor networks—a survey of requirements, protocols, and challenges. IEEE Commun. Surv. Tutorials **16**(3), 1391–1412 (2014)
5. AlMheiri, S.M., AlQamzi, H.S.: Data link layer security protocols in wireless sensor networks: a survey. In: 2013 10th IEEE International Conference on Networking, Sensing and Control (ICNSC), Evry, pp. 312–317 (2013)
6. Khedim, F., Labraoui, N., Lehsaini, M.: Dishonest recommendation attacks in wireless sensor networks: a survey. In: 2015 12th International Symposium on Programming and Systems (ISPS), Algiers, pp. 1–10 (2015)
7. Elgenaidi, W., Newe, T., O'Connell, E., Mathur, A., Toal, D., Dooly, G.: Reconfiguration of neighbouring nodes in coastal monitoring wireless sensor networks based on leader node recommendation. In: 2017 4th International Conference on Control, Decision and Information Technologies (CoDIT), Barcelona, pp. 0348–0352 (2017)
8. Tomić, I., McCann, J.A.: A survey of potential security issues in existing wireless sensor network protocols. IEEE Internet Things J. **4**(6), 1910–1923 (2017)
9. Grover, J., Sharma, S.: Security issues in wireless sensor network—a review. In: 2016 5th International Conference Reliability, Infocom Technologies and Optimization (Trends and Future Directions) (ICRITO), pp. 397–404 (2016)
10. Yegireddi, R., Kumar, R.K.: A survey on conventional encryption algorithms of cryptography. In: 2016 International Conference on ICT in Business Industry & Government (ICTBIG), Indore, pp. 1–4 (2016)
11. Aissani, S., Omar, M., Tari, A., Bouakkaz, F.: μKMS: micro key management system for WSNs. IET Wirel. Sens. Syst. **8**(2), 87–97 (2018)
12. Burg, A., Chattopadhyay, A., Lam, K.Y.: Wireless communication and security issues for cyber-physical systems and the Internet-of-Things. Proc. IEEE **106**(1), 38–60 (2018)
13. Gandino, F., Ferrero, R., Montrucchio, B., Rebaudengo, M.: Fast hierarchical key management scheme with transitory master key for wireless sensor networks. IEEE Internet of Things J. **3**(6), 1334–1345 (2016)
14. Gopalakrishnan, B., Bhagyaveni, M.A.: Anti-jamming communication for body area network using chaotic frequency hopping. Healthc. Technol. Lett. **4**(6), 233–237 (2017)
15. Gope, P., Hwang, T.: A realistic lightweight anonymous authentication protocol for securing real-time application data access in wireless sensor networks. IEEE Trans. Ind. Electron. **63** (11), 7124–7132 (2016)
16. Sayad Haghighi, M., Wen, S., Xiang, Y., Quinn, B., Zhou, W.: On the race of worms and patches: modeling the spread of information in wireless sensor networks. IEEE Trans. Inf. Forensics Secur. **11**(12), 2854–2865 (2016)
17. Huang, M., Yu, B., Li, S.: PUF-assisted group key distribution scheme for software-defined wireless sensor networks. IEEE Commun. Lett. **22**(2), 404–407 (2018)
18. Kesavan, V.T., Radhakrishnan, S.: Cluster based secure dynamic keying technique for heterogeneous mobile wireless sensor networks. China Commun. **13**(6), 178–194 (2016)
19. Li, Z., Wang, H., Fang, H.: Group-based cooperation on symmetric key generation for wireless body area networks. IEEE Internet of Things J. **4**(6), 1955–1963 (2017)
20. Moara-Nkwe, K., Shi, Q., Lee, G.M., Eiza, M.H.: A novel physical layer secure key generation and refreshment scheme for wireless sensor networks. IEEE Access **6**, 11374–11387 (2018)
21. Oliveira, L.M.L., Rodrigues, J.J.P.C., de Sousa, A.F., Denisov, V.M.: Network admission control solution for 6LoWPAN networks based on symmetric key mechanisms. IEEE Trans. Ind. Inf. **12**(6), 2186–2195 (2016)
22. Shim, K.A.: BASIS: a practical multi-user broadcast authentication scheme in wireless sensor networks. IEEE Trans. Inf. Forensics Secur. **12**(7), 1545–1554 (2017)

23. Shin, S., Kwon, T.: Two-factor authenticated key agreement supporting unlinkability in 5G-integrated wireless sensor networks. IEEE Access **6**, 11229–11241 (2018)
24. Tian, F., et al.: Secrecy rate optimization in wireless multi-hop full duplex networks. IEEE Access **6**, 5695–5704 (2018)
25. Al-Turjman, F., Kirsal Ever, Y., Ever, E., Nguyen, H.X., David, D.B.: Seamless key agreement framework for mobile-sink in iot based cloud-centric secured public safety sensor networks. IEEE Access **5**, 24617–24631 (2017)
26. Wu, J., Ota, K., Dong, M., Li, C.: A hierarchical security framework for defending against sophisticated attacks on wireless sensor networks in smart cities. IEEE Access **4**, 416–424 (2016)
27. Yildiz, H.U., Ciftler, B.S., Tavli, B., Bicakci, K., Incebacak, D.: The Impact of Incomplete Secure Connectivity on the Lifetime of Wireless Sensor Networks. IEEE Syst. J. **12**(1), 1042–1046 (2018)
28. Zhang, J., Tao, X., Wu, H., Zhang, X.: Secure transmission in SWIPT-powered two-way untrusted relay networks. IEEE Access **6**, 10508–10519 (2018)
29. Zhao, J.: Topological properties of secure wireless sensor networks under the q-composite key predistribution scheme with unreliable links. IEEE/ACM Trans. Netw. **25**(3), 1789–1802 (2017)
30. Zheng, G., et al.: Multiple ECG fiducial points-based random binary sequence generation for securing wireless body area networks. IEEE J. Biomed. Health Inf. **21**(3), 655–663 (2017)
31. Tejashwini, N., Shashikumar, D.R., Satyanarayan Reddy, K.: Mobile communication security using Galios field in elliptic curve cryptography. In: 2015 International Conference on Emerging Research in Electronics, Computer Science and Technology (ICERECT), Mandya, pp. 251–256 (2015)
32. Oliveira, L.B., Wong, H.C., Bern, M., Dahab, R., Loureiro, A.A.F.: SecLEACH - a random key distribution solution for securing clustered sensor networks. In: Fifth IEEE International Symposium on Network Computing and Applications (NCA 2006), Cambridge, MA, pp. 145–154 (2006)

Game-Decision Model for Isolating Intruder and Bridging Tradeoff Between Energy and Security

Bhagyashree Ambore(✉) and L. Suresh

Department of Computer Science and Engineering, CiTech, Bengaluru, India
ambore.bhagyashree@gmail.com, suriakls@gmail.com

Abstract. Irrespective of the type and degree of capability, any adversaries' results in significant loss of energy among the physical devices connected in Internet-of-Things (IoT). After reviewing existing trends of research-based solution, it was seen that various complex cryptographic solution has been put forward to claim potential security but they were not found to claim if their solution could stop unwanted energy depletion too during the process of safe-guarding the physical devices. Therefore, this paper presents a game-decision making model that is capable of assessing the level of vulnerability as well as legitimacy of the IoT devices during the communication process that finally leads to identification followed by isolating the compromised or malicious IoT devices. The study outcome was found to offer better identification of the threats for multi-staged games in contrast to existing methods.

Keywords: Internet-of-Things · Attacks · Energy · Battery · Encryption · Security · Sensors · Game · Vulnerability

1 Introduction

Internet-of-Things (IoT) is not only about connectivity among different physical devices to form a large network but it is also about significant amount of generated data that is highly valuable [1, 2]. As it forms connectivity of different number of heterogeneous devices working over different routing strategy, therefore, designing a uniform security solution is not possible in IoT. At present, the security system of IoT focuses on securing either application layer, or transportation layer, or perception layer [3]. There are also different review studies carried out towards addressing security protocols in IoT [4–9], however, there are various questions that are yet left unsolved from the approaches in existing security solution. The first question will be –is there any good alternative for strong encryption mechanism? The second question will be why the existing security solutions are so attack specific. Owing to the novel nature of the technology, answers to such question are yet to be explored. If this answers were ever found, than then next question will be why the researchers have not emphasized on their solution by considering energy factor. The IoT devices are usually wireless and low-powered hardware which cannot execute complex security protocols. Hence, existing attacks e.g. denial of service, Sybil attack, routing attack, as well as many other

R. Silhavy (Ed.): CSOC 2019, AISC 986, pp. 11–20, 2019.
https://doi.org/10.1007/978-3-030-19813-8_2

unknown attacks too cost the network resource as well as node battery just to resist it. Moreover, there are various types of attacks that are only meant for energy depletion [10]. Hence, this paper presents a novel technique where game theory has been used for modeling a simple decision making framework with capability of isolating compromised IoT devices. Section 2 discusses about the existing research work followed by problem identification in Sect. 3. Section 4 discusses about proposed methodology followed by elaborated discussion of algorithm implementation in Sect. 5. Comparative analysis of accomplished result is discussed under Sect. 6 followed by conclusion in Sect. 7.

2 Related Work

This section discusses the research work carried out towards security in IoT. Existing literatures has been witnessed mainly theoretical discussion emphasizing over the theoretical aspect of it (Bertino et al. [11], Bhattarai and Wang [12], Burg et al. [13], Nurse et al. [14], Singh et al. [15], Szymanski [16], Wolf and Serpanos [17]). Existing literatures have also presented studies towards securing physical layer in IoT (Hu et al. [18]) by adding artificial noise. Security problems could also be solved using software defined network where integer linear programming was proven to be best approach to solve the problem (Liu et al. [19], Villari et al. [20]). Security-based connectivity between IoT and upcoming industry 4.0 is quite high. A recent study shows that Hidden Markov Model could be used for constructing intelligence to resist security breaches in IoT applications working on Industry 4.0 (Moustafa et al. [21]). Such security features could be further upgraded by enhancing conventional digital signature (Mughal et al. [22]). Apart from digital signature, symmetric encryption, other hashing mechanism, and public key encryption are also found helpful in resisting low end threats over IoT devices (Pereira et al. [23], Raza et al. [24], Xiao and Yu [25]). The most advanced version in cryptography called as block chain is recently investigated by many researchers and were claimed to offer potential resistance for IoT devices (Qu et al. [26]). It was also seen that usage of homomorphic encryption could increase the privacy feature in IoT devices (Song et al. [27]). Apart from encryption, it also improves performance of re-encryption too. Usage of game theory has been reported to assist in modeling solution towards learning and resisting threat (Wu and Wang [28]). A unique study has been presented by Xu et al. [29] where ontology has been used for modeling network threats over IoT. The study has also formulated various rules and reasoning mechanism to resisting security threats in IoT. Another unique study was presented by Zhang et al. [30] where potential of public key encryption has been claimed with lesser size of secret key. The study outcomes of above discussed literatures have presented solution for some of the potential threats and it was found claiming its successful operation using numerical validation. However, apart from advantages, there are certain associated limitations that are briefly highlighted in next section.

3 Problem Description

A closer look at all the existing security approaches shows that none of the solutions have actually identified that owing to different forms of attack, the IoT devices significantly depletes energy which adversely affect the data forwarding performance of itself as well as its neighboring nodes. Existing system uses potentially sophisticated encryption to resist attack without even considering the fact that IoT nodes (e.g. sensors) couldn't actually process complex encryption protocols. Hence, it can be said that none of the existing security solutions have actually considered the energy problems among the IoT nodes. Hence the problem statement is "*Designing a non-cryptographic solution that bridges the trade-off between security and energy efficiency among the IoT nodes is a computationally challenging task*". The next sections briefs of proposed solution.

4 Proposed Methodology

The core goal of proposed system is to resist all sorts of attack in IoT that makes the devices deplete its energy. The proposed system aims for introducing a novel, simple and yet robust framework that is capable of identifying and isolating the compromised IoT devices considering the fact that there is no predefined information about the threat (Fig. 1).

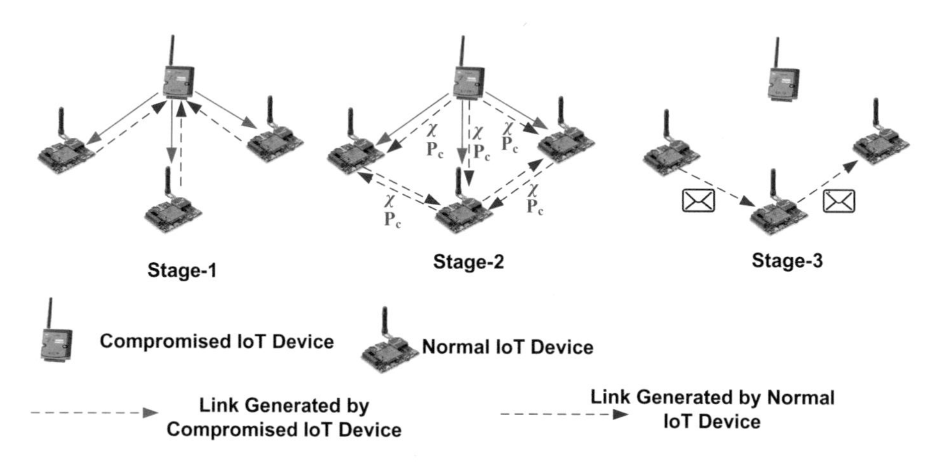

Fig. 1. Schematic architecture of AEOC

The proposed system implements game theory concept that allows the IoT device to perform certain vulnerability calculation from its neighboring node, assuming that it doesn't know the intention of its neighbor node. Using probability concept and depending upon the extracted information of vulnerability as well as legitimacy, the proposed system makes a decision to isolate all the active connection from any compromised IoT device. By doing this, only the necessary amount of energy is spent to cater up data packet forwarding process. Hence, proposed system is capable of resisting any forms of threats towards active communication process in IoT. The next section briefs about the algorithm implementation for this process.

5 Algorithm Implementation

The prime motive of the proposed algorithm design is to offer identification of the any forms of attacks or intrusion among the network of IoT devices those results in energy depletion. The algorithm basically targets to bridge the gap between the energy efficiency and security in communication of IoT. The complete ideology of the proposed study is that there is no predefined idea of any form of presence of intruder as well as energy is one of the prime targets for every attack directly or indirectly. Hence, the algorithm is essentially meant for balancing both the security needs as well as energy efficiency by establishing a relationship between them. The algorithm is meant to address resist any form of attack that directly or indirectly depletes the power of IoT device. The steps of the algorithm are as follows:

Algorithm for Identification of compromised IoT device
Input: *C* (cut-off vulnerability), *H* (vulnerability cost)
Output: identification of Compromised IoT device
Start
1. **While** $H<C$ **Do**
2. **If** $P_c<f(A_1)$
3. flag $A_1(p_1)$
4. **Else**
5. flag $A_1(p_2)$
6. **EndIf**
7. Update *A* & compute χ
8. Return(Node-A→compromised IoT device)
End

The discussion of the algorithmic steps is as follows: The algorithm uses probability theory in its entire formulation and therefore C will represents a user-defined cut-off vulnerability factor. The initialization of this value could be carried out using statistical approach. The first step of this algorithm is to compute a decision matrix *H* that represents possible decision to be carried out by the system to capture the precise information about the legitimacy of the IoT device. The computation of the *H* is carried out mathematically by scalar multiplication of probability of compromised IoT device with the non-vulnerability parameter (Fig. 2).

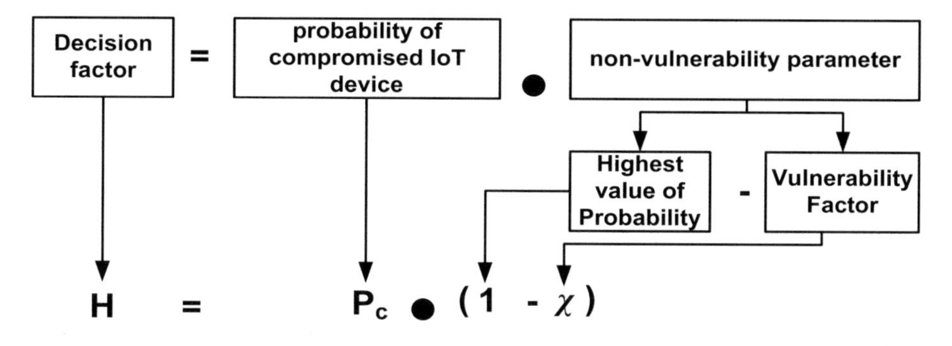

Fig. 2. Formulation of decision factor

The proposed system performs computation of vulnerability factor χ as $\chi \rightarrow f(B_1, B_2, c)$, where B_1 and B_2 represents behavioral parameter of IoT devices. B_1 directly represents total number of packets being transmitted by node while B_2 represents total number of packets being dropped by a node. A closer look will show that just by obtaining B_1 and B_2, it is not feasible to determine the legitimacy of the node as such behaviour could be exhibited by both normal or compromised IoT device. The variable c represents network coefficient (Fig. 3).

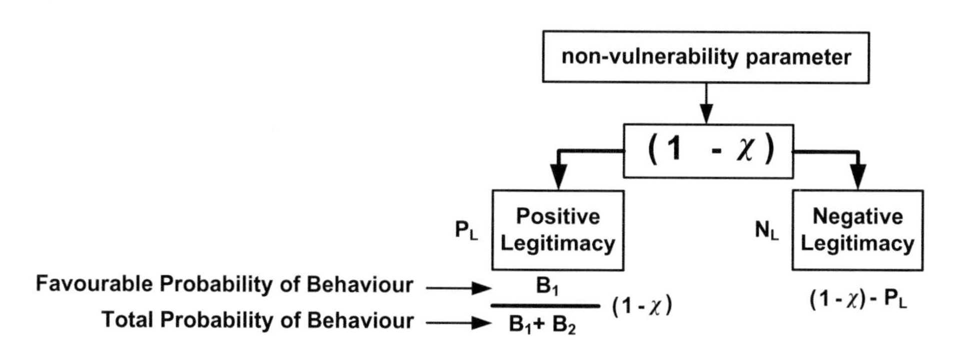

Fig. 3. Probability computation of legitimacy of IoT device

Hence, Line-1 represents the situation when there is good probability of capturing the information associated with attackers that could possibly deplete energy. The next part of the algorithm checks for the situation when the probability of compromised IoT device is found to be less than anticipated cost function of forwarding the packets A_1. The algorithm formulates a set of actions for IoT device as $A = \{A_1, A_2, A_3, A_4\}$, where A_1 represents action towards packet transmission, A_2 represents actions towards dropping the packet, A_3 represents action towards updating the topology information of IoT device along with legitimacy report, and A_4 represents actions towards launching attacks in IoT. A closer look into this action formulation will show that a compromised IoT node could exhibit A_1, A_2, and A_4 actions while it will not execute A_3 attack as it would get caught. Hence, a normal node could also exhibit A_1 and A_2 actions but it would definitely execute A_3 actions. Hence, A_1 and A_2 are common actions for normal and compromised nodes while A_3 action is executed by normal IoT device only while A4 action is exhibited by compromised IoT nodes only (Fig. 4).

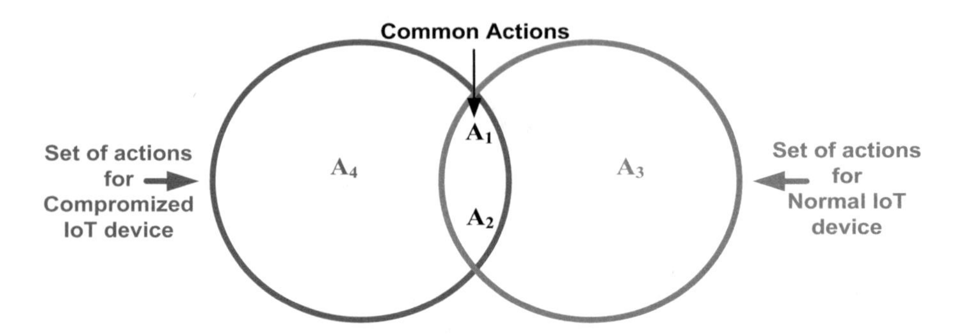

Fig. 4. Classifications of actions adopted by IoT nodes

Therefore, Line-2 represents a condition when an IoT device chooses to perform transmission of the packet. In such case, the condition represent to that of both normal as well as compromised IoT nodes. It is because of the reason that A_1 as well as A_2 are common actions for both normal and compromised IoT device. It is a research challenge to make a proper identification of legitimacy of the IoT device that is carried out by two ways. The algorithm either appoints the higher probability value p_1 (Line-3) or recomputed the new probability value i.e. p_2 (Line-5). According to the algorithmic steps, the violation of Line-2 represents the case when packet transmission is carried out for the purpose out for harming the network i.e. by compromised IoT device. Therefore, if the packet is transmitted for better intension (that can be tracked by lower value of P_c in Line-2) than it represents normal IoT device where probability pc is assigned the highest possible value i.e. 1 (Line-3) otherwise it just represent a presence of compromised IoT device (as Pc value is always more for compromised IoT device). In such condition, the probability p_2 is further updated as amount of resources involved in initiating an attack event divided by total profit owned during the attack. The total profit could be further initialized and investigated for multiple behavioral case study of attack in IoT. The final outcome of the proposed algorithm is to finally update A followed by computation of vulnerability parameter χ (Line-7) to state that Node A is compromised IoT device (Line-8). The complete algorithm considers that a transmitting node A could be either normal or malicious (with no predefined information about its legitimacy) while receiver node-B is always normal IoT device.

6 Results Discussion

The proposed logic discussed in prior section has been implemented with 500 IoT nodes dispersed in 1000 × 1200 m^2 simulation area. The complete analysis was recorded for 600 simulation rounds on increasing stages of games. The analysis has been carried out considering the performance parameter of legitimacy and vulnerability factor with respect to probability of compromised IoT device. Discussion of process of calculating legitimacy as well as vulnerability factor has been briefed in prior section. For an effective analysis, the study outcome has been compared with work carried out by Agah et al. [31] and Hamdi et al. [32]. Agah et al. [31] has presented a definitive technique where each node (normal and compromised) can increase its respective capability based on its type (normal node protects and malicious node attacks). Similarly, work of Hamdi et al. [32] is slightly enhanced version of Agah et al. [31] where the system allocates probability to each node working on definitive strategy. In order to perform comparative analysis, only the core aspect of the algorithm has being implemented over the similar test-environment where the proposed system was investigated. A closer look into the outcome shows that proposed system offers reduced threats to IoT devices in increasing staged games as compared to existing system (Fig. 5). The prime reason behind this is existing mechanism calls for one round of check for all the nodes in communication in order to ascertain the facts of legitimacy of the node, but proposed system performs progressive strategy to monitor the malicious behaviour of compromised IoT device by using empirical value of vulnerability. Therefore, in a long run of multi-staged games, the proposed system will always reduce threat level (reduction in P_c).

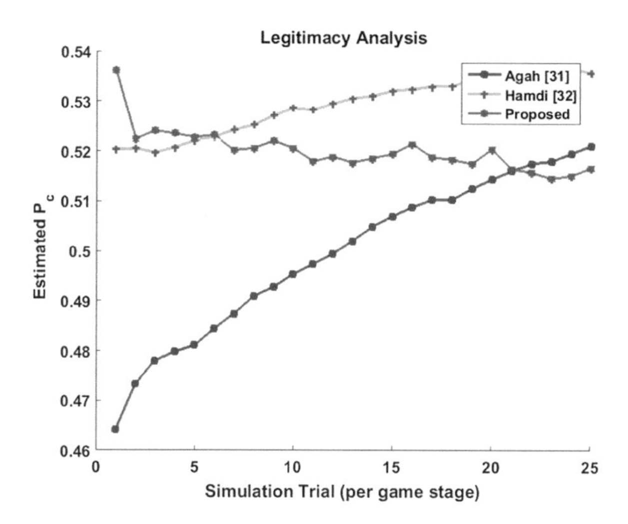

Fig. 5. Comparative analysis of legitimacy

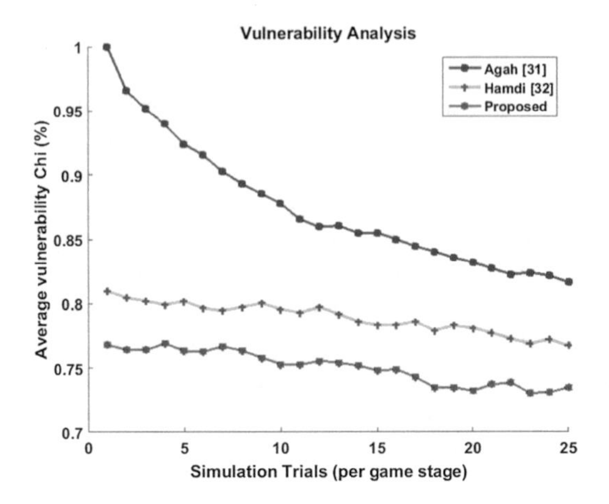

Fig. 6. Comparative analysis of vulnerability

A distinct performance of lowering threat level by proposed system can be seen in Fig. 6 in comparison to existing system. With increasing simulation trials, the vulnerability spontaneously minimizes and hence it can be also said that the energy dissipation also minimizes too. As the proposed system considers energy being dissipated owing to energy-based attackers, so the reduction in threat level is equivalent to minimization of unwanted (or illegitimate) energy depletion. Therefore, a good balance is maintained for energy depletion due to security problems in IoT.

7 Conclusion

This paper has presented a discussion of one of the simple and yet unique solution to resist intrusion in communication system affecting performance of IoT devices. The contribution of the proposed study are as follows: (i) without using any form of encryption, the proposed study offers simple and practical solution of security, (ii) the solution present is resistive against majority of the reported threats that directly or indirectly causes the node to deplete energy thereby acting as one solution towards multiple security breaches, (iii) it offers a capability to identify and isolate the attacker or the compromised IoT nodes without affecting the ongoing communication, (iv) the data packet is only send when the complete nodes involved in the process is proven legitimate or non-vulnerable. The study outcome has proven with good capability to identify the threats in contrast to existing system. Our future work will be focused on further optimizing the security performance.

References

1. Hwang, K., Chen, M.: Big-Data Analytics for Cloud, IoT and Cognitive Computing. Wiley, Hoboken (2017)
2. Krishna Prasad, A.V.: Exploring the Convergence of Big Data and the Internet of Things. IGI Global, Hershey (2017)
3. Bhadoria, R.S., Chaudhari, N., Tomar, G.S., Singh, S.: Exploring Enterprise Service Bus in the Service-Oriented Architecture Paradigm. IGI Global, Hershey (2017)
4. Humayed, A., Lin, J., Li, F., Luo, B.: Cyber-physical systems security—a survey. IEEE Internet of Things J. **4**(6), 1802–1831 (2017)
5. Benzarti, S., Triki, B., Korbaa, O.: A survey on attacks in Internet of Things based networks. In: 2017 International Conference on Engineering & MIS (ICEMIS), Monastir, pp. 1–7 (2017)
6. Mendez, D.M., Papapanagiotou, I., Yang, B.: Internet of things: survey on security. J. Inf. Secur. J.: Global Persp. **27**(3), 162–182 (2018)
7. Sain, M., Kang, Y.J., Lee, H.J.: Survey on security in Internet of Things: state of the art and challenges. In: 2017 19th International Conference on Advanced Communication Technology (ICACT), Bongpyeong, pp. 699–704 (2017)
8. Benabdessalem, R., Hamdi, M., Kim, T.H.: A survey on security models, techniques, and tools for the Internet of Things. In: 2014 7th International Conference on Advanced Software Engineering and Its Applications, Haikou, pp. 44–48 (2014)
9. Dragomir, D., Gheorghe, L., Costea, S., Radovici, A.: A survey on secure communication protocols for IoT systems. In: 2016 International Workshop on Secure Internet of Things (SIoT), Heraklion, pp. 47–62 (2016)
10. Shakhov, V., Koo, I., Rodionov, A.: Energy exhaustion attacks in wireless networks. In: 2017 International Multi-Conference on Engineering, Computer and Information Sciences (SIBIRCON), Novosibirsk, pp. 1–3 (2017)
11. Bertino, E., Islam, N.: Botnets and Internet of Things security. Computer **50**(2), 76–79 (2017)
12. Bhattarai, S., Wang, Y.: End-to-end trust and security for Internet of Things applications. Computer **51**(4), 20–27 (2018)
13. Burg, A., Chattopadhyay, A., Lam, K.Y.: Wireless communication and security issues for cyber-physical systems and the Internet-of-Things. Proc. IEEE **106**(1), 38–60 (2018)
14. Nurse, J.R.C., Creese, S., De Roure, D.: Security risk assessment in Internet of Things systems. IT Prof. **19**(5), 20–26 (2017)
15. Singh, J., et al.: Twenty security considerations for cloud-supported Internet of Things (2015)
16. Szymanski, T.H.: Security and privacy for a green Internet of Things. IT Prof. **19**(5), 34–41 (2017)
17. Wolf, M., Serpanos, D.: Safety and security in cyber-physical systems and Internet-of-Things systems. Proc. IEEE **106**(1), 9–20 (2018)
18. Hu, L., et al.: Cooperative jamming for physical layer security enhancement in Internet of Things. IEEE Internet of Things J. **5**(1), 219–228 (2018)
19. Liu, Y., Kuang, Y., Xiao, Y., Xu, G.: SDN-based data transfer security for Internet of Things. IEEE Internet of Things J. **5**(1), 257–268 (2018)
20. Villari, M., et al.: Software defined membrane: policy-driven edge and Internet of Things security. IEEE Cloud Comput. **4**(4), 92–99 (2017)
21. Moustafa, N., Adi, E., Turnbull, B., Hu, J.: A new threat intelligence scheme for safeguarding industry 4.0 systems. IEEE Access **6**, 32910–32924 (2018)

22. Mughal, M.A., Luo, X., Ullah, A., Ullah, S., Mahmood, Z.: A lightweight digital signature based security scheme for human-centered Internet of Things. IEEE Access **6**, 31630–31643 (2018)
23. Pereira, G.C.C.F., et al.: Performance evaluation of cryptographic algorithms over IoT platforms and operating systems. Secur. Commun. Netw. **2017**, 16 (2017)
24. Raza, S., et al.: S3K: scalable security with symmetric keys—DTLS key establishment for the Internet of things. IEEE Trans. Autom. Sci. Eng. **13**(3), 1270–1280 (2016)
25. Xiao, D., Yu, Y.: Cryptanalysis of compact-LWE and related lightweight public key encryption. Secur. Commun. Netw. **2018**, 9 (2018)
26. Qu, C., et al.: Blockchain based credibility verification method for IoT entities. Secur. Commun. Netw. **2018**, 11 (2018)
27. Song, W.-T., Hu, B., Zhao, X.-F.: Privacy protection of IoT based on fully homomorphic encryption. Wirel. Commun. Mob. Comput. **2018**, 7 (2018)
28. Wu, H., Wang, W.: A game theory based collaborative security detection method for Internet of Things systems. IEEE Trans. Inf. Forensics Secur. **13**(6), 1432–1445 (2018)
29. Xu, G., et al.: Network security situation awareness based on semantic ontology and user-defined rules for Internet of Things. IEEE Access **5**, 21046–21056 (2017)
30. Zhang, M., et al.: Tolerating sensitive-leakage with larger plaintext-space and higher leakage-rate in privacy-aware Internet-of-Things. IEEE Access **6**, 33859–33870 (2018)
31. Agah, A., Das, S.K., Basu, K.: A game theory based approach for security in wireless sensor networks. In: IEEE International Conference on Performance, Computing, and Communications, pp. 259–263 (2004)
32. Hamdi, M., Abie, H.: Game-based adaptive security in the Internet of Things for eHealth. In: 2014 IEEE International Conference on Communications (ICC), Sydney, NSW, pp. 920–925 (2014)

Simplified Framework for Resisting Lethal Incoming Threats from Polluting in Wireless Sensor Network

Somu Parande(✉) and Jayashree D. Mallapur

Department of Electronics and Communication,
Basaveshwar Engineering College, Bagalkot, Karnataka, India
somuparande63@gmail.com

Abstract. Accomplishing enhanced security in a resource-constrained sensor node is one of the most challenging aspects in the area of Wireless Sensor Network (WSN). Review of current research approaches using shows that still there is a big gap in the evolving security practices and what is needed in reality for WSN. Therefore, the proposed system introduces a novel security scheme that is capable of identifying the security threats that usually comes in the deceiving form of incoming request. The analytical scheme presented by proposed system is simple and progressive manner where malicious requests after being identified are diverted as a mechanism of resistance. The simulated outcome of the proposed system shows that it offers good energy retention as well as better packet delivery performance in contrast to existing secure and energy efficient protocols and is in-line with practical security need of WSN.

Keywords: Wireless Sensor Network · Security · Intrusion · Attacks · Network security · Secure routing

1 Introduction

The wireless sensor network (WSN) is introduced by MEMS technology, communication, and network system. Initially, the WSN application is employed for security purpose in small scale and later it is come up with IoT application and remote monitoring applications [1]. A WSN can be defined as a spatial dispersed device with sensor which having capability to sense the physical surroundings and then collets relevant data according to application requirement and transmit data to sink node via wireless media [2]. Different types of sensor are used in WSN such as Ultra sonic sensor [3], temperature sensor, proximity sensor, accelerometer, light sensor, IR sensor, pressure sensor and smoke or gas sensor [4]. The WSN are highly employed by many application area such as healthcare, military, security and surveillance, industrial process control, and environmental monitoring [5]. The WSN are more useful where the wired set up is not able to deploying like forest, under the water or ground. Also, it is used in harsh and hostile environment. The main advantage of WSN is the system is scalable and flexible. The lack of computation power and communication resources (limited sensing range, limited storage capacity and limited energy) is the prime

R. Silhavy (Ed.): CSOC 2019, AISC 986, pp. 21–30, 2019.
https://doi.org/10.1007/978-3-030-19813-8_3

disadvantage of the WSN [6]. Due to sensor node resource constraints nature, and deployment in unfriendly environment, WSN become central attraction for the various security vulnerabilities. There lots of researches efforts have been made to tackle the security issues associated to WSN [7–10]. Hence this paper introduces an optimal technique for security in wireless sensor networks. Section 2 discusses about the existing research work followed by problem identification in Sect. 3. Section 4 discusses about proposed methodology followed by elaborated discussion of algorithm implementation in Sect. 5. Comparative analysis of accomplished result is discussed under Sect. 6 followed by conclusion in Sect. 7.

2 Related Work

This section discusses about the existing research work taken in effort of WSN security. The work carried out by Jiu Cui et al. [11], presents a novel encryption algorithm for data confidentiality and to provide secure data transmission to end-user. Kemedi et al. [12] adopted a power preserving key generation system in perception layer for providing forward/backward security. Similarly, Kong et al. [13] presented a bidirectional relay network security protocol using network coding and key generation method in order to resist wiretap attacks. Osanaiye et al. [14] discusses different techniques for DoS attacks in various layer of WSN. Shah and Jinwala [15] presented a data aggregation scheme by with novel encryption technique to provide privacy, resilience, and avoidance capacity against different attacks with less key storage usage. Luo et al. [16] presented Access control model for data security in cross domain settings of the IoT. Wu et al. [17] presents a certificate-less signature technique for medical field on cloud environment to protect from intruder. Li et al. [18] discusses three authentications approaches for security improvement in WSN using cryptanalysis. Hajji et al. [19] presents a robust routing protocol for improving the life time of the WSN. Meena and Sharma [20] adopted a FLSO protocol method in WSN which use to manage the energy consumption and security problem by using re-keying technique. Lu and Sun [21] proposed a data aggregation algorithm under space-time correlation factor for avoiding additive attacks in WSN. Tayebi et al. [22] presents a advance chaotic depended technique for enhancing the defensive property of WSN. Boudia et al. [23] employs a data aggregation method with hop by hop verification technique for early detection of attacks in WSN. Sun et al. [24] presents an intrusion detection model with V- detector algorithm for tackling various attacks and for reducing the computation, data storage etc. Qin et al. [25] adopts a trusted secure routing mechanism for handling the common attacks appears in WSN. Hatzivasilis et al. [26] presents a safe routing protocol system for securing routing in cyber systems and IoT based applications. Umar et al. [27] presents a trust worthy-cross layer protocol based fuzzy logic system for mitigating the security threats in the sensor network. Meng et al. [28] presents a intrusion identification system based on sampling prediction to improve trust-key management in WSN. Shafiee et al. [29] presents a comparative analysis for cognitive spectrum sensing algorithms in order to make efficient parameter for WSN. Mehmood et al. [30], adapts knowledge-based Context Aware intrusion detection for malicious node identification in WSN.

3 Problem Description

The existing approaches discussed in prior section have offered solution but their applicability is more towards specific forms of attacks or the problems. Another significant problem is that it doesn't not consider various resource-based constrains of the sensors that also reduces the practicability of the sensors. Existing security approaches are also found to be more dependent on the encryption-based approach to claim higher degree of resistance from lethal attacks. However, such security solution has always been evolved at the cost of network lifetime that also has indirect impact of data delivery performance. Therefore, the statement of problem is "It is quite a difficult task to offer simple and potential security solutions to resist lethal threats in WSN using cost effective computational approach". The next section briefs of proposed solution.

4 Proposed Methodology

The core ideology of the proposed study is to evolve up with a simple and progressive strategy to capture all forms of request generated by the adversary as the first step to stop intrusion over WSN. Adopting analytical research methodology, the proposed system presents a simple scheme shown in Fig. 1.

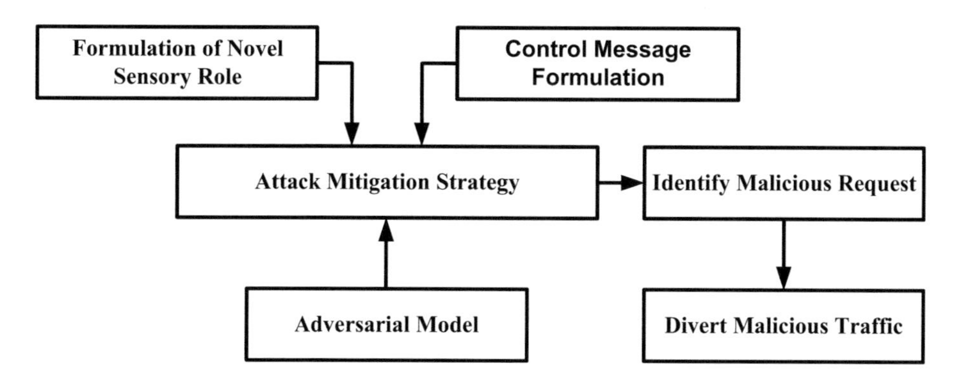

Fig. 1. Analytical scheme of proposed methodology

The proposed system introduces a novel coordinator node system that is meant for minimizing the network overhead by assisting in enhancing the data transmission rate. The control message has been worked upon to split the operation of route request for bearing the specific information associated with updates using routing matrix. A new adversarial model has been introduced by generalizing majority of the lethal threats in WSN who are more targeting to attack network with higher links and connectivity. The proposed system introduces a mechanism where the malicious request is identified and then diverted into different direction thereby wasting resources of the adversarial node. The significant contribution of the proposed solution is that it offers capability to stop the intrusion in its first attempt itself. The next section elaborates about the system design.

5 System Design

In order to develop a system that offers higher resistance towards any form of security threats, it is essential to first validate the form of the request that are incoming towards the cluster head during data aggregation in WSN. However, this is absolutely not an easy task and hence the proposed system introduces a new secure routing system for addressing this problem. The contribution of the system design of the proposed system (Fig. 2):

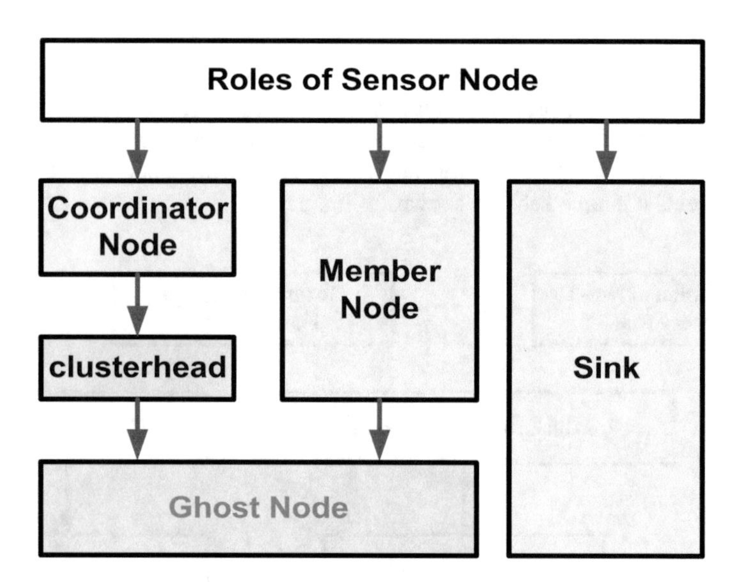

Fig. 2. Taxonomies of sensory role

- *Novel Sensory Role*: Unlike conventional role of a sensor node in WSN, the proposed routing scheme doesn't only use cluster head and member node but it introduces two more different roles i.e. ghost node and coordinator node. Basically, a coordinator node is one out of many clusters hear that undertakes decision of forwarding the aggregated data to the sink or perform re-transmission or send back then. It will also mean that proposed system offers a supportability of multihop routing with these sensors.
- *Novel Control Message*: Majority of the existing routing scheme in WSN uses only four types of control message i.e. route_request, route_reply, route acknowledgement, and route_error. However, the proposed system makes slight changes in route_reply by forwarding the reply of controlling the node's action as consequences of data aggregation. The route_request message performs declaration of the

information related to neighboring nodes of a target node. The severity of the attack is judged by the intrusion level over the route_request control message. The broadcasting of the route_reply is carried out by coordinator node periodically and such messages are only forwarded via coordinator node thereby minimizing the overhead.

- *Adversarial Model*: It was seen from the existing approaches that they are highly specific to a single form of attack which renders inapplicability of the existing approaches. Therefore, a new adversarial model is designed in such a way that it incorporates malicious behavior of majority of the lethal threats in WSN. In this process, the vulnerable node will try to perform less broadcasting in order to ensure that they are not much exposing their location and resources to too many of the sensors. The attacker node, on the other hand, will try to extract more information from such vulnerable node in order to obtain information of large chain of networks which exists in multi-hop. For this purpose, the attacker node broadcast a falsified control message of route_request just to show that they are available to act as a coordinator node for the target source node. It is because once they become coordinator node than they will gain rights to use route_reply and start controlling the topology maliciously. Figure 3 shows the sequence of different phases of initiating and launching the attacks by the adversary.
- *Attack Mitigation Strategy*: The attack mitigation strategy of the proposed system is very unique. Referring Fig. 4, the first level of protection involves formulating a routing table that retains information of diversified hops for active connections only. After receiving the malicious request of an adversary, the source node compares it with that of local updates, which will never match. In such case, the source node appoints a non-existing node using IP masking called as primary ghost node who forwards a fake profile of IP of node that never exists in the existing network. As the primary ghost node forwards information in such a way that adversary believes the information forwarded by ghost node and allocates its resources to achieve it in second level of protection. In third level of protection, the adversary makes communication with secondary ghost node that is again virtually developed by primary ghost node.

One interesting point to be observed is that proposed system doesn't implement any form of encryption mechanism and it performs simple heuristic-based comparisons to find if the incoming request is legitimate or not. In case it is found illegitimate than it appoints a non-existing set of nodes that is means for generating a confusive and non-existing network that leads to degradation of resources by the attacker and thereby discourage them from any further attacks.

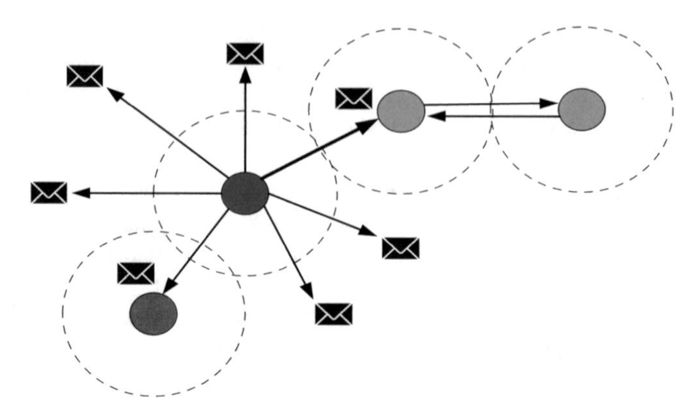

Phase 1: attacker receives route_request

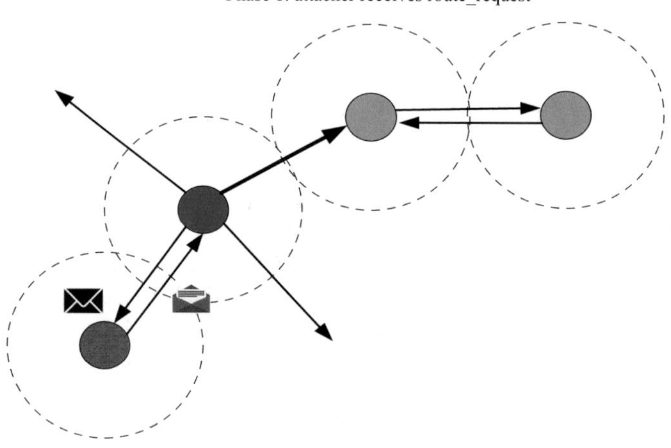

Phase-2: attacker forward route_reply

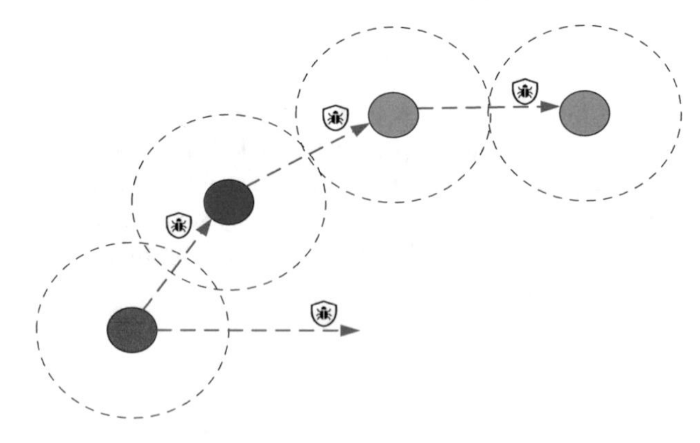

Phase-2: attacker forward route_reply

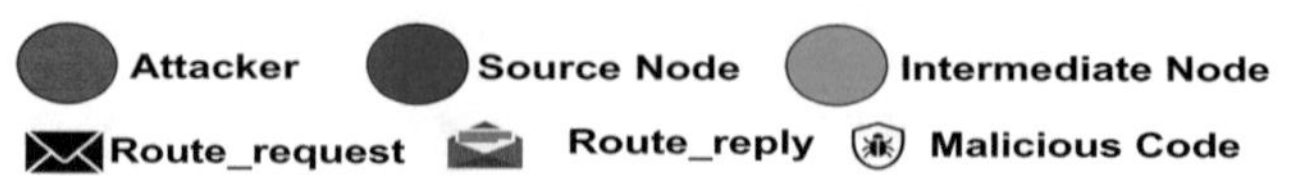

Fig. 3. Adversarial model strategies

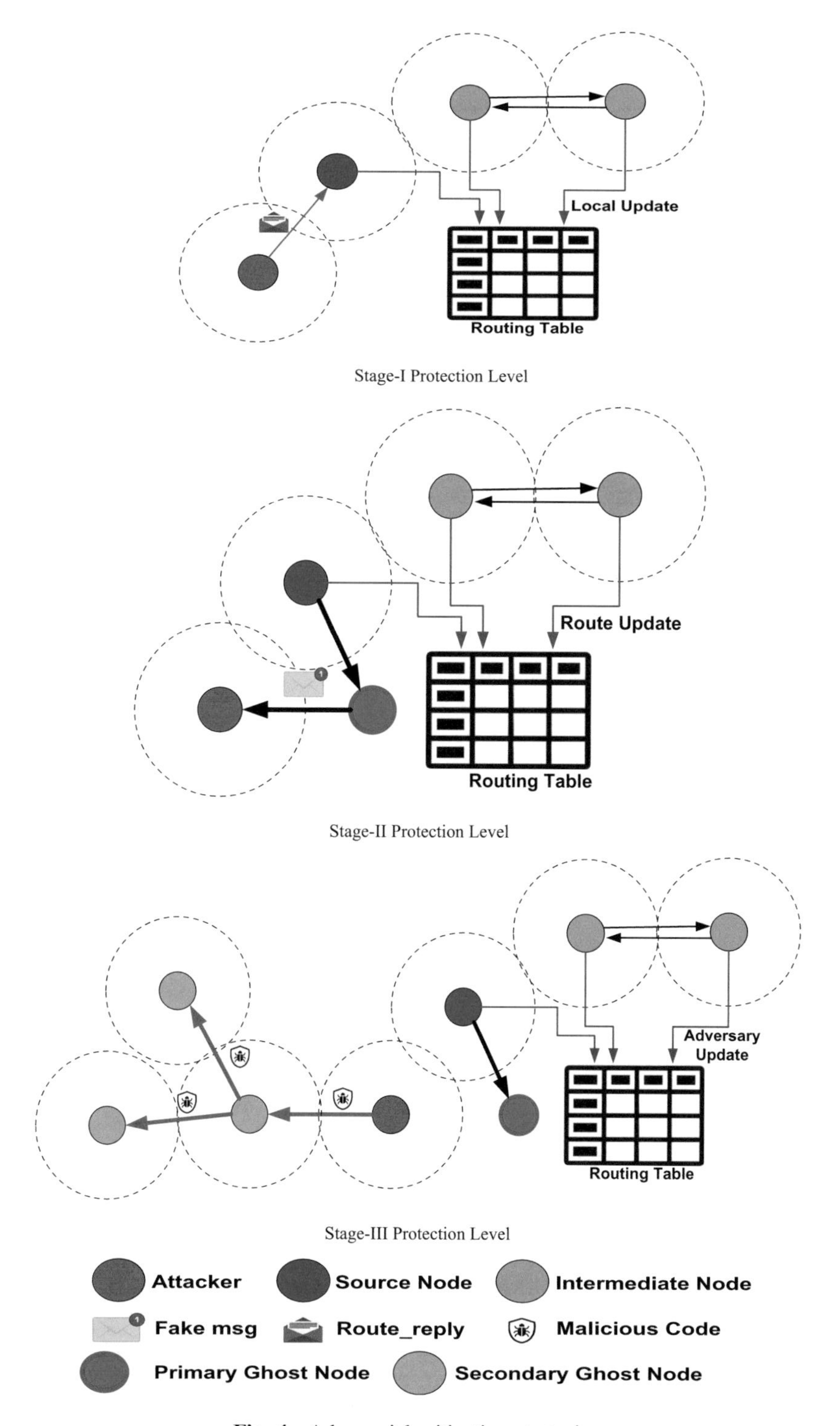

Fig. 4. Adversarial mitigation strategies

6 Results Discussion

As the proposed study introduces a novel mitigation strategy for security, hence the analysis of security approach is carried out using some practical parameters e.g. residual energy and throughput. The simulation has been carried out considering 500–600 sensors bearing MicaZ mote property randomly spread over 1000×1100 m^2 simulation area.

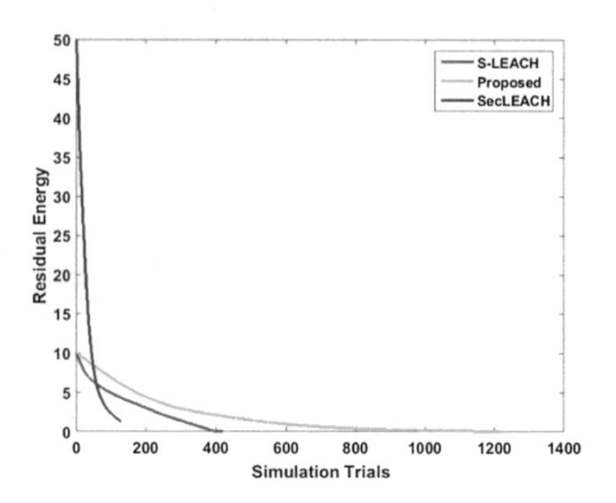

Fig. 5. Comparative analysis of residual energy

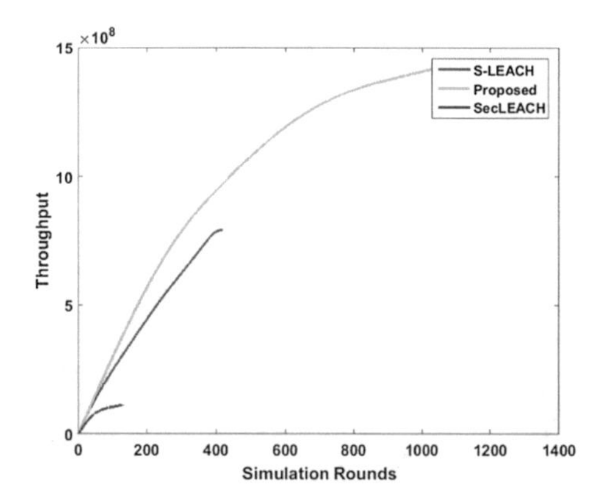

Fig. 6. Comparative analysis of throughput

The outcome shows that proposed system offers better energy retention capability (Fig. 5) as well as better form of data delivery services (Fig. 6). As SecLEACH [31] carry out random key distribution; hence it undergoes multiple checks just to authenticate the malicious request. Similarly, S-LEACH [32] speed up this process using pair wise key distribution, however, when the attacks are dynamic, S-LEACH slowly drains energy. For similar reason, performance of throughput is affected by existing approaches. Although, they are claimed to be preventive against various threats, such prevention cost is more than performance anticipated for existing system. The proposed system uses highly progressive approach of confirming the legitimacy of incoming request to understand the threat level and uses coordinator node which reduces lot of work load during data aggregation. At the same time, proposed system takes 0.22361 s of processing time while the existing system needs 1.72111 s.

7 Conclusion

Majority of the attacks in WSN starts by compromising any of the weaker sensor nodes. Normally that starts from broadcasting the forged information within the network. The proposed system is all building a secure level of trust from the heuristics of successfully created routes and then it compares the entire incoming request for network participation. The proposed design principle after finding the presence of malicious request diverts the request using a ghost sensor. The benefit of this process is that an attacker has no other option but to rely on the forged information passed on to them by a ghost node. Hence, without using any form of encryption technique, the proposed system offers good capability to resist malicious traffic.

References

1. Kamila, N.K.: Handbook of Research on Wireless Sensor Network Trends, Technologies, and
2. Kocakulak, M., Butun, I.: An overview of wireless sensor networks towards Internet of Things. In: 2017 IEEE 7th Annual Computing and Communication Workshop and Conference (CCWC), Las Vegas, NV, pp. 1–6 (2017)
3. Mehta, K., Pal, R.: Energy efficient routing protocols for wireless sensor networks: a survey. Energy 165 3 (2017)
4. Li, S.E., et al.: Kalman filter-based tracking of moving objects using linear ultrasonic sensor array for road vehicles. Mech. Syst. Sig. Process. **98**, 173–189 (2018)
5. Borges, L.M., Velez, F.J., Lebres, A.S.: Survey on the characterization and classification of wireless sensor network applications. In: IEEE Communications Surveys and Tutorials, vol. 16, no. 4, pp. 1860–1890 (2014)
6. Nayak, P., Devulapalli, A.: A fuzzy logic-based clustering algorithm for WSN to extend the network lifetime. IEEE Sens. J. **16**(1), 137–144 (2016)
7. Tomić, I., McCann, J.A.: A survey of potential security issues in existing wireless sensor network protocols. IEEE Internet Things J. **4**(6), 1910–1923 (2017)
8. Butun, I., Morgera, S.D., Sankar, R.: A survey of intrusion detection systems in wireless sensor networks. In: IEEE Communications Surveys and Tutorials, vol. 16, no. 1, pp. 266–282 (2014)
9. Xu, J., Yang, G., Chen, Z., Wang, Q.: A survey on the privacy-preserving data aggregation in wireless sensor networks. China Commun. **12**(5), 162–180 (2015)
10. Feng, W., Yan, Z., Zhang, H., Zeng, K., Xiao, Y., Hou, Y.T.: A survey on security, privacy, and trust in mobile crowdsourcing. IEEE Internet Things J. **5**(4), 2971–2992 (2018)
11. Cui, J., et al.: Data aggregation with end-to-end confidentiality and integrity for large-scale wireless sensor networks. Peer Peer Netw. Appl. **11**(5), 1022–1037 (2018)

12. Moara-Nkwe, K., et al.: A novel physical layer secure key generation and refreshment scheme for wireless sensor networks. IEEE Access **6**, 11374–11387 (2018)
13. Kong, Y., et al.: The security network coding system with physical layer key generation in two-way relay networks. IEEE Access **6**, 40673–40681 (2018)
14. Osanaiye, O.A., Alfa, A.S., Hancke, G.P.: Denial of service defence for resource availability in wireless sensor networks. IEEE Access **6**, 6975–7004 (2018)
15. Shah, K.A., Jinwala, D.C.: Privacy preserving, verifiable and resilient data aggregation in grid-based networks. Comput. J. **61**(4), 614–628 (2018)
16. Luo, M., Luo, Y., Wan, Y., Wang, Z.: Secure and efficient access control scheme for wireless sensor networks in the cross-domain context of the IoT. Secur. Commun. Netw. (2018)
17. Wu, L., Xu, Z., He, D., Wang, X.: New certificateless aggregate signature scheme for healthcare multimedia social network on cloud environment. Secur. Commun. Netw. (2018)
18. Li, W., Li, B., Zhao, Y., Wang, P., Wei, F.: Cryptanalysis and security enhancement of three authentication schemes in wireless sensor networks. Wirel. Commun. Mob. Comput. (2018)
19. El Hajji, F., Leghris, C., Douzi, K.: Adaptive routing protocol for lifetime maximization in multi-constraint wireless sensor networks. J. Commun. Inf. Netw. **3**(1), 67–83 (2018)
20. Meena, U., Sharma, A.: Secure key agreement with rekeying using FLSO routing protocol in wireless sensor network. Wirel. Pers. Commun. **101**, 1–23 (2018)
21. Lu, Y., Sun, N.A.: Resilient data aggregation method based on spatio-temporal correlation for wireless sensor networks. EURASIP J. Wirel. Commun. Netw. **2018**(1), 157 (2018)
22. Tayebi, A., Berber, S., Swain, A.: Security enhancement of fix chaotic-DSSS in WSNs. IEEE Commun. Lett. **22**(4), 816–819 (2018)
23. Merad Boudia, O.R., Senouci, S.M., Feham, M.: Secure and efficient verification for data aggregation in wireless sensor networks. Int. J. Netw. Manag. **28**(1), e2000 (2018)
24. Sun, Z., Xu, Y., Liang, G., Zhou, Z.: An intrusion detection model for wireless sensor networks with an improved V-detector algorithm. IEEE Sens. J. **18**(5), 1971–1984 (2018)
25. Qin, D., Yang, S., Jia, S., Zhang, Y., Ma, J., Ding, Q.: Research on trust sensing based secure routing mechanism for wireless sensor network
26. Hatzivasilis, G., Papaefstathiou, I., Manifavas, C.: SCOTRES: secure routing for IoT and CPS. IEEE Internet Things J. **4**(6), 2129–2141 (2017)
27. Umar, I.A., Hanapi, Z.M., Sali, A., Zulkarnain, Z.A.: TruFIX: a configurable trust-based cross-layer protocol for wireless sensor networks. IEEE Access **5**, 2550–2562 (2017)
28. Meng, W., Li, W., Su, C., Zhou, J., Lu, R.: Enhancing trust management for wireless intrusion detection via traffic sampling in the era of big data. IEEE Access **6**, 7234–7243 (2018)
29. Shafiee, M., Vakili, V.T.: Comparative evaluation approach for spectrum sensing in cognitive wireless sensor networks (C-WSNs). Can. J. Electr. Comput. Eng. **41**(2), 77–86 (2018)
30. Mehmood, A., Khanan, A., Umar, M.M., Abdullah, S., Ariffin, K.A., Song, H.: Secure knowledge and cluster-based intrusion detection mechanism for smart wireless sensor networks. IEEE Access **6**, 5688–5694 (2018)
31. Oliveira, L.B., Wong, H.C., Bern, M., Dahab, R. Loureiro, A.A.F.: SecLEACH - a random key distribution solution for securing clustered sensor networks. In: Fifth IEEE International Symposium on Network Computing and Applications (NCA 2006), Cambridge, MA, pp. 145–154 (2006)
32. El_Saadawy, M., Shaaban, E.: Enhancing S-LEACH security for wireless sensor networks. In: 2012 IEEE International Conference on Electro/Information Technology, Indianapolis, IN, pp. 1–6 (2012)

An Efficient Approach Towards Formal Verification of Mixed Signals Using Feed-Forward Neural Network

D. S. Vidhya(✉) and Manjunath Ramachandra

Assam Don Bosco University, Guwahati, India
dsvidhya0770@gmail.com, manju_r_99@yahoo.com

Abstract. The world of integrated has seen vast advancement and most of the semiconductor industries were focusing on analog and mixed signals circuits as it is cost effective solutions on a single chip i.e., system on chip (SoC). This advancement provides analog, digital, and essential mixed signal circuits are integrated on a semiconductor device, which is used to build any modern consumer electronic applications including smart devices. In order to verify or test the mixed signals traditional techniques are not encouraging because of cost, performance and production time. The formal verification is technique which provides the evident of conscious algorithms in a system with respect to formal methods. The demand of formal verification in system on chip (SoC) designs in context of both software and hardware is high because of its cost and accuracy. In this paper, the design of formal verification of mixed signals using Feed Forward Neural Network (FFNN) is analyzed. The design includes mixed signal module, FFNN implemented design, and formal verification for analyze the accuracy of the trained network. The training of FFNN network is performed by using Verilog HDL over Xilinx platform on low cost Artix-7 FPGA. From outcomes it is found that formal verification achieves 94.00% of accuracy with less processing time.

Keywords: Feed forward · Formal verification · Mixed signal · Neural network · Back propagation · Training module · FPGA

1 Introduction

The recent advancement in the domain of integrated system design has witnessed lot of alters in the computer aided design (CAD) process. The prime interest in this domain has got much focus in hardware description language (HDL) based System on Chip designs (SoCs) to analyze the behavior of mixed signals (MS) [1]. Recently, some of the techniques and procedures were presented in recent past to design system for analyzing the behavior of mixed signals. However, with available techniques it is tough task to analyze the signal behavior as it needs proper knowledge of analog and mixed signals (AMS) [2] and [3]. During verification of circuits, various challenges may take place in all the steps of design. Also, various differences in circuits design can be found in different classes of AMS designs [4]. The verification process of AMS designs at different adaption level of the analog circuits can be defined with respect to

R. Silhavy (Ed.): CSOC 2019, AISC 986, pp. 31–39, 2019.
https://doi.org/10.1007/978-3-030-19813-8_4

uninterrupted signal quantities [5]. Also, the problem associated with the proper functionalities of integrated designs required to meet the system performance with respect to energy uses and area [6]. Thus, the paper presents a formal verification approach to perform the verification of the mixed signals by using Feed forward neural network (FFNN). The entire paper is framed as: Sect. 2 highlights the related work towards the formal verification of mixed using different training methods. The problem Identification is addressed with research gaps in Sect. 3. Section 4 explains the proposed methodology in detail with architecture of Training module for formal verification. The Results are analyzed for the proposed method in Sect. 6. Finally, the overall proposed work is summarized in the Sect. 7.

2 Related Work

Most recent researches towards verification of analog and mixed signals is presented in this section where some of them are selected from different publications. The survey of [7] discussed various techniques involved in verification of analog and mixed signals where it is observed that the verification is the critical step of the design that helps to assure the circuit design preciseness and its behavior with respect to system design. Also, some of the issues in formal verification were addressed with existing research gaps in verification of AMS circuits. The extended work of [8] gives the framework based on differential learning approach to perform formal verification by considering mixed signals. The technique has presented an analytical modeling approach that considers an algorithm to generate the multiple mixed signals meant for feasible operation of the mixed signal circuits, algorithm for training and verification. From outcomes analysis it was found that [8] achieved 98.7% accuracy with better speed of response than other machine learning algorithms. Similar research addressing verification of SoC with mixed signal is found in [9] by using Verilog model. The design of [9] builds an equivalent model of high-level radio frequency (HLRF) by integrating Verilog language with mixed signals. The verification analysis is performed by using digital methods like random data generation, manual verification methodology, assertion-based verification and coverage-driven verification. From the analysis it is been found that [9] yields faster verification than other existing approaches. A modelling approach for analog/mixed signals is found in [10] which uses Verilog coding. The work [10] adopted a top-level methodology which brings integration of simulation and modelling in Cadence environment. Through this methodology, [10] is able to achieve efficient and accurate model. The work of [11] presented an automated verification model for mixed signal SoCs designed for modern application by integrating analog, digital and mixed signals. The verification is performed at different levels and at different modes. The research problem considered in this paper is described in below section.

3 Problem Statement

From the previous section, it has been noticed that an amount of work carried on formal verification of mixed signal is quite good as a software approach. But very few amounts of work are carried as a hardware approach. The hardware approaches are having certain relevance and research issues which are yet to assemble. The identification of research gaps from the existing related work is explained. Very less work towards analog and mixed signals for formal verification in hardware architecture view point from past seven years of the existing works. Most of the research works on mixed signals towards on industrial applications point of view. Hence there is a research gap for complete formal verification model for mixed signals. Very few works done analytically for formal verification of mixed signals, which are not efficient computational models or standard benchmark models. Most of the existing works on hardware viewpoint, the formal verification is done for small examples like E-XOR operations, FIFO Models etc. not worked for complex examples like mixed signals training, etc. Most of the existing work towards the formal verification of mixed signals is facing less cost-effective solutions and optimization techniques to improve the performance over SOC. In proposed modelling we are trying to overcome the above-mentioned problems. The next section gives the overview of the proposed work as a methodology to fix the above problems.

4 Research Methodology

The proposed design aims to design a Hardware architecture model for formal Verification of mixed signal using FFNN. The feed forward neural network is designed employing Back-propagation Training (BPT) algorithm which is utilized to enhance the accuracy by performing itself. The FFNN is having perceptron layer (PL) architecture which is trained by Back-propagation to update the weights and map the input-outputs. The major objective for implementing hardware architecture of FFNN for formal verification method to improve the accuracy in real time scenario. The Complete design flow of proposed methodology is shown in the Fig. 1.

The proposed Methodology of Formal verification of mixed signals is represented below. The model mainly having two designs which includes one is implementation design and secondly, reference design. The implementation design includes mixed signals generation model using Matlab followed by Hardware architecture of FFNN using Back propagation training model. The reference design includes Random mixed signal generation using Matlab followed by Reference design using IP-Core based output generation. The both designs outcome placed to perform the formal verification using equivalence checking. Find-out the performance analysis of proposed design in terms of accuracy based on the error finding using counter method. The next section explains the detailed proposed design approach towards formal verification.

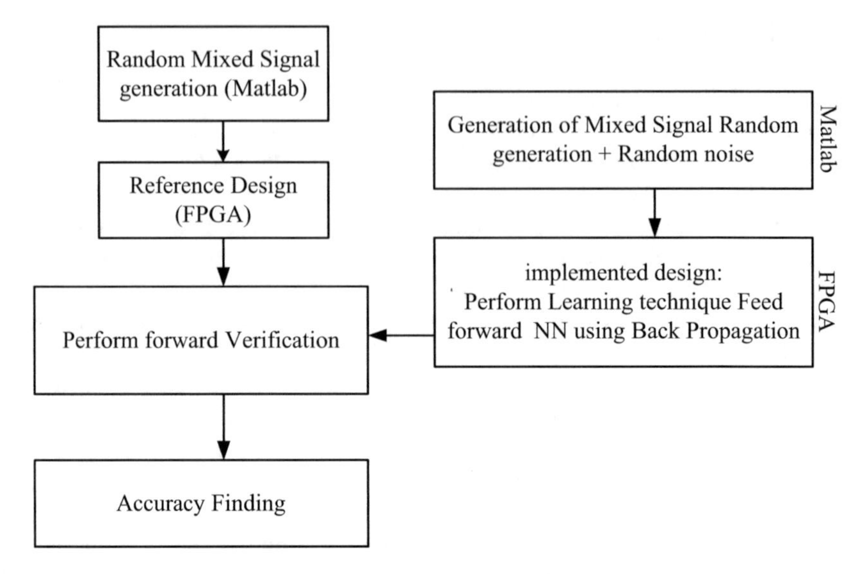

Fig. 1. Research methodology of the proposed design

5 Proposed Design Approach

The proposed design for formal verification of mixed signals is explained in detail. The proposed design includes mainly three modules namely, Generation of mixed signal, Training Module and Formal Verification module.

5.1 Generation of Mixed Signal

The generation of multiple mixed signals is used to test the formal verification under FFNN training. The mixed signal is generated using Matlab tool. The mixed signal generation includes analog and digital form of signal which is having higher or lower current and voltage factors. The mixed signal is stable or unstable is based on voltage or current variation. Consider the sine wave with defined sampling frequency 'Fs' and defined interval time 't' with varying amplitude 'a' to form a data signal. Add the pseudo random noise to the data signal to form a mixed signal. Change the 'a' and 'Fs' for multiple generation of mixed signals. These signals are used to train the network FFNN which is followed by Formal verification.

5.2 Training Module

The training module includes the FFNN using Back-propagation Training (BPT) algorithm is represented in the Fig. 2. The Training module considers the 2-3-1 network structure for FFNN. The Network structure is having two input layers, three hidden layers and one output layer along with back propagation module. Each input layer receives the inputs from mixed signal and performs the multiplication with initial defined weights; later the weights will be updated. The two input layers have din1 and din2 8-bit inputs and 8-bit w1, w2, w3 are weights are multiplied using individual six multipliers and generates the six different outputs namely il1, il2, il3, il4, il5, and il6, which is inputs to hidden layer. In hidden layer, three different Linear neuron model are used to generate the three outputs namely hl1, hl2, and hl3. The linear Neuron module contains two inputs il1 and il2 along with weight wh1 are multiplied to get the two products and which are added using Carry ahead adder. The added results are followed by linear function using LUT to obtain the hl1 output. The output layer is having 3 inputs are multiplied by three weights wo1, wo2, wo3, the three multiplied outcomes are added which are followed by sigmoid function to generate the final output 'out'.

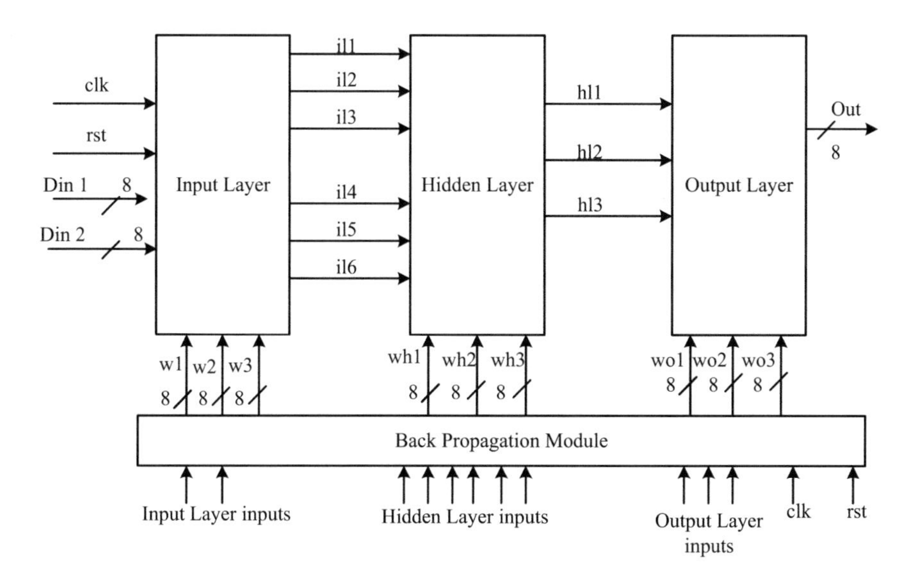

Fig. 2. Feed forward neural network employing BPT

The Back-propagation algorithm is used to generate the weights which are in updated form. The updated weight for each layer is defined as below:

$$\text{Updated Weight} = I_w + (L_r * D_i * D_o * O_p * (1 - O_p)) \tag{1}$$

Where the I_w is the initial weights, L_r is the learning rate which is set to 1, Di is difference between output layer output O_p and actual output A_o. The actual output A_o is stored in memory locations. D_o is the different layers input/output which depends on the layers. The O_p is the output layer output.

5.3 Formal Verification Module

The complete formal verification model includes instantiation of Implemented Design and reference design which results the accuracy. In the formal verification Model, The Implemented Design is tested through test cases, which is having the mixed signals it generates the 4000 samples, and each is 8-bit, which is fed to complete 2-3-1 FFNN network Results the trained data. The reference design includes the mixed signal generation using random data and pseudo random noise which is followed by IP Core based Block RAM which is having Referred output signal. This trained data and reference data from the reference design are tested by equivalence checking, if both the data are matched equally, then formal verification is "successful" otherwise it is Unsuccessful. To find out the error in the design, counter method is used. If the both the trained data and reference data are equal, counter is zero, if not equal, counter starts counting, wherever both the data are not matching till end of the samples. Based on the count, the accuracy will be calculated for formal verification.

6 Results and Analysis

The proposed Feed forward Neural Network of 2-3-1 network type for formal verification (FV) of different types of mixed signals results are analyzed in the below section. The proposed work is modeled over Xilinx ISE 14.7 environment using Hardware language Verilog-HDL and waveform simulated using Modelsim 6.5 simulator and implemented on FPGA includes Artix-7 family, XC7A100T-3 device, with package CSG324.

The synthesized results of proposed Design with both implemented and reference design for formal verification includes the 195-Slice registers, 357-Slice LUT's, two-Block RAM/FIFO Module used in references design for mixed signal generation and 39 DSP Elements which are tabulated in Table 1 and shows the available and used area resources constraints and percentage of utilization.

Table 1. Device utilization of the proposed deign for formal verification

FPGA Device: Artix-7: XC7A100T-3CSG324			
Logic utilization	Available	Used	Utilization
Number of slice registers	126800	195	0%
Number of Slice LUTs	63400	357	0%
Number of fully used LUT-FF pairs	430	122	28%
Number of bonded IOBs	210	48	22%
Number of Block RAM/FIFO	135	2	1%
Number of BUFG/BUFGCTRLs	32	1	3%
Number of DSP48E1s	240	39	16%

The power utilization of Feed forward Neural Network for formal verification of mixed signals is analyzed using X-Power analyzer tool from Xilinx ISE. The Total power consumed by the network is 0.117 W which includes the dynamic power of 0.035 W.

The proposed method targets to work for formal verification of mixed signals using trained FFNN apart from the hardware resource constraints results, accuracy is major performance parameter for formal verification. The analysis of formal verification is done using equivalence checking which includes two designs, the implemented design (RTL) and reference design (RTL) and it finds both the design outputs are equivalent or different. If Different, find out the errors using counter method to get the exact accuracy. The analysis of the proposed work is carried out over different test environments by changing the network type or number of samples which feed to the input layers. The implemented design includes complete hardware architecture of FFNN which is fed by mixed signals as an input to the network, followed by three hidden layer and single output layer gives the final results. The reference design includes Matlab generated mixed signals as fed to Block RAM IP cores which gives reference model.

The formal verification outcomes with an accuracy of multiple mixed signals are represented in Fig. 3. The Table 2 shows the three mixed signals with their accuracies after formal verification of the FFNN network. The study is focused on multiple mixed signals. The Proposed technique using FFNN along with Back propagation Training (BPT) analysis found with a higher accuracy rate mixed signal-3 with a 95.95% as compared to other two mixed signals of formal verification.

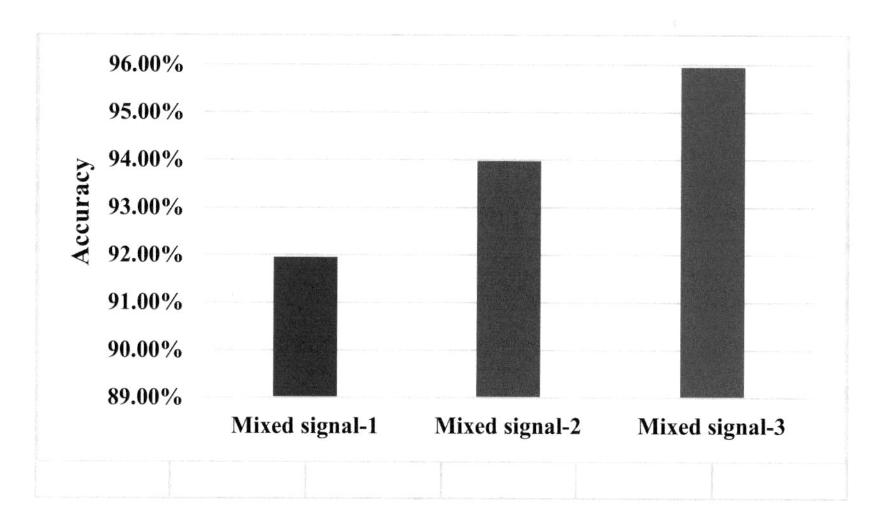

Fig. 3. Formal verification outcomes with an accuracy of multiple mixed signals

The average of all the three mixed signal accuracy is nearly 94.00% for formal verification of FFNN model. The outcome is satisfactory on System on Chip Level, also with digital modeling and its verification with good accuracy.

The processing time of the proposed model which is obtained to be 0.080 ms which is an average of multiple mixed signals? Still can improve the accuracy, by updating the weights properly in Back propagation training. The complete formal verification of proposed design is applicable to any mixed signals type which includes both data and noise, and it is proven from the implementation.

Table 2. Trained network outcomes for FV of multiple mixed signals

Feed Forward Neural Network (2-3-1)		
Sl. No	Signal type	Accuracy
1	Mixed signal-1	91.95%
2	Mixed signal-2	93.98%
3	Mixed signal-3	95.95%
Average accuracy		**94.00%**

7 Conclusion

This paper has presented a cost-effective solution to formal verification of mixed signals using FFNN over FPGA based systems or SOC. The usage rate of mixed signaling in industry is increasing drastically and it is one of the part of SOC Design design, which increases the demand in formal verification. Improper verification cost us with more timing. Hence it is compulsory to performance the formal verification in optimized manner to improve the error rate with better accuracy. The proposed Formal verification model includes Mixed signal generation, Implemented design using FFNN, Reference Design and formal verification using equivalence checking. The RTL-RTL verification is done. The synthesized area utilization is tabulated with less resources. The proposed design consumes standard accuracy rate around 94.00% using FFNN and less 0.080 ms processing time. The processing time will be vary which depends on the number of samples. In Future, to improve the accuracy and speed of the training network in formal verification of mixed signals, the Differentially fed neural network (DFNN) will be designed on hardware platform with low cost.

References

1. Bryant, R.E., et al.: Limitations and challenges of computer-aided design technology for CMOS VLSI. In: Proceedings of the IEEE 89.3, pp. 341–365 (2001)
2. Erden, M.S., et al.: A review of function modeling: approaches and applications. Ai Edam **22** (2), 147–169 (2008)
3. Zaki, M.H., Tahar, S., Bois, G.: Formal verification of analog and mixed-signal designs: a survey. Microelectron. J. **39**(12), 1395–1404 (2008)
4. Wielinga, B., Schreiber, G.: Configuration-design problem solving. IEEE Expert **12**(2), 49–56 (1997)
5. Gupta, S., Krogh, B.H., Rutenbar, R.A.: Towards formal verification of analog designs. In: Proceedings of the 2004 IEEE/ACM International Conference on Computer-Aided Design. IEEE Computer Society (2004)
6. Kundert, K., Chang, H.: Verifying all of an SOC-analog circuitry included. IEEE Solid-State Circ. Mag. **1**(4), 26–32 (2009)
7. Vidhya, D.S., Manjunath, R.: Research trends in formal verification process for analog and mixed signal design. Int. J. Comput. Appl. (0975 – 8887) **109**(11), 1–6 (2015)

8. Vidhya, D.S., Ramachandra, M.: A novel design in formal verification corresponding to mixed signals by differential learning. In: Computer Science On-line Conference. Springer, Cham (2017)
9. Yang, X., et al.: Mixed-signal system-on-a-chip (SoC) verification based on systemVerilog model. In: 2013 45th Southeastern Symposium on System Theory (SSST). IEEE (2013)
10. Gang, P.: Behavioral modeling and simulation of analog/mixed-signal systems using Verilog-AMS. IEEE Youth Conference on Information, Computing and Telecommunication, YC-ICT 2009. IEEE (2009)
11. Harinarayan, G.S., et al.: Automated Full Chip SPICE simulations with self-checking assertions for last mile verification & first pass Silicon of mixed signal SoCs. In: 2016 29th IEEE International System-on-Chip Conference (SOCC). IEEE (2016)

PCCV: Parallel Cancellation and Convolution Viterbi Encoding/Decoding Approach for MIMO-OFDMA for Efficient Resource Allocation and Power Consumption

B. Archana[1(✉)] and T. P. Surekha[2]

[1] Department of Electronics and Communication Engineering, GSSS Institute of Engineering and Technology for Women, Mysore, India
archanab.research@gmail.com

[2] Department of Electronics and Communication Engineering, Vidyavardhaka College of Engineering, Mysore, India

Abstract. The advancement in wireless communication with devices like smart mobile and tablets have significantly improved the necessary throughput of wireless communication schemes. However, achieving better speed data packets amenities is a major concern for the quality of various multimedia communications applications. The recent technique OFDM offers many of the significant features, but it has got some interference at carriers, i.e., inter-carrier interference (ICi). In order to tackle these interferences, the existing techniques adapted cyclic prefix which again leads to redundancy in transmitted data. Also, the transmission of this redundant data may lead to consuming some extra power and bandwidth. These limitations can be eliminated by using parallel cancellation mechanism. The MIMO-OFDMA is the effective scheme to achieve this high data rate which also associations with benefits of both the architectures of MIMO and OFDMA modulation approach. Thus, the MIMO-OFDMA system is analyzed with three different approaches like performance analysis without parallel cancellation, OFDMA with parallel cancellation (OFDMA-PC) and proposed system having OFDMA with parallel cancellation and Convolution Viterbi encoding/decoding for 4×4 transmitter and receiver. From performance analysis, it is observed that the proposed system in terms of power and resource allocation (bandwidth) is achieved with high data rate by minimized BER rate and also with that the power consumption with least BER is reduced.

Keywords: BER · Convolution encoder · Parallel cancellation · Power consumption · Resource allocation SNR · Viterbi decoding

1 Introduction

The speedy growth in wireless communication in today's world with devices like smart mobile and tablets have significantly improved the necessary throughput of wireless communication schemes [1]. However, achieving better speed data packets amenities is a major concern for the quality of various multimedia communications applications [2]. The MIMO-OFDMA is the effective scheme to achieve this high data rate which also

R. Silhavy (Ed.): CSOC 2019, AISC 986, pp. 40–49, 2019.
https://doi.org/10.1007/978-3-030-19813-8_5

associations with benefits of both the architectures of MIMO and OFDMA modulation approaches [3, 4]. Together these approaches can help to get better performance and high data rate in wireless communication mechanisms [5]. The core execution problems in high difficulty MIMO recognitions for various sub-carrier symbols in the MIMO-OFDMA receiver systems [6]. In a recent trend, a huge number of high data rated devices were facing the problem of intersymbol interference (ISI). In order to compute the desired channel at the receiver end, the channel estimation mechanisms can be used and which also helps to improve device capacity of OFDM system [7]. The mechanisms of channel estimation are mainly focused on the conceptuality of the favorable multiple path channels [8]. However, the existing approaches have presented the concepts with the real-time implementation of the multiple paths channels having a wide bandwidth which leads with the sparse in OFDMA architecture [9, 10]. These kinds of multiple path communication channels acquire some propagation delays. With the conventional approaches, most of the elements of impulse response are of zero have some noise floor with least number of delayed path elements, and it idealizes that the multiple path channels exhibit the sparse architecture [11, 12]. Hence, no such mechanism is available to compute the sparse data. Thus, the conventional computational mechanism does not consider this sparse signal of the communication channel. However, with available MIMO-OFDMA transceiver system, it is quite taught task to achieve reduced BER for the receive data signal generated by a device. This paper aims to build a MIMO-OFDMA system of 4×4 transmitter and receiver to achieve better resource allocation and power consumption in terms of BER and SNR with better signal quality. The paper is divided into various subsections like Background of MIMO and OFDMA (Sect. 2), Problem statement (Sect. 3), Research Methodology (Sect. 4), Algorithm implementation (Sect. 5), Performance analysis (Sect. 6) and Conclusion (Sect. 7).

2 Related Work

The existing researches considering MIMO-OFDMA system performance in terms of resource allocation and power consumption are discussed by focusing on inter-carrier interference (ICi) and resolved by adopting different approaches. The initial work of Archana and Sureka [13] presented a research survey discussing the current state of the art in the research domain of MIMO-OFDMA system for efficient communication. The work towards mitigating the inter cell interference is found in Lopez-Perez et al. [14], in which self-organizing algorithm is used for enhancing the system level performance. The limitation of [14] is that it limits with inter cell communication. Similar kind of research is found in Also Haily et al. [15], where user access is analyzed for multi-radio. The resource allocation among the devices is considered in Huang et al. [16], and a protocol based algorithm is presented on the basis of Nash equilibrium and attained better gain and sum rate. The work of Xiao et al. [17] used a computationally efficient optimal algorithm for improving the quality of service by performing better resource allocation, but it is not compared with any existing technique. The efficient scheduling mechanism for resource allocation in MIMO is found in Cao et al. [18] where the two-step method is used to bring the balance to computational complexity and resource allocation. The previous work of Archana and Sureka [19] presented a compressive sensing based channel estimator for MIMO-OFDMA. With the numerical outcome of [19] suggests that the system was a low-cost system to have better MIMO-OFDMA system. From the above surveys, it was observed that very rare works were incorporated with the minimization of BER, resource allocation, and power consumption.

3 Problem Statement

The OFDM exhibits many of the significant features which make it as good modulation techniques, but it has got some interference at carriers, i.e., inter-carrier interference (ICi). In order to tackle these interferences, some of the existing techniques adapt cyclic prefix which again leads to redundancy in transmitted data. Further transmission of this redundant data may lead to consuming some extra power and bandwidth. These limitations can be eliminated by using parallel cancellation mechanism. Over the period of research, many of the practices suggested that many real-time multi-path channels with a broad bandwidth may lead to sparse in architecture. During this, the communication channel of multipath collects some propagation delays. The conventional approaches provide impulse response elements of zero under a noisy condition with few delayed path elements which indicate sparse architecture of multipath channel. Hence no conventional mechanism performs the estimation of sparse data where these mechanisms ignore the sparse signal to minimize the BER of received data. Thus, there is a need of the system which considers the above pitfalls and comes up with energy efficient and better resource allocation in MIMO OFDMA.

4 Research Methodology

In order to overcome the above-discussed problems, the MIMO-OFDMA system is analyzed with three different approaches like performance analysis without parallel cancellation, OFDMA with parallel cancellation (OFDMA-PC) and proposed system having OFDMA with parallel cancellation and Convolution Viterbi encoding/decoding for 4×4 transmitter and receiver. In the proposed system, MIMO-OFDMA concept is considered to minimize BER ratio under high noise condition. The MIMO works on the basis of optimal binary search tree (OBST) scheme. The transmitter side of OFDMA uses IFFT operation while the receiver side uses FFT operation. However, to overcome Inter-Carrier Interference (ICI) issues few transmitters data is processed by using IFFT operation in the MIMO system. In the proposed system, instead of the parallel operation of IFFT and FFT, all the transmitter data are processed through FFT operation at transmitter side while IFFT operation at a receiver side. The interchanging of both FFT and IFFT at transmitter and receiver the zero value of BER can be achieved at 5 dB in traditional MIMO-OFDMA system while in proposed MIMO-OFDMA system zero BER can be achieved at −5 dB. In this paper also, a method of forwarding error correction (FEC) and detection approaches like Convolution encoding and Viterbi decoding utilized before the modulation of transmitted data and after demodulation of the received data respectively. The FEC mechanism also provides an additional protection layer for data in terms of detection and correction of errors in received data by a 1 dB improvement. Further to augment OFDM systems, some notations on OBST-OFDMA were analyzed with the 4×4 OBST-OFDM codeword matrix which offers diversity order 2 having a coding rate of 1. For the complex transmission, this orthogonality can't be achieved with the code rate of 1 during the transmit diversity is exceeding 2. There exist some of the studies to offer a framework for non-orthogonal complex transmission with a coding rate of 1 and having diversity order equal to the number of transmit antennas. Thus, the 4×4OBST-OFDM system is considered with a non-orthogonal transmission matrix, a coding rate of 1, and transmission diversity of 4. Also, the system of orthogonal transmission matrix is analyzed with a code rate of 0.5. The Following Fig. 1 gives the architectural model of the proposed system.

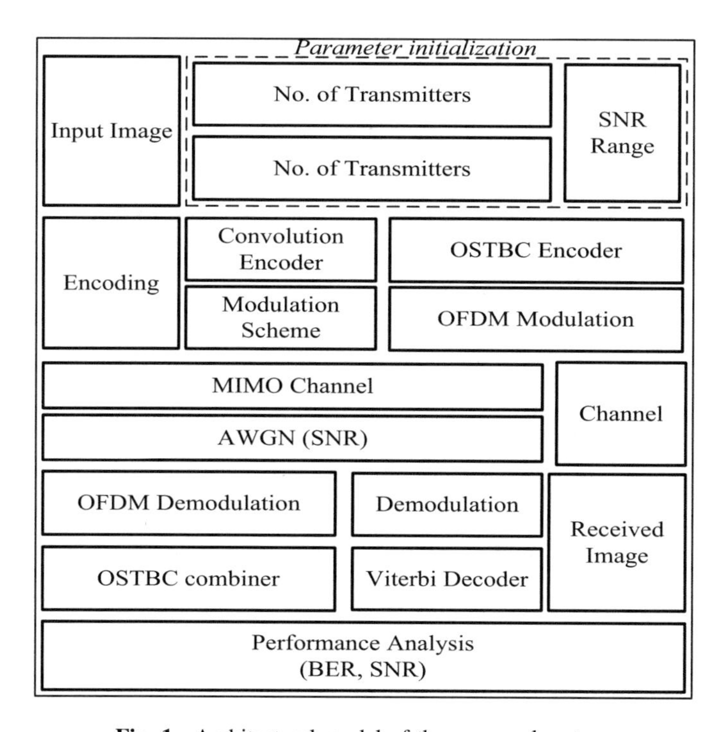

Fig. 1. Architectural model of the proposed system

In the proposed system an image is considered and is then converted into a binary format. On the same binary data pre-processing operations can be performed to get good quality of input signal. Further, the convolution encoding approach is adapted over the input signal. The same signal is modulated by using the QPSK modulation technique. This modulated signal is encoded using the OSTBC encoding technique depending upon the number of the transmitter are used in this framework. At the transmitter end, the OFDM modulation will be done along with the parallel cancellation scheme. Then the signal is converted time domain to frequency domain. These signals are transmitted through the communication channel.

The channel includes the MIMO channel along with AWGN noise. This signal to noise ratio is adjusted using signal range. Then it is transmitted. The received signal is demodulated using the OFDM Demodulating scheme along with parallel cancellation method. This received, and the demodulated signal is applied with OSTBC combining scheme depending upon the number of receiver devices are used in this framework. Then, this signal is demodulated using the QPSK demodulating method. The Viterbi decoding scheme is used to filter all the noise present in the received signal. In the end, the received data is converted to the original format. Then, the performance analysis of the proposed framework is performed using bit error rate (BER). The BER is compared with 3 different techniques, such as without parallel cancellation, OFDMA with parallel cancellation (OFDMA-PC) and proposed system having OFDMA with parallel cancellation and Convolution Viterbi encoding/decoding for 4×4 transmitter and receiver.

5 Algorithm Implementation

This section gives the implementation of the algorithm to achieve enhanced resource allocation and minimized power consumption. The algorithm is given as explain.

Algorithm: to enhance resource allocation and minimize power consumption

Input: I, I_1, Cd, Qmd, α, β, SNRr, λ,δ τ, η, φ, ς, Φ
Output: Reduced BER

Start

1. **Initialize** λ =1e6, δ= [0 2e-6], τ = [0 -10], η =30.
2. resize (I_{org}, 0.1)←I_{org}← read (I)
3. data←I_1 ← binary (I_{org})
4. t← trellis structure(I_1)
5. Cd ←Conv_encode (data, t)
6. Qmd ←QPSK_Mod (bits, symbol, value)
7. modData←step (Qmd, data, Cd)
8. [E_{data}, n_2] ←OSTBC_encoder (Qmd, modData, β (1))
9. [E_{Cd} ,N_1,n_2]←ofdm_mod(E_{data});
10. Channel←MIMOChannel(λ, δ, τ, η, ς, β, α, φ);
11. Qmd_{AWGN}←$AWGN_{Channel}$(Noise, SNRr, Φ,1)
12. [α, G_p]←Ofdm_demod($\alpha_{\_Cd}$, num α,n_2 , φ_{Cd});
13. C_{data}←OSTBC $_{Combining}$(α_{Signal}, α_{signal_Cd}, numβ, numα, φ,n_{2_Cd});
14. Qmd_{demod}←$QPSK_{demodv}$(Symbol,bits)
15. R_{data}← B_{i2}(R_{data})
16. **If** (SNRr>1)
    SNR ← [2: end];
    else
    SNR ← [];
17. **If** (α>1)
    SNR ← [1: end];
    else
    SNR← [];
18. **If** (SNRr>1 && β >1 && α >1)
    Apply MIMO_OFDM (SNR, β, α, I_{org})
    else
    NOP
19. **end if;**
20. R_{data}← Apply Viterbi(R_{cd}, t, t_b,'trunc','hard');
21. Reshape bits → original data format

End

During first step of the algorithm, the initialization of sampling rate (λ = 1e6), path delay (δ = [0 2e−6]), average path gain (τ = [0−10]), maximum Doppler shift (η = 30) is performed. Later, an image from the disk is selected and is resized with a multiplication factor of 0.1 (step-2). The resized image is converted into binary form and is stored as data (step-3). Using MATLAB, trellis structure (t) of the stored binary form of data is formed for convolution encoder which provides every possibility of input to the encoder subjected with both outputs and state transition of an encoder having states of binary form 00, 01, 10 and 11 which yields 2-bit output from 1-bit input (step-4). Further, considering both the factors data and t the convolution encoder is applied which gives a convoluted data (Cd) Step-5. The modulation scheme QPSK is applied to the signal to yield the modulated signal (Step-6), and then the encoding is performed by using orthogonal space-time block code (OSTBC) encoder block that encodes the input symbol sequence using OSTBC. The use of OSTBC along with MIMO helps to yield high SNR and low BER (Step-7). The OSTBC can be adapted with the system having feedback to the transmitter from the receiver. Later, the signal is demodulated by using the QPSK demodulation approach. For the same signal, the SNR values will be calculated and then apply the MIMO-OFDMA technique by considering original data, SNR value, transmission range and receiver range (Step-8 to 17). Then, the Viterbi decoding mechanism is adapted to filter the noise in the received signal. Finally, the received signal is converted/reshaped into the original format. The notations used in the algorithms are given in Table 1.

Table 1. Description of notations

SI. no	Notation	Description
1	I	Input image
2	I_1	Binary image
3	Cd	Convoluted data
4	Qmd	QPSK modulated data
5	α	Receiver range
6	β	Transmitter range
7	SNRr	SNR range
8	λ	Sampling rate
9	δ	Path delay
10	τ	Average path gain
11	η	Maximum doppler shift
12	φ	Path gain
13	ς	Spatial correlation
14	Φ	Signal power

6 Results and Analysis

The proposed MIMO-OFDMA system of 4×4 transmitter and receiver is analyzed to achieve better resource allocation and power consumption in terms of BER and SNR with better signal quality under noisy conditions. The simulation of the method is performed by using MATLAB. The performance is analyzed by comparing with different approaches like without Parallel cancellation, OFDMA with parallel cancellation (OFDMA-PC) and proposed system having OFDMA with parallel cancellation and Convolution Viterbi (CV) encoding/decoding for 4×4 transmitter and receiver. In the proposed system, the parallel operation of FFT and IFFT were replaced at both transmitter and receiver which help to reduce BER. The proposed system introduced Forward error correction (FEC) and detection mechanisms like convolutional encoding before modulating the transmitted data and Viterbi decoding after demodulating the received data. This significance of the FEC mechanism is that it offers an additional protection layer for the data in terms of detection and correction of errors in received data by a 1db improvement. The following section follows with the outcomes accomplished with respect to different approaches for communication.

The comparative analysis is considered with 4-transmitter and 4-receiver for all the different approaches like without Parallel cancellation, OFDMA with parallel cancellation (OFDMA-PC) and proposed system having OFDMA with parallel cancellation and CV encoding/decoding for 4×4 transmitter and receiver. The following Fig. 2 gives the outcomes attained from MATLAB simulation for all the approaches at different SNR values. The numerical data obtained for BER were represented in Table 2.

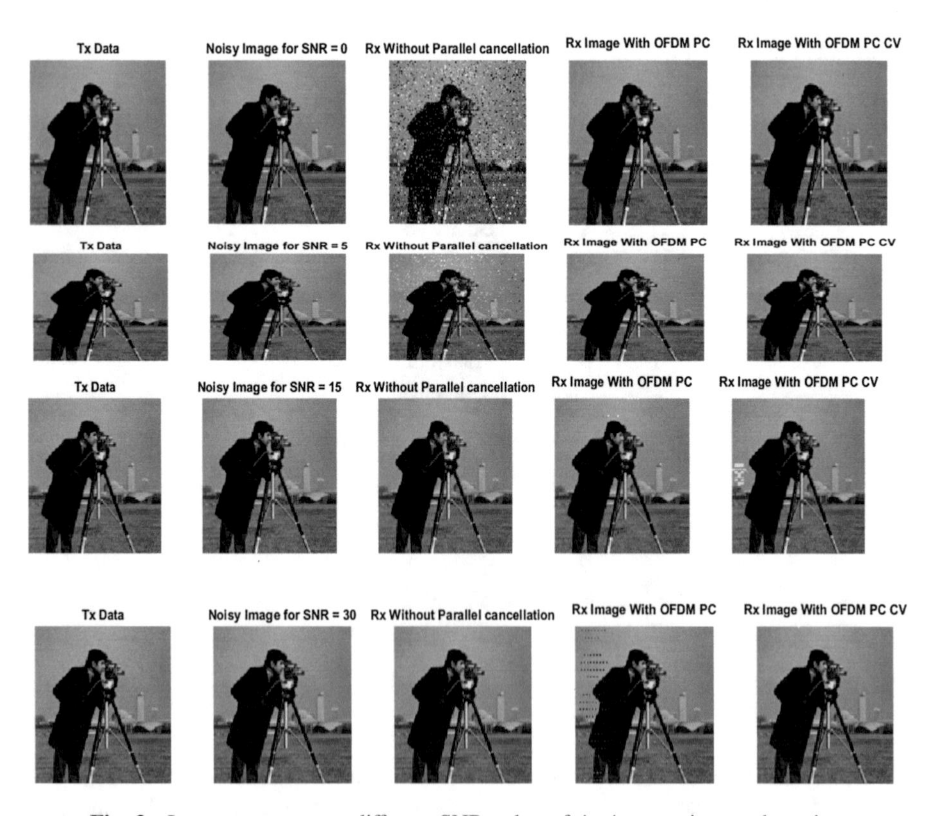

Fig. 2. Image outcomes at different SNR value of 4×4 transmitter and receiver

Table 2. Outcomes of BER at different SNR

SNR (dB)	BER		
	Without parallel cancellation	With parallel cancellation	With parallel cancellation and convolutional coding
0	0.0065	0.0008	0.0007
5	0.0025	0.0011	0.0006
10	0.0012	0.0009	0.0002
15	0.0013	0.0012	0.0001
20	0.0063	0.0003	0.0001
25	0.0002	0.0002	0.0000
30	0.0000	0.0001	0.0000

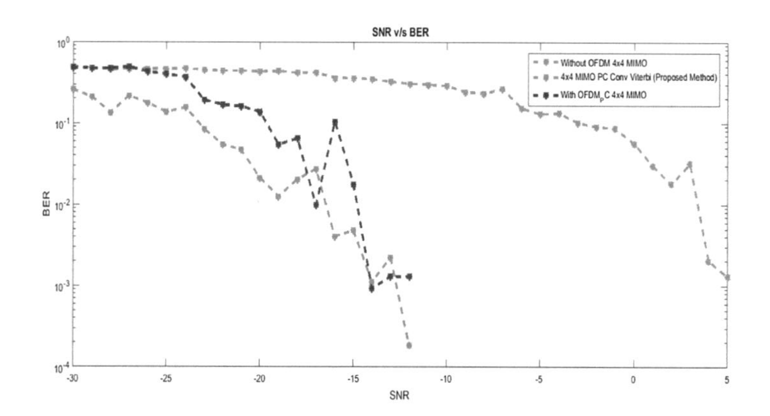

Fig. 3. Comparison of SNR vs. BER for different schemes

The above Fig. 3 indicates the plot of comparative analysis of the different schemes which have adopted different approaches like without parallel cancellation, with parallel cancellation and the combination of parallel cancellation & convolution Viterbi encoding/decoding. From Fig. 3, it is observed that the bit error rate (BER) is increased with a decrease in the Signal to Noise Ratio (SNR) value. The SNR with least value indicates the signal is having more unwanted noise while least value of BER indicates the least error occurred during signal transmission. In general sense, the BER is reciprocal to SNR. Thus, from all the approaches it is found that the OFDMA system without parallel cancellation has highest BER than other approaches having OFDMA system with parallel cancellation and proposed system (OFDMA system with parallel cancellation & convolution Viterbi encoding and decoding). This indicates that the error rate in the OFDMA system without parallel cancellation is high as it does not reduce the unwanted noise. In comparison with the OFDMA system having a parallel cancellation and the OFDMA system having a combination of parallel cancellation and Viterbi encoding and decoding (Proposed system), the proposed system has least error rate than other approaches. Thus it can be said that the proposed system in terms of power and resource allocation (bandwidth) is achieved with high data rate by minimized BER rate and also with that the power consumption with least BER is reduced.

7 Conclusion

This paper presents a MIMO-OFDMA system of 4×4 transmitter and receiver by using parallel cancellation and convolution Viterbi encoding and decoding method to achieve better resource allocation and power consumption in terms of BER and SNR with better signal quality. Through this system, the computational complexity of the system, as well as the cost of the system, is achieved. The proposed OFDMA system having a combination of parallel cancellation and Viterbi encoding and decoding (Proposed system) achieved less BER than other approaches. From this, it can be concluded that the system attained high data rate transmission by minimizing the BER rate and power consumption.

The proposed system can be considered in future research to fulfill the security concerns of MIMO-OFDMA having parallel cancellation and Viterbi encoding or decoding approach.

References

1. Checko, A., et al.: Cloud RAN for mobile networks—a technology overview. IEEE Commun. Surv. Tutorials **17**(1), 405–426 (2015)
2. Khalifa, T., Naik, K., Nayak, A.: A survey of communication protocols for automatic meter reading applications. IEEE Commun. Surv. Tutorials **13**(2), 168–182 (2011)
3. Feng, D., et al.: A survey of energy-efficient wireless communications. IEEE Commun. Surv. Tutorials **15**(1), 167–178 (2013)
4. Harjula, I., et al.: Practical issues in the combining of MIMO techniques and RoF in OFDM/a systems. In: Proceedings of the 7th WSEAS International Conference on Electronics, Hardware, Wireless, and Optical Conference (2008)
5. Nosratinia, A., Hunter, T.E., Hedayat, A.: Cooperative communication in wireless networks. IEEE Commun. Mag. **42**(10), 74–80 (2004)
6. Olwal, T.O., Djouani, K., Kurien, A.M.: A survey of resource management toward 5G radio access networks. IEEE Commun. Surv. Tutorials **18**(3), 1656–1686 (2016)
7. Akyildiz, I.F., Pompili, D., Melodia, T.: State-of-the-art in protocol research for underwater acoustic sensor networks. In: Proceedings of the 1st ACM international workshop on Underwater networks. ACM (2006)
8. Gkelias, A., Leung, K.K.: Multiple antenna techniques for wireless mesh networks. In: Wireless Mesh Networks. Springer, Boston, pp. 277–307 (2008)
9. Rawat, P., et al.: Wireless sensor networks: a survey on recent developments and potential synergies. J. Supercomput. **68**(1), 1–48 (2014)
10. Hughes, L., Wang, X., Chen, T.: A review of protocol implementations and energy efficient cross-layer design for wireless body area networks. Sensors **12**(11), 14730–14773 (2012)
11. Tse, D., Viswanath, P.: Fundamentals of Wireless Communication. Cambridge University Press, Cambridge (2005)
12. Bajwa, W.U., et al.: Compressed channel sensing: a new approach to estimating sparse multipath channels. Proc. IEEE **98**(6), 1058–1076 (2010)

13. Archana, B., Surekha, T.P.: Resource allocation in LTE: an extensive review on methods, challenges, and future scope. In: Communications on Applied Electronics (CAE). Foundation of Computer Science FCS, New York, USA, vol. 3, no. 2, October 2015. ISSN: 2394-4714
14. Lopez-Perez, D., Chu, X., Vasilakos, A.V., Claussen, H.: Power minimization based resource allocation for interference mitigation in OFDMA femtocell networks. IEEE J. Selected Areas Commun. **32**(2), 333–344 (2014)
15. Alsohaily, A., Sousa, E.S.: On the utilization of multi- mode user equipment in multi-radio access technology cellular communication systems. IEEE Access **3**, 787–792 (2015)
16. Huang, J., Yin, Y., Zhao, Y., Duan, Q., Wang, W., Yu, S.: A game-theoretic resource allocation approach for intercell device-to-device communications in cellular networks. IEEE Trans. Emerg. Topics Comput. **PP**(99), 1 (2014)
17. Xiao, X., Tao, X., Lu, J.: Energy-efficient resource allocation in LTE-based MIMO-OFDMA systems with user rate constraints. IEEE Trans. Veh. Technol. **64**(1), 185–197 (2015)
18. Cao, H., Cai, J., Alfa, A., Zhao, Z.: Efficient resource allocation scheduling for MIMO-OFDMA-CR downlink systems. In: 2016 8th International Conference on Wireless Communications & Signal Processing (WCSP), Yangzhou, pp. 1–5 (2016)
19. Archana, B., Surekha, T.P.: A compressive sensing based channel estimator and detection system for MIMO-OFDMA system. In: Proceedings of the Computational Methods in Systems and Software. Springer, Cham, pp. 22–31 (2018)

Streaming Analytics—Real-Time Customer Satisfaction in Brick-and-Mortar Retailing

Felix Weber(✉)

University of Duisburg-Essen, Essen, Germany
felix.weber@icb.uni-due.de

Abstract. The manifold changes in retailing in recent years has led to the scenario where competition is mainly driven by price. However, this one-sided focus on price as the only competitive instrument has become a significant problem for many retailers due to the increase in competition and reduction in scope for price differentiation. For brick-and-mortar stores in particular, however, customer satisfaction within the store is also a decisive factor, although this can currently only be assessed manually by employees as there are no analytical processes in place. Active evaluation and control of overarching measures is technically and economically not yet feasible. The aim of this research is to sketch a sentiment analytics model to analyze customer satisfaction for brick-and-mortar retailing. Using the presented Customer Satisfaction Streaming Index (CSSI), a mathematical model is developed that is tailored to the characteristics of the available data sources. In a second step, a framework for conducting big data analyses based on a standard retail system architecture is demonstrated, and a prototypical implementation is demonstrated. The preliminary results show that this is a suitable method for brick-and-mortar retailers. As the quality of social media sources might not be fully sufficient, alternate resources are discussed.

Keywords: Streaming analytics · Sentiment analysis · Retail · Customer satisfaction

1 Introduction

1.1 Retail

Changes in the retail trade in recent years have been manifold: globalization has led to international competitors gaining access to the markets, and they are now able to offer products of the same quality at lower prices. In addition, other market participants, particularly those from e-commerce, are entering the already highly competitive market, thereby intensifying competition. This and other factors have led to growing predatory competition, not only due to overcapacity but also to stagnating overall market volumes. The scope for price differentiation has declined overall, as "technical-functional harmonization" has taken place in many product categories. Due to the similarities among product ranges and business types in the retail sector—especially in grocery retailing—price is largely the only remaining competitive instrument. Increased price transparency on the part of consumers is based on the increased

R. Silhavy (Ed.): CSOC 2019, AISC 986, pp. 50–59, 2019.
https://doi.org/10.1007/978-3-030-19813-8_6

transparency brought about by the omnipresent availability of the Internet and price search engines and comparisons. The (positive) connection between customer satisfaction and the willingness to buy is almost undisputed scientifically [1–5] and could be used as a further competitive instrument in German food retailing in addition to the focus on price. The changed customer behavior and the wider availability of information must not only benefit the customers, but retailers should also use this to their own advantage. The presented research project presents a case study of stationary retail using an existing retail system architecture based on standard software.

1.2 Customer Satisfaction

There have been various studies on customer satisfaction in retail and its key drivers. Although generalization across all retail sectors and types of distribution is not permissible, the performance criteria are weighted differently but always include the following factors: price, availability, and advice [6, 7].

The deviation between reality, real customer satisfaction, and perceived reality means that the assumptions made by the retail companies about customer satisfaction cannot deviate further. For example, a study conducted by the largest German retail association [8] showed that the gap between consumer demand and current fulfilment by retailers is large for many product categories. In many cases, as a glance at the young consumer category "Smart Natives" showed, it will become even bigger in the future if retailers do not act. The low attention and deviation in stationary retail is due, among other things, to the fact that most customer satisfaction measurement and evaluation methods known to date are based on direct customer surveys [9], which are difficult and costly to carry out operationally in stationary retail. Today, customer satisfaction in stationary retailing is mainly recorded for benchmarking and company comparisons, and it is determined by external service providers on an individual or aggregated level. Due to this approach, however, the results of these surveys and evaluations are only given to the service provider with great delay. In addition, employees in the stores are currently the only contact persons who can receive customer feedback and react to it. However, they usually do not have the necessary time—or at least this is not explicitly regulated for most retailers—to actively receive and pass on customer feedback. In most cases, the reaction to feedback from a customer depends on the situation, but a structured and analytical process is lost due to the daily workflow and effort.

In contrast, Amazon, as the omnipresent example for online retail and a major uprising competitor for stationary retail, has a strong focus on the customer, and customer feedback is collected and analyzed at every possible point of customer interaction [10]. Besides the approach of collecting customer feedback and sentiment data, the retail giant also relies on a large knowledge base and capability to analyze this data. As the line between online and offline retailing is becoming blurred due to the implementation of omnichannel scenarios, this might be a major disadvantage for traditional retailers.

1.3 Increased Technology Adoption and Omnichannel Allows Broad Access to Data

The technological changes relevant for the retail sector are mainly driven by the broad adoption of the Internet and new technologies, such as smart mobile devices (handhelds, smartphones, and tablets) and corresponding software applications (mobile apps, mobile payments, e-marketing, or location-based services). This increased adoption of mobile devices and the ubiquitous access to the Internet does not only increase the transparency over prices and information for customers but also enables retail companies to connect with their customers. Besides providing the possibility for stationary retailers to individually identify their customers, these technological advances can be used as a data source to collect customer sentiment-related data. Social media, mobile applications, and the Internet of Things (IoT) enable the collection of relevant data.

In recent years, the number of participants and, at the same time, the intensity of use of **social media** has increased significantly. More than 500 million tweets (6,000 per second) are created every day. This mass adaptation is an opportunity for marketing and consumer research. There is now a whole field of research called "Social Media Analytics". In addition to data volume and diversity, the velocity, known as the three "Vs", characterizes the term "Big Data". In recent years, Big Data Analytics, which until now has functioned predominantly through offline data processing, has also expanded to include batch processing, online processing, and streaming. In general, the control of social media networks is impossible and therefore, the best strategy is to use the networks for one's own benefit [11]. There are two main strategies: active use as a marketing tool or passive use as a data source. This data can be used for a wide variety of purposes: alerting, text analysis, sentiment analysis, social network analysis, trend analysis, attribute analysis, association analysis, and text mining. If these use cases are executed by Big Data analyses in real time, they are referred to as Streaming Analytics.

With the rise of **omnichannel retailing, mobile devices** are taking a key role in all available channels [12]. Not only are personal mobile devices, smartphones, and tablets relevant but handheld-based shopping assistants and self-scanners or self-checkout devices are also considerable sources of customer data. Combined personal and non-personal mobile devices cover all parts of human interactions [13] in retailing. Their possible uses are broad: shopping list applications, shopping assistant, navigation, payments or self-checkout. All of these possible use-cases allow the integration of the collection of active and passive customer sentiments and feedback data.

Fig. 1. Example of a hardware-based feedback system by HappyOrNot [14]

Another possible way of collecting the needed data is through the use and integration of **IoT** and hardware at the **point-of-sale (POS)**. Hardware-based feedback systems include modified POS-hardware that allows each customer to provide instant feedback in at the checkout. Also, specialized hardware with the single use-case of collecting feedback (see Fig. 1) can be applied.

2 Methods

2.1 Customer Satisfaction Streaming Index (CSSI)

In empirical marketing and behavior research, a large number of different definitions and models of customer satisfaction have developed over time. However, there is an abstract agreement on the fact that customer satisfaction is not a directly observable and hypothetical construct. A simple explanatory approach in the sense of the basic model, that can be derived without straining deeper theoretical and psychological backgrounds at this point is the "confirmation/disconfirmation paradigm" [15–17]. In this paradigm, customer satisfaction is the result of a psychological construct based on a comparison between actual experience with a service or product (perceived performance) and a specific standard of comparison (expectations). If the comparison corresponds to the perceived performance with the expected target performance, there is confirmation. Depending on the operationalization of this construct, an evaluation at the level of a specific situation, branch, assortment, product, or the entire shopping experience is made possible. Based on this scientifically accepted paradigm [18], an evaluation of customer satisfaction in retail can be developed using data from social media, IoT sources, and mobile devices (Fig. 2).

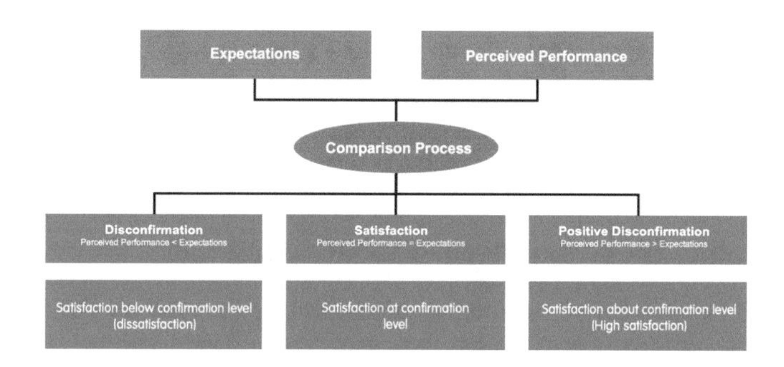

Fig. 2. Schematic representation of the confirmation/disconfirmation paradigm

There are already various methods and procedures that are used in retail to determine customer satisfaction, the evaluation of stores, or the price image [19], but they involve operationalized questionnaires or direct customer surveys, which cannot be mapped in real-time. Since no specific questions can be asked but depend on the active input of customers, neither a standardization nor an operationalization in the sense of the classical questioning instruments is given. Therefore, only a basic tendency can be derived, and the developed index orients itself at the well-known procedure of the customer satisfaction index. Analytically, all indices work under the same basic structure: using a set (n) of evaluation characteristics (k) for a set (m) of asked customers (i), the weighted ($W_{k,\ i}$) average value of the single evaluations (Z) is formed. In order to do justice to the characteristics of the source medium, the importance levels of the individual rating points ($W_{k,\ i}$) are weighted over time using the adjusted exponential function of the time difference. Timeliness is a decisive variable in the short-lived social media environment, and therefore older contributions have lower weightings.

$$\mathrm{CSSI} = \sum_{k-m} \sum_{i=1-n} \left(\mathrm{Z}_{\mathrm{k,i}} * \left(\mathrm{W}_{\mathrm{k,i}} * \frac{1}{\mathrm{e}^{t_\mathrm{n} - t_{n-1}}}\right)\right)$$

n = Number of evaluation characteristics
m = Number of customers
Z = Satisfaction value per evaluation point and customer
W = Importance per evaluation point and customer
k = Evaluation characteristic
i = Customer
t_n= Evaluation date
t_{n-1}= Time of entry of the evaluation

Fig. 3. Customer Satisfaction Streaming Index (CSSI)

This procedure makes it possible to evaluate a current result with maximum importance (e.g., a branch that needs a clean-up), but to adjust it over time based on its decreasing importance (branch was already cleaned, but the negative impression remains). A normalization is carried out on a few variables analogous to the variables of the sentiment analysis ("strong positive" to "strong negative"). The CSSI (see Fig. 3) contains a geographical delimitation and is not surveyed in a comprehensive manner but rather per branch and geographical region.

3 Results

3.1 Prototypical Implementation in a Retail Architecture

Some events and therefore their data representation, can be characterized as an "endless" stream of parallel or consecutive instances. To perform batch processing on this type of data, the data must be stored, the collection stopped, and the data processed. This process is then restarted for the next batch run. In contrast, stream processing is performed on data streams as an uninterrupted chain of events in near real-time. Over the years, numerous streaming platforms have been developed that follow the stream processing paradigm. This is based on the goal of querying a continuous data stream and recognizing conditions within a small period of time. These platforms are all suitable for the objective of determining the customer sentiment index described above. However, the information obtained in this way is only of limited use if it is considered singularly. Much more decisive than the pure key figure itself is the cause-and-effect relationship that leads to it. For this reason, cause and effect analysis in the BI environment is usually the downstream and still completely manual follow-up process for all activities. Only with a possible cause link that is domain- and application-specific does every type of real-time analysis make sense. In the case described, a decreasing CSSI cannot be used for a reaction without possible reasons (see Fig. 4).

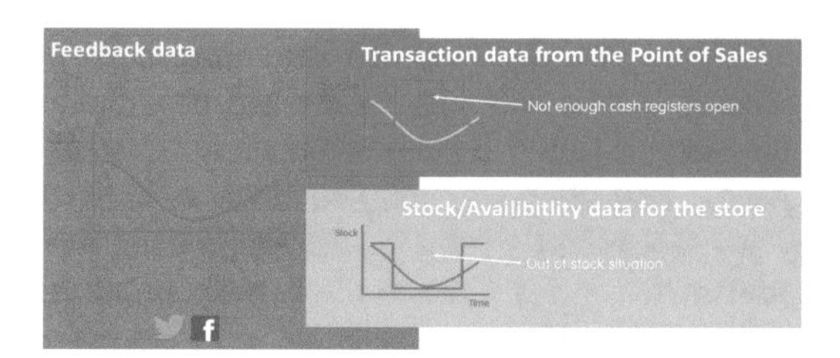

Fig. 4. CSSI put into the business context

Therefore, integration into the existing system architecture in retail is required. In order to map the above-mentioned decisive influencing factors of customer satisfaction in retail, product availability, and price, integration into the existing merchandise management processes and systems is required. The existing system architectures today are still technologically and architecturally designed for batch processes [20] and are a challenge not to be underestimated due to the enormous volume, for example, of POS data at peak times in food retailing with several cash registers in several thousand branches.

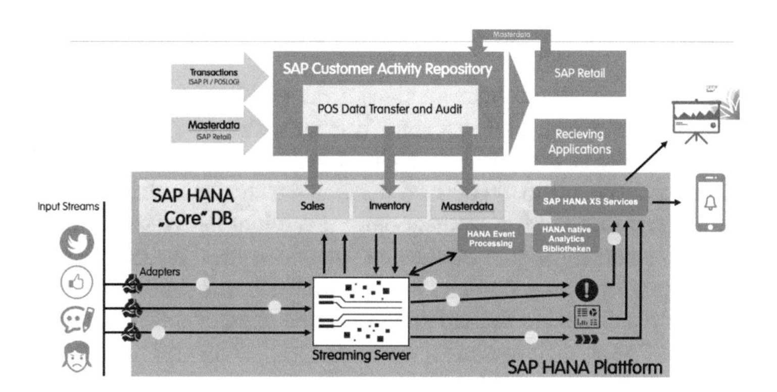

Fig. 5. System architecture based on the reference architecture for the retail sector

Implementation in a system architecture must be appropriate to the existing requirements and conditions of the retail trade. The standard software provider SAP lists "The SAP Model Company for Omnichannel Retail" [21] as a reference architecture for retail companies. The SAP Customer Activity Repository offers harmonized data models and functions for all multichannel transaction data in real-time for internal use or for consuming applications. SAP HANA is the subordinate in-memory database management system [22] that is used for all applications. It serves primarily as a database server and was also designed to allow extended analyses and contains an application server. The technical implementation takes place in the mentioned system architecture (Fig. 5). At the technical level, streaming analytics have been implemented with the use and adaptation of "Smart Data Integration" on the SAP Streaming Server integrated into the platform. In a first step, the incoming social media contributions, tweets, are sorted according to suitable keywords and geographical assignment to a branch. The tweets are then subjected to a sentiment analysis, which delivers the results of the sentiment, events, organizations, and situations mentioned. In the case of a decisive event (very negative mood, wrong articles, importance of the originator, etc.), this information is transmitted in real-time to the corresponding recipients via a push notification. All events also flow into the calculation of the branch-specific CSSI-index. A more in-depth analysis is possible, analogous to a drilldown. In this way, each event is enriched with appropriate in-depth information (out-of-stock, price changes or duration of the cashing processes at the time of the event) from the SAP CAR.

3.2 Preliminary Results

This case study involves one of Germany's largest retailers. The German retail sector is characterized by an oligopolistic market with strong intra-competition between existing retailers and rising inter-competition between traditional and new "pure" digital players [23]. With the withdrawal of Kaiser's Tengelmann, one of the last major independent retail chains, in 2016 and the looming market entry of Amazon with Amazon Fresh, this competition has intensified further. On top of that, a waning scope for differentiation between operating types [24], increased costs, an overall increased price awareness [25], and influence of the company's price image on the customer's choice for a retail chain can be observed.

The retail group examined here operates a range of different distribution chains, hypermarkets, supermarkets, and discount formats with wide assortment widths and depths: 8000 products at the discount chain, around 35,000 products in the regular supermarket chain, and up to 80,000 products at the hypermarket chain. Moreover, each store serves up to 10,000 customers daily. An initial test of the above-mentioned prototypical implementation was conducted, in which around 250.000 tweets were analyzed over a period of 11 months in the year of 2017. The sentiment analysis within the SAP HANA database and the streaming service examined 70% of the assessable messages. Overall, 32% of all messages were rated as "strong positive", 20% as "strong negative", 12% as "weak negative", 3% as "minor problem", another 3% as "major problem", and the rest as a "weak negative" or "weak positive". The final set of messages were assigned a rating of "neutral". Based on this data, the CSSI was calculated and provided to the business users. The resulting information about customer satisfaction can be accessed. This can be done via a dashboard (see Fig. 6) or a direct push notification for the end user. Depending on the user group, there are different possible applications. In the branch itself, the alert function is the main point of interaction for the branch management. However, access to the dashboard can also be used at any time to obtain regular evaluations or to obtain more detailed information in the event of a warning. At a higher organizational level, it is possible to compare the different branches with each other.

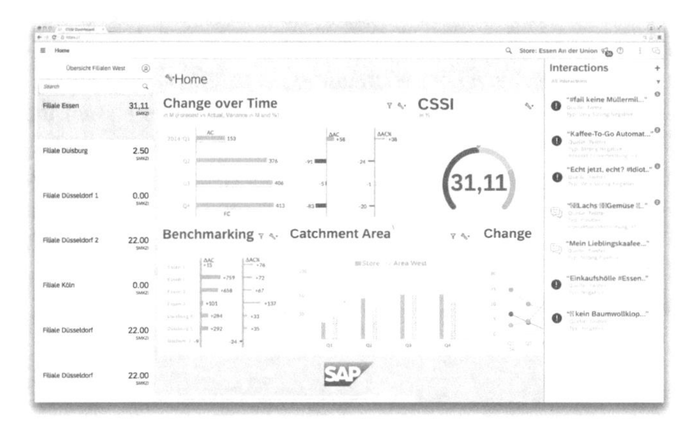

Fig. 6. Dashboard with CSSI in branch view

4 Discussion

The prototypical implementation of the newly formulated Customer Satisfaction Streaming Index (CSSI) is a general analytical approach to deal with customer feedback. The presented approach is a suitable way for unaliasing the input generated from customers on different social media sources. The SMCSI, as a general benchmark for assessing the customer, works very well and may provide the possibility to close the information gap between online and offline retailers, at least partly. The gathered data can also be seen as a step to counteract the current low penetration of Artificial Intelligence and Machine Learning within the domain of retailing [26]. An integration to Machine Learning models could presents a suitable use-case to enhance marketing atomization as an example. The idea of a direct notification to the sales personnel is partially hard to handle. Firstly, the required technical equipment is not available. Secondly, the distribution of mobile devices to sales personnel in retailing has been traditionally low or non-existent at the case study company. Reviewing the initial processed data, it seems likely that the content and frequency tweets would allow real-time monitoring for single store locations, but the overall general sentiment is very well pictured. However, due to the low usage of tweets with a location tag in Germany, matching to a unique store is a huge challenge. A text analysis based on machine learning could help here, but it is not clear whether this process is feasible. Another drawback is the fact that Facebook restricted the access to a wide range of search APIs after the so-called "Cambridge Analytica scandal". The much larger market share of Facebook over Twitter in the German market presents a major challenge to data availability for this project. This leaves the use of instore IoT devices as the only option for collecting customer feedback. An empirical study needs to be conducted in a second research step. A simple button at the checkout would be the easiest way to gather the relevant data.

With the presented adaptation of the well-known metrics for measuring customer satisfaction to the CSSI and the integration of the existing information from the transaction and master data systems in retail, the retailer only has the option of using the CSSI as a KPI for operational control. It enables reactions to events to occur in real-time and can be used to reveal basic trends over time. Benchmarking between the different branches is also possible. Initial practical tests have shown that KPI calculations work well, but an empirical test of the proposed real-time alerting needs to be done to uncover a number of previously undiscovered problems. How the CSSI developed here behaves in relation to a proven method of marketing and behavioral research still has to be tested with empirical studies. What is certain, however, is that the CSSI and the above implementations for the stationary retail trade offer a unique opportunity to respond to customer feedback in real-time.

References

1. Hallowell, R.: The relationships of customer satisfaction, customer loyalty, and profitability: an empirical study. Int. J. Serv. Ind. Manag. **7**(4), 27–42 (1996)
2. Homburg, C., Koschate, N., Hoyer, W.D.: Do satisfied customers really pay more? A study of the relationship between customer satisfaction and willingness to pay. J. Mark. **69**(2), 84–96 (2005)
3. Francioni, B., Savelli, E., Cioppi, M.: Store satisfaction and store loyalty: the moderating role of store atmosphere. J. Retail. Consum. Serv. **43**, 333–341 (2018)
4. Kumar, V., Anand, A., Song, H.: Future of retailer profitability: an organizing framework. J. Retail. **93**(1), 96–119 (2017)
5. Anderson, E.W.: Customer satisfaction and price tolerance. Mark. Lett. **7**(3), 265–274 (1996)
6. Renker, C., Maiwald, F.: Vorteilsstrategien des stationären Einzelhandels im Wettbewerb mit dem Online-Handel. In: Binckebanck, L., Elste, R. (eds.) Digitalisierung im Vertrieb: Strategien zum Einsatz neuer Technologien in Vertriebsorganisationen, pp. 85–104, Springer Fachmedien Wiesbaden, Wiesbaden (2016)
7. Fleer, J.: Kundenzufriedenheit und Kundenloyalität in Multikanalsystemen des Einzelhandels: Eine kaufprozessphasenübergreifende Untersuchung. Springer, Wiesbaden (2016)
8. IFH. Catch me if you can - Wie der stationäre Handel seine Kunden einfangen kann (2017). https://www.cisco.com/c/dam/m/digital/de_emear/1260500/IFH_Kurzstudie_EH_digital_Web.pdf. Accessed 23 July 2018
9. Töpfer, A.: Konzeptionelle Grundlagen und Messkonzepte für den Kundenzufriedenheitsindex (KZI/ CSI) und den Kundenbindungsindex (KBI/ CRI). In: Töpfer, A (ed.) Handbuch Kundenmanagement: Anforderungen Prozesse, Zufriedenheit, Bindung und Wert von Kunden, pp. 309–382. Springer, Berlin (2008)
10. Anders, G.: Inside Amazon's Idea Machine: How Bezos Decodes Customers (2012). https://www.forbes.com/sites/georgeanders/2012/04/04/inside-amazon/#73807ee56199. Accessed 20 May 2018
11. Constantinides, E., Romero, C.L., Boria, M.A.G.: Social Media: A New Frontier for Retailers?. In: European Retail Research, pp. 1–28 (2008)

12. Piotrowicz, W., Cuthbertson, R.: Introduction to the special issue information technology in retail: toward omnichannel retailing. Int. J. Electron. Commer. **18**(4), 5–16 (2014)
13. Evanschitzky, H., et al.: Consumer trial, continuous use, and economic benefits of a retail service innovation: the case of the personal shopping assistant. J. Prod. Innov. Manag. **32**(3), 459–475 (2015)
14. HappyOrNot. Case Study: Elkjøp Leading Consumer Electronics Industry with Excellent Customer Service (2013). https://www.happy-or-not.com/en/case-studies/elkjop-leading-consumer-electronics-industry-with-excellent-customer-service/. Accessed 22 Nov 2017
15. Oliver, R.L.: Effect of expectation and disconfirmation on postexposure product evaluations: an alternative interpretation. J. Appl. Psychol. **62**(4), 480 (1977)
16. Bösener, K.: Kundenzufriedenheit, Kundenbegeisterung und Kundenpreisverhalten: Empirische Studien zur Untersuchung der Wirkungszusammenhänge. Springer, Berlin (2014)
17. Simon, A., et al.: Safety and usability evaluation of a web-based insulin self-titration system for patients with type 2 diabetes mellitus. Artif. Intell. Med. **59**(1), 23–31 (2013)
18. Fornell, C., et al.: The American customer satisfaction index: nature, purpose, and findings. J. Mark. **60**(4), 7–18 (1996)
19. Becker, J., Schütte, R.: Handelsinformationssysteme domänenorientierte Einführung in die Wirtschaftsinformatik, 2nd edn. Redline-Wirtschaft, Frankfurt/M (2004)
20. Schütte, R.: Analyse des Einsatzpotenzials von In-Memory-Technologien in Handelsinformationssystemen. In: IMDM (2011)
21. Woesner, I.: Retail Omnichannel Commerce – Model Company (2016). https://blogs.sap.com/2016/12/22/retail-omnichannel-commerce-model-company/. Accessed 01 July 2017
22. Plattner, H., Leukert, B.: The In-Memory Revolution: How SAP HANA Enables Business of the Future. Springer, Berlin (2015)
23. Schütte, R., Vetter, T.: Analyse des Digitalisierungspotentials von Handelsunternehmen. In: Handel 4.0. Springer, Berlin, pp. 75–113 (2017)
24. Meffert, H., Burmann, C., Kirchgeorg, M.: Marketing: Grundlagen marktorientierter Unternehmensführung Konzepte - Instrumente - Praxisbeispiele, 12th edn, pp. 357–768. Springer Fachmedien Wiesbaden, Wiesbaden (2015)
25. Daurer, S., Molitor, D., Spann, M.: Digitalisierung und Konvergenz von Online-und Offline-Welt. Zeitschrift für Betriebswirtschaft **82**(4), 3–23 (2012)
26. Weber, F., Schütte, R.: A domain-oriented analysis of the impact of machine learning—the case of retailing. Big Data Cogn. Comput. **3**(1), 11 (2019)

An Efficient Mechanism to Improve the Complexity and System Performance in OFDM Using Switched Beam Smart Antenna (SSA)

T. G. Shivapanchakshari[1,2(✉)] and H. S. Aravinda[3]

[1] Department of ECE, Cambridge Institute of Technology, Bangalore, India
tgsresearch2013@gmail.com

[2] Visvesvaraya Technological University, Belagavi, India

[3] Department of ECE, JSSATE, Bangalore, India

Abstract. The recent past has witnessed the requirement of high-speed data communication in wireless network. In fulfilling these needs various modulation techniques like Phase-Shift Keying (PSK), Frequency-Shift Keying (FSK), Amplitude-Shift Keying (ASK), Quadrature Amplitude Modulation (QAM), Frequency-Division Multiplexing (FDM) etc. The Orthogonal Frequency Division Multiplex (OFDM) is a modulation technique and significant air link technologies to achieve future wireless communications. In order to bring the efficiency in bandwidth, system performance, smart antenna and adaptive resource allocation techniques are significantly implemented in OFDM. But, the utilization of fully adaptive beam forming in an OFDM system enhances the complexity of medium access control layer design and which affect the implementation of adaptive resource allocation (ARA). This paper introduces an efficient mechanism of Cross-layer ARA (CARA) with Hybrid Adaptive Array (HAA) and Switched-beam Smart Antenna (SSA) for OFDM systems. By using various smart antenna (SA) mechanisms based on requirements of different Quality of Service (QoS), the proposed system significantly minimizes the complexity of ARA in OFDM system and maintains the better system performance.

Keywords: Complexity · Modulation techniques · Orthogonal Frequency Division Multiplex (OFDM) · Quality of Service (QoS) · Smart antennas · Wireless communication

1 Introduction

The futuristic wireless communication needs an efficient way of data or information transmission with higher transmission rate and bandwidth efficiency. Due to existence of limited channel bandwidth (LCB), the adaptive resource allocation (ARA) has becomes very important. In order to improve the performance of the overall system, an integrated design of Media Access Control (MAC) and physical layer (PL) is most suitable. The technique in current days called orthogonal frequency division multiplex (OFDM) has become a most significant air-link technique for the futuristic wireless

R. Silhavy (Ed.): CSOC 2019, AISC 986, pp. 60–66, 2019.
https://doi.org/10.1007/978-3-030-19813-8_7

communications in different application categories [1]. The application of OFDM is widespread in today's world for digital media broadcasting applications. Also, the OFDM has been a primary PL architecture in IEEE802.11a/g, HIPERLAN/2 and recent Long-Term Evolutions (LTE) like 3GPP and 3GPP2 air-interface evolution [2]. The main functionality of the OFDM is that it splits the high rated data stream as parts of low rated parallel data sub streams and will be processed simultaneously over various sub-carriers.

Traditionally, the mobile communication systems use the Smart Antenna (SA) techniques to overcome the challenges of the LCB and help to fulfill the upcoming demands of the wireless communications [3]. The intelligent utilization of the SA can provide the improvement in the system performance with enhancement in the spectrum efficiency and channel capacity and also with extension in the coverage range. The SAs can be of two categories [4] like adaptive array SA (AASA) and Switched-beam Smart Antenna (SSA). The AASA can track every user within the given cell with a separate adaptive beam pattern while the SSA chooses a beam pattern among all the patterns for every user based on the user location [4]. The AASA yields the better performance than the SSA but the implementation of AASA is very complex and which need to be relying on the exact Channel Estimation (CE) and beam forming algorithms convergence speed [5, 6]. Thus, the application of AASA leads to biggest challenge in implementation of MAC layer.

This paper efficient mechanism of Cross-layer ARA (CARA) with Hybrid Adaptive Array (HAA) and Switched-beam Smart Antenna (SSA) for OFDM systems is presented. The proposed hybrid SA (HSA) system integrates both the AASA and SSA to attain the higher improvement in the performance than existing system. In this paper, various SA schemes are used as per the Quality of Service (QoS) requirement of the different users. The prime intension of this paper is to explore the significances of SAs over OFDM system when ARA is subjected to minimize the complexity of ARA.

2 Problem Statement

The use of SAs will enhance the throughput of the OFDM system, which offers reuse of the same subcarriers over different beams for spatially separated users [7]. The beam forming process can enhance the signal-to-interference-plus-noise ratio (SINR) of users which unnecessarily utilizes the higher modulation schemes. In OFDM system without ARA, the fully AASA for every user can provide the significant performance with higher implementation cost than SSA [8]. In case the ARA is applied, the implementation of AASA at base station (BS) needs to cover every user which is not a best idea because of following reasons:

The beam forming process of AASA relies on training symbols or direction of arrival (DOA). The significances of the DOA estimation or training symbols put lots of load over the PL and MAC layers. Most of the users in a cell are mainly of low data rate (LDR) or voice users [9]. Even though the LDR users holds only one or two subcarriers but when the system is fully loaded and the users of LDR becomes higher than the need of separate adaptive beam forming at every user become very high because the implementation of the process must be performed separately for different users [10]. Thus, for the support of AASA in OFDM system with huge number of users, the ARA implementation may lead to complicated optimization issue. Under this condition, the system can become unstable means the adaptive algorithm cannot attain the convergence [2, 11].

Thus, there is a need of suboptimal and realistic SA scheme for its implementation in OFDM with the application of ARA. The following section describes the design and implementation of the proposed system.

3 Proposed System

The architectural diagram of the proposed system is represented in Fig. 1. The proposed system is named as "efficient mechanism of Cross-layer ARA (CARA) with Hybrid Adaptive Array (HAA) and Switched-beam Smart Antenna (SSA) for OFDM systems".

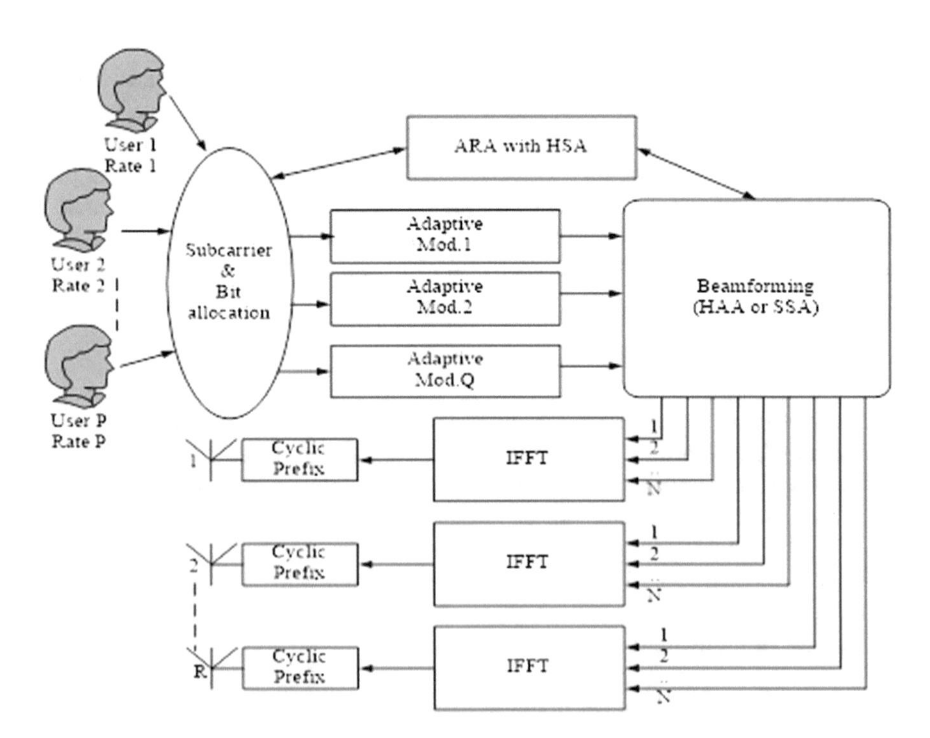

Fig. 1. Implementation diagram of proposed system

The proposed system various processing units, first the different users serial bit streams will be subjected to the subcarriers and bit allocation unit (BAU). In this unit, different user's bits are loaded over different subcarriers. Then, the data stream outputs of the BAU are given to the Q-adaptive modulator and appropriate modulation (appropriate mod.) levels are selected for different subcarriers. Later, the adaptive modulators output is forwarded to the beam forming module where the SA schemes can be chosen for different needs of users QoS. To perform the real video data transmission of high data rate users demanding the cluster of subcarriers, AASA can be utilized. In order to process the low data rate users like voice users this consumes very rare subcarriers. The module of ARA consisting of HSA helps to control the entire processing. Once the data is transformed as time domain samples by inverse fast Fourier transform (IFFT) and later the Cyclic Prefix (CP) will be added, the streams will be forwarded through antennas (R).

The Pseudo code implementation for the proposed system is presented in following algorithm:

```
Pseudo code for OFDM system performance enhancement pro-
cess
Start:
1: Generate →Pk
2: Cr & Ba ←perform
3: Apply→ Am
4: Perform→Bf
5: Apply→ Chc
6: Perform→ODm
7: Sig←Extract
8: Apply→ADm
9: Receive←Pk
10: Analyze ← Results
End
```

The collected data from the different users is used to generate the packets (Pk) and then performed the carrier (Cr) and Bit Allocation (Ba). Later, the obtained output is subjected to adaptive modulation (Am) and then Beam forming (Bf) is performed. Then the Channel condition (Chc) is applied. The OFDM De-modulation (ODm) is performed at receiver side to obtain the generated Pk through Signal (Sig) extraction and Adaptive De-modulation (ADm). Finally, the Voice User data traffic (VUdt) packet degradation ratio, fixed bit rate (FBR) degradation ratio and Average Delay per data traffic (ADdt) is extracted and analyzed for computation of Total packets Transmitted (TpT). The following section discusses the obtained results analysis.

4 Results Analysis

The entire OFDM system is designed by using MATLAB and selected the III beams and VI beams antennas. For the performance analysis the proposed method used three different SAs like Hybrid SA, Perfect SA and Switched beam SA and the numbers of users are set randomly between 35 to 210. The following Figures give the outcomes of the proposed system. The Figs. 2, 3 and 4 represents the respective plots of the Voice User data traffic (VUdt) packet degradation ratio, fixed bit rate (FBR) degradation ratio and Average Delay per data traffic (ADdt) and Total packets Transmitted (TpT) corresponding to Number of Users (Nu). By observing the Figs. 2 and 4 found that the perfect SA provides te better performance than hybrid SA and Switched beam SA for both the VUdt and ADdt terminals. This happens due to the implementation of the Switched beam SA in Hybrid SA and Switched beam SA schemes for VUdt and ADdt terminals while the perfect SA considered precisely towards the VUdt and ADdt terminals. However, the AASA are subjected to the FBR terminals in Hybrid SA, the FBR packet degradation ratio of the hybrid SA is lower than the Switched beam SA scheme and is observed in Fig. 3.

It is also examined that the Hybrid SAs FBR packet degradation ratio almost at the level of Perfect SA scheme. Also found that every FBR terminal accesses more subcarrier than the VUdt terminal and hence the FBR holds the significant role if the TpT maximization is considered. Also, the FBR terminal can utilize the modulation (high order) schemes in both the hybrid SA and Perfect SA scheme where AASA are implemented for FBR terminals. The Fig. 5, indicated the TpT of hybrid SA nearly closer to Perfect SA. In both the Figs. 2 and 4, the VUdt and ADdt increases with increase in Nu. Because with less Nu the user can choose the significant subcarriers and beams while at high Nu the user can choose the suboptimal subcariers and beams for satisfactory QoS. Under the condition with the limited subcarriers and beams the TpT converges to desired upbound if the Nu is increases.

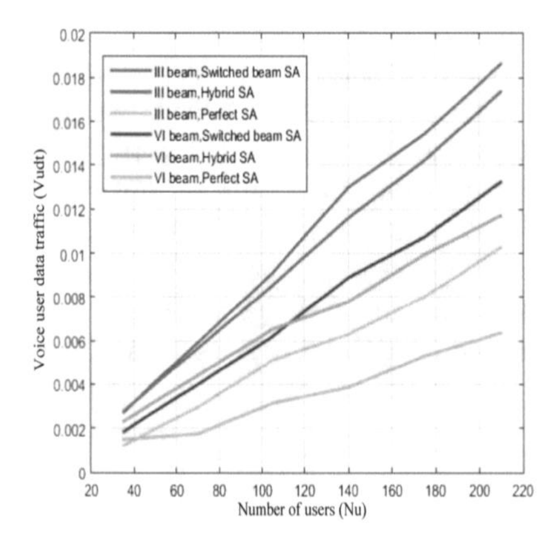

Fig. 2. Plot of VUdt vs Nu

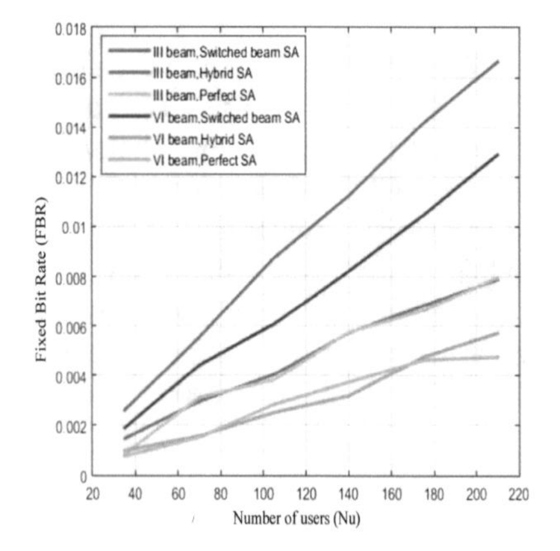

Fig. 3. Plot of FBR vs Nu

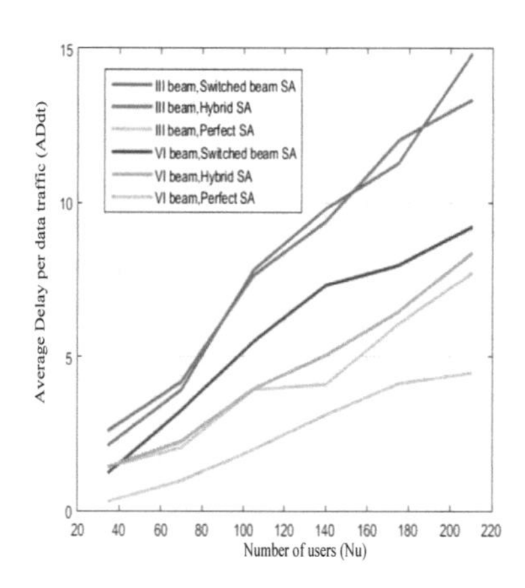

Fig. 4. Plot of ADdt vs Nu

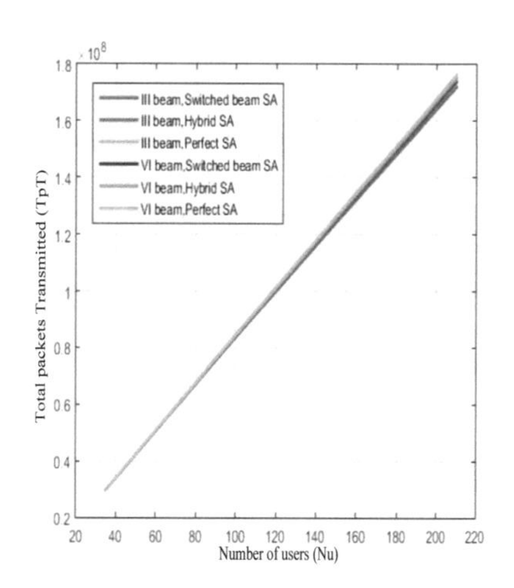

Fig. 5. Plot of TpT vs Nu

From all the outcomes it is been observed that the VI beam case outforms than of III beam case because the VI beam offers high spatial channels for users than III beam. The scheme complexity is also an major concern, in AASA the beamforming computational complexity and tracking algorithms are of about 0(M2) in which M is the training sequence length. The tracking process of AASA can change the interference scenario to all the co-channel users and hence complexity increases in the AASA implementation. The implementation of switched beam SA is very easy where the system can scan the beam outputs and selects the significant beam for transmission. The proposed Hybrid SA scheme exhibits low complexity than the Perfect SA scheme because the Nu of high data rate are very small. Also, hybrid SA exhibit the high throughput with better TpT than Perfect SA and Switched beam SA. The final outcomes measured for VUdt, FBR, ADdt and TpT are 6.56×10^{-3}, 5.13×10^{-3}, 6.56×10^{-3} and $1.76 \times 10^{+8}$ respectively. Hence the Proposed system gives better solution of Performance enhancement and system complexity.

5 Conclusion

This paper presented an efficient work towards fulfilling the recent requirement of high-speed data communication in wireless network. The proposed system addressed the efficient mechanism of Cross-layer ARA (CARA) with Hybrid Adaptive Array (HAA) and Switched-beam Smart Antenna (SSA) for OFDM systems is presented. The proposed hybrid SA (HSA) system integrates both the AASA and SSA to attain the higher improvement in the performance than existing system. In this paper, various SA schemes are used as per the Quality of Service (QoS) requirement of the different users. The perfect SA, switched beam SA and Hybrid SA are considered with III beam or VI beam antennas with adaptive resource allocation (ARA). The Nu for the system is randomly selected from 35 to 210. The outcomes (Voice User data traffic (VUdt) packet degradation ratio, fixed bit rate (FBR) degradation ratio and Average Delay per data traffic (ADdt) and Total packets Transmitted (TpT)) measured corresponding to Nu found more significant. From the above outcomes it can be concluded that the proposed system offers the better performance with reasonable complexity level.

Further the proposed system can be considered for futuristic improvement in the OFDM system performance with other smart antennas under different user traffic scenario. Even the proposed system can be implemented for the security concerns of OFDM system and performance analysis with other metrics.

References

1. Rohling, H., May, T.: Comparison of PSK and DPSK modulation in a coded OFDM system. In: 1997 IEEE 47th Vehicular Technology Conference, vol. 2. IEEE (1997)
2. Hu, H., Guo, K., Weckerle, M.: Hybrid smart antennas for OFDM systems-a cross-layer approach. In: 2005 IEEE 16th International Symposium on Personal, Indoor and Mobile Radio Communications, PIMRC 2005, vol. 4. IEEE (2005)
3. Balanis, C.A., Ioannides, P.I.: Introduction to smart antennas. Synth. Lect. Antennas **2**(1), 1–175 (2007)
4. Alexiou, A., Haardt, M.: Smart antenna technologies for future wireless systems: trends and challenges. IEEE Commun. Mag. **42**(9), 90–97 (2004)
5. Shivapanchakshari, T.G., Aravinda, H.S.: Review of research techniques to improve system performance of smart antenna. Open J. Antennas Propag. **5**(02), 83 (2017)
6. Rezk, M., et al.: Performance comparison of a novel hybrid smart antenna system versus the fully adaptive and switched beam antenna arrays. IEEE Antennas Wirel. Propag. Lett. **4**(1), 285–288 (2005)
7. Teja, K.R., Chopra, S.R.: Review of massive MIMO, filter bank multi carrier and orthogonal frequency division multiplexing (2017)
8. Hadzi-Velkov, Z., Gavrilovska, L.: Performance of the IEEE802.11 wireless LAN under influence of hidden terminals and Pareto distributed packet traffic. In: Proceedings of the IEEE ICPWC 1999, pp. 221–225, February 1999
9. Jakes, W.C.: Microwave Mobile Communications, 2nd edn. IEEE Press, New York (1993)
10. Hu, H.L., Zhu, J.K.: Performance analysis of distributed-antenna communication systems using beam-hopping under strong directional interference. Wirel. Pers. Commun. **32**(1), 89–105 (2005)
11. Razavilar, J., Rashid-Farrokhi, F., Liu, K.J.R.: Software radio architecture with smart antennas: a tutorial on algorithms and complexity. IEEE J. Sel. Areas Commun. **17**(4), 662–676 (1999)

Modeling the Engineering Process as a Thinging Machine: A Case Study of Chip Manufacturing

Sabah Al-Fedaghi(✉) and Aya Hassouneh

Computer Engineering Department, Kuwait University,
P.O. Box 5969, 13060 Safat, Kuwait
sabah.alfedaghi@ku.edu.kw,
ayah.hasoneh@grad.ku.edu.kw

Abstract. Engineering processes often involve serious potential consequences, such as damage to the environment and injury to people. Modeling such processes is aimed at understanding internal workings by prescribing procedures to be followed and making predictions about the project. Models are typically represented using diagrams (e.g., flowcharts, manual schematics UML diagrams, circuits, and technical sketches). In these diagrams, technical notions are mixed with implementation issues, which leads to fragmental representation and difficulties in creating appropriate documentation. A conceptual framework suitable for managing engineering processes is currently lacking. Without loss of generality, we focus on the VLSI manufacturing process to exemplify these processes and propose a new technique called a thinging machine (TM) to develop conceptual modeling in this context. We demonstrate the ability of a TM to provide an environment for managing the manufacturing of integrated circuits and supporting high-level control policies.

Keywords: Engineering process · VLSI manufacturing · Conceptual model · Diagrammatic representation · Management project

1 Introduction

Engineering processes often involve serious potential consequences, such as damage to the environment and injury to people. A process model is aimed at understanding the internal workings of an engineering project by prescribing procedures to be followed and making predictions about project outcomes. It is also aimed at providing an overview of the mechanisms used in system operations for decision making and control. This paper builds a *conceptual model* in the domain of *engineering systems*. The term *model* refers to an abstract representation of some part of the real world, developed as a means of communication among stakeholders when a complex system is built. Desirable features in such models include the ability to capture relevant and significant aspects of a real phenomenon concisely, with understandability and complete specification of activities. A conceptual model uses diagrammatic notations as an abstract picture of critical concepts and their interrelationships in reality. Its purpose is to express a universal description without technological aspects and to guide the

R. Silhavy (Ed.): CSOC 2019, AISC 986, pp. 67–77, 2019.
https://doi.org/10.1007/978-3-030-19813-8_8

subsequent design phase. In this context, modeling methodologies such as Unified Modeling Language (UML) [1] and System Modeling Language (SysML) [2] have been utilized for object-oriented software design and conceptual modeling. However, further research is needed to develop a modeling apparatus that integrates constituents and permits the assemblage of applications from shared processes in diverse areas of engineering. For example, UML's origin in software engineering may limit its appropriateness for conceptual modeling [3]. Dori [4] noted that "as the inherent complexity and interdisciplinary nature of systems increases, the need for a universal modeling, engineering, and lifecycle support approach becomes ever more essential. The unnecessary complexity and software orientation of UML – the current standard language – calls for a simpler, formal, generic paradigm for systems development" [4].

Without loss of generality, we specifically concentrate on the very-large-scale-integration (VLSI) process of manufacturing integrated circuits and propose a conceptual model for it based on a new modeling technique called a thinging machine (TM). A TM is constructed from five operations (i.e., creating, processing, receiving, releasing, and transferring) performed on things (sand, blank wafers, silicon ingots, etc.).

The VLSI manufacturing process is an iterative process that involves an engineering scheme with several levels of abstraction [5]. The manufacturing scheme is critical to the cost of computer chips [6]. From specification to fabrication, it involves a sequence of processes in which an integrated circuit is created. Abstract requirements are converted into a register transfer description and then moved to circuit representation. The chip is finally shaped using geometric forms representing circuit elements and interconnections. This process requires various tools (e.g., VLSI CAD) and management apparatuses: "As a design is processed, it must be passed from tool to tool. For example, the designer may use a schematic capture tool for initial input, then they wish for their design to be minimized and finally simulated" [7]. End-to-end management of the design process is very important. According to Hodges and Rounce [7], it is too easy to make changes to a design without documenting them or without keeping backup versions, in case these changes need to be referenced at a later date. It is equally possible to accidentally delete files with no means of retrieval. Problems also arise when building up a hierarchy of blocks, with each level making use of the blocks in the levels below. The various tools often do not support this system of hierarchy directly [7].

Designers have used workflow-based methods to synchronize the execution of multiple tasks and activities in VLSI manufacturing processes [8, 9]. Designers also incorporate database systems [10] to provide an environment for data management. They use several diagrammatic techniques (e.g., a weighted directed graph to represent the flow of work, in which nodes represent tasks and edges depict the sequencing of tasks). Signal flow graphs [11] and petri-nets [12] are other examples of graphing techniques used to analyze such processes. According to Xiu [13], system-level description does not involve implementation details, since it is aimed at viewing a chip from a big-picture perspective. Full details in a system-level description become abstracted, retaining enough features to validate the embodied functions in the process.

2 Related Works

This paper formulates a conceptual approach to system modeling using the VLSI design process as an example. Heavily theoretical approaches to control systems analysis have "caused a decline in the intuitive understanding of engineering systems and the way in which they behave" [14]. The VLSI production process is analogous to the design tasks involved in building construction [15]. First, we produce an architectural drawing based on requirements and specifications. We lay out floor plans and define constraints based on the connectivity, accessibility, and size of internal spaces. We follow this step by depicting streams of flow (e.g., electrical wiring and water pipes) in the building, including the requirements, synchronization, and triggering among different flows (e.g., of supplied power and the uniformity of flow). Maintaining integrity within spaces by specifying partitions and shields is also an issue [15]. The VLSI production process similarly grows from a blueprint drawn on the basis of specifications. A designer bases VLSI "floor plans" on the connectivity/space of components, with constraints on placement of the blocks. The designer plans electrical flows in a power-grid topology that includes power requirement supplied over the topology for uniform distribution across all parts of the chip and standard cells. Then the synchronization phase involves timing and region analyses to place cells (circuits) that share data near each other, to minimize timing and reduce wire lengths.

According to Sherwani [16], the production process of a VLSI chip is a complex human-resource management project. As a result, managers partition it along functional lines, and different teams design different units. Therefore, a given unit may not be at the same level of production as another given unit; one unit may be in the logic-planning phase, while another unit may be completing its physical design phase. Designers must work with partial data at the chip level [16]. A great share of the complexity involved is related to the management of transformations of representations. An overall conceptual structure for managing representations is currently missing. Designers typically use sketches of stages and phases in the form of flowchart-like depictions, which are an example of a design style that "shows the physical design cycle with emphasis on timing" [16]. Using such flowchart-like diagrams to represent a complex process has many well-known limitations, as shown in the software development cycle.

A need exists to develop a more sophisticated documentation tool that does not require technical expertise. This is the objective of applying a TM in this paper and demonstrating its viability in the manufacturing course for integrated circuits. A TM (also called a flowthing machine) is a new diagrammatic language. In the next section, we will briefly review TM as a purely conceptual framework that specifies flows of artifacts in a process [17, 18]. It provides an environment for managing the design process and supporting high-level control policies. Managers can use the resultant conceptual picture in documentation, explanation, education, and control.

3 Thinging Machine

The aim of conceptual modeling is to obtain a description of a system and its behavior. Despite the differences in the conceptual notions used to describe systems, certain fundamental similarities make the modeling task possible. A suitable unifying concept in this context is *thinging*, which refers to defining a boundary around some portion of reality, separating it from everything else, and then labeling that portion of reality as a named thing [19]. Heidegger [20] introduced the term *thinging* to refer to the ontological the existence, presence, or being of a thing. Al-Fedaghi [17, 18] deviated from Heideggerian thought by considering a thing as an abstract machine. A TM generates and handles a thing and its constituent subthings. A TM can craft things and is defined in terms of its five functions: to create, process (change), receive, release, and transfer things, as shown in Fig. 1. The TM model additionally utilizes triggering (denoted by a dashed arrow) to establish connections with other machines. A machine that crafts (handles) things is itself a thing that is crafted.

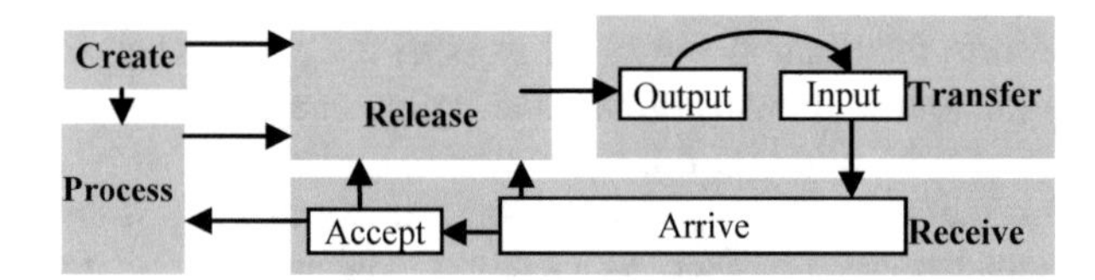

Fig. 1. Thinging machine.

Example of TM Modeling. Water is formed (created) by combining molecular hydrogen (H_2) and oxygen (O_2). This process is typically described as $2H_2 - O_2 \rightarrow 2H_2O$. A water machine processes an oxygen thing and two hydrogen things to create a water thing (Fig. 2). The dashed arrow indicates triggering, and the solid arrows indicate flows. The number of hydrogen atoms received will be discussed when introducing the notion of events.

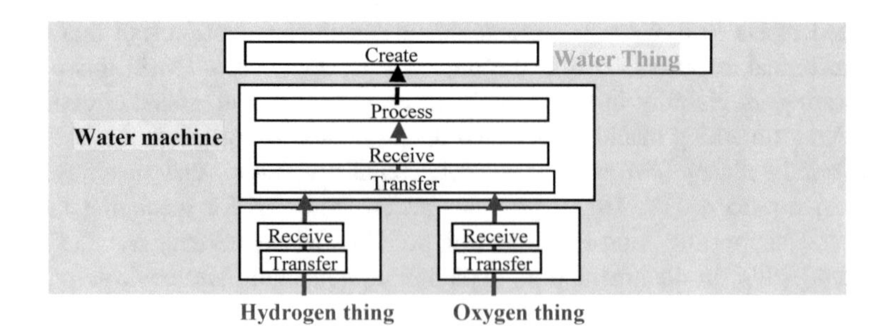

Fig. 2. A water machine processing things.

The behavior of a TM as a grand machine is specified in terms of events. An event is a thing (i.e., it can be created, processed, released, transferred, and/or received). Figure 3 shows meaningful events in the example of a water machine. Each event involves a time machine, a region machine, and the events themselves as a machine. Figure 4 shows the chronology of events that form water.

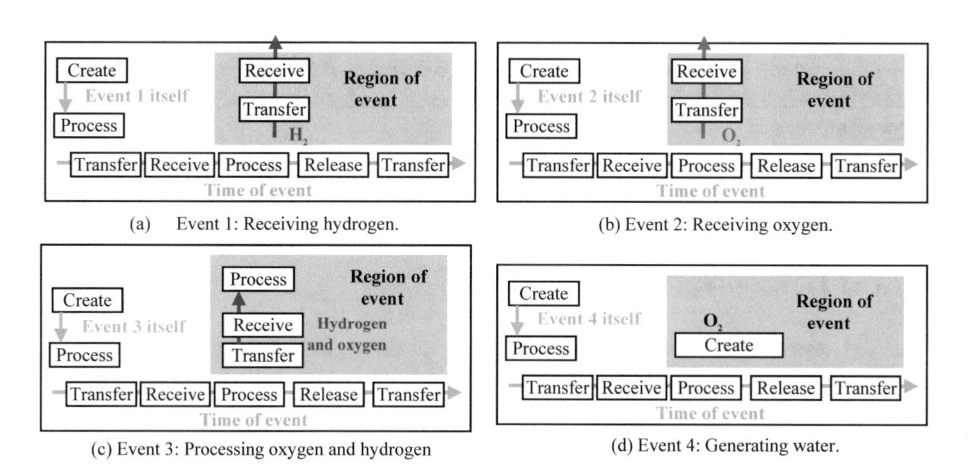

Fig. 3. Events in the water machine from Fig. 2.

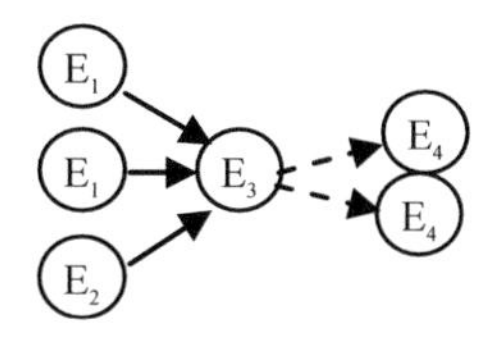

Fig. 4. The behavior of the machine in Fig. 2.

4 Modeling VLSI Manufacturing

Figure 5 partially shows the manufacturing process for integrated circuits.

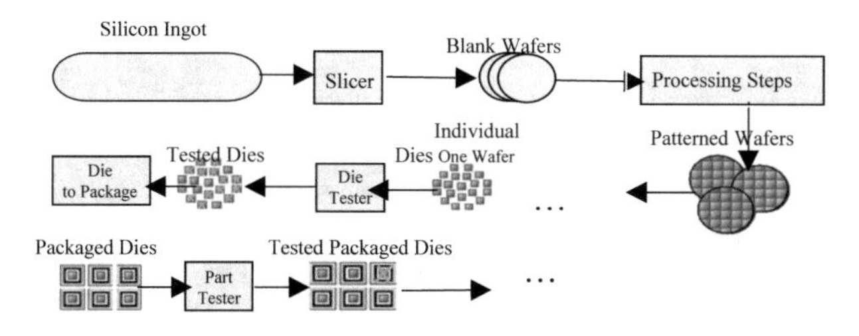

Fig. 5. The manufacturing process for integrated circuits (adapted from [6])

According to Patterson and Hennessy [6], the process starts with a silicon crystal ingot, which is sliced into wafers. The wafers go through a series of processing steps, during which patterns of chemicals are placed on each wafer, creating the transistors, conductors, and insulators. Integrated circuits contain only one layer of transistors but may have up to eight levels of metal conductors, separated by layers of insulators. The patterned wafer is then diced into chips. A wafer containing microprocessors can be diced to discard only those sections that contain flaws rather than the whole wafer. The cost of an integrated circuit rises quickly as the die size increases. Using this next-generation process shrinks a large die and uses smaller sizes for transistors and wires, to reduce the cost.

Next, as a demonstration of applying TM to the VLSI engineering process, we model Patterson and Hennessy's [6] manufacturing process from Fig. 5.

5 TM Description

The TM representation consists of diagrams for three phases of modeling (i.e., the static, dynamic, and control stages), as will be described in the following three sections.

5.1 Static Specification

As shown in Fig. 6, the chip-manufacturing process starts with the arrival of sand (circle 1) to be processed (2), which triggers the creation of a silicon crystal ingot (3).

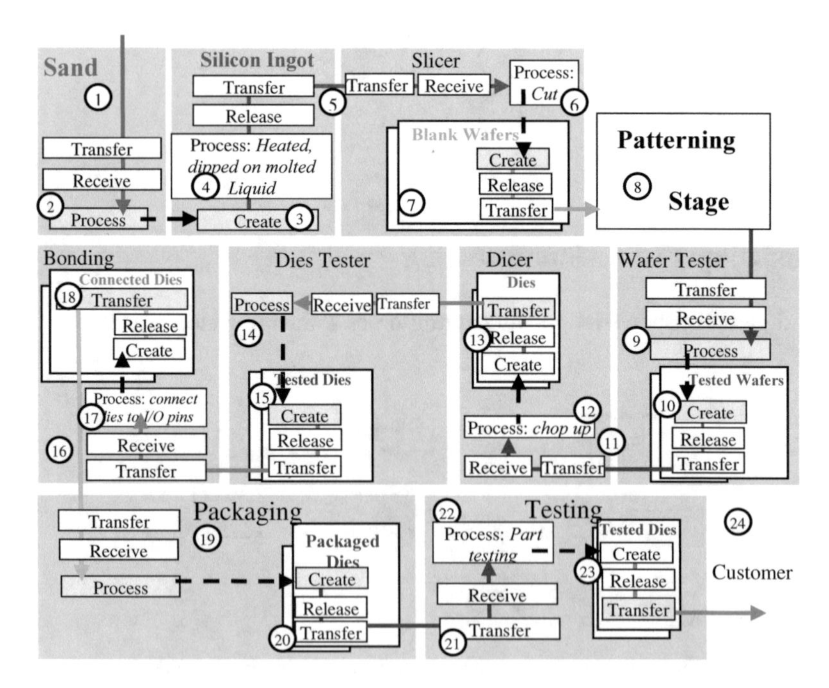

Fig. 6. The manufacturing process for integrated circuits.

The ingot is processed (4) and then flows to the slicer (5). In the slicer (5), it is processed and cut (6) to produce blank wafers (7). Then, each blank wafer (7) flows to the patterning system (8). These patterned wafers then flow to the wafer tester (9) to be tested, to make a map of their good parts (10). Then, many independent components are placed on a single wafer to cope with its defects. The wafers (10) flow to the dicer (11) to be chopped up (12) into dies (13). After the dies are tested, dies that contain flaws are discarded (14). The good dies (15) flow to the bonding stage (16), to be connected (17) to the input/output pins of a package (18). In the packaging stage (19), the packaged dies (20) flow to the testing stage (21) to be tested (22) a final time. Finally, the tested dies (23) are shipped to the customers (24).

The patterning stage (circle 8 in Fig. 6) includes eight phases: wafer fabrication, thermal oxidation, masking, etching, doping, metallization, passivation, and electrical testing. The phases can be modeled similarly to in Fig. 6; however, space limitations do not permit including them in this paper.

5.2 Behavior

The TM representation in Fig. 6 is a static description of the activities in the involved processes. Perinbanayagam [21] stated that activities "can become an object of analysis, communication, and record only to the extent that such processes are apprehended and arrested in presumptively static forms" [21]. Figure 6 as a static form leads to the behavior of things being modeled during events. The chronology of activities can be extracted by orchestrating the sequence of events within the interacting processes.

An *event* in a TM is a thing that can be created, processed, released, transferred, and received. The thing becomes "alive" in events. Note that the process stage of an event means that an event has run its course. Accordingly, the static arrangement of things flowing in machines can reveal the choreography of the execution of events. Modeling of the dynamic behavior occurs in a succeeding phase that is separate from the static picture of production and involves an "events space" in which events happen [22]. Events are the "trains" that run over the networks of flow tracks in the static specification. In a TM, an event is specified by the spatial region, time, and substages. Note that a conceptual event refers to sets of (momentary) events in space and time that form a meaningful event in the context of the involved model; however, not "every happening [in history] deserves the title of [historical] event" [23].

Consider the meaningful events in the slicer machine in Fig. 6: Processing Ingot (circle 6) and Creating Blank Wafer (7) as modeled in Fig. 7 and Include the Time (3) and the Regions of Events (4 and 5; sub diagrams of Fig. 6). Figure 8 shows the chronology of the ingot being processed and cut to produce many blank wafers.

Accordingly, we generalize this capturing of behavior and apply it to the entire manufacturing process in Fig. 6. First the events are specified as follows.

- Event 1 (E_1): Sand flows to the system.
- Event 2 (E_2): The sand is processed.
- Event 3 (E_3): A silicon ingot is generated.
- Event 4 (E_4): The silicon crystal ingot is processed under heat and melted.
- Event 5 (E_5): The silicon crystal is sent to the slicer.

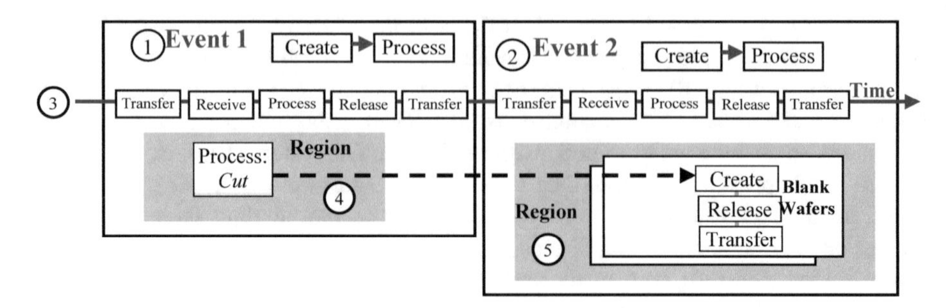

Fig. 7. The two events *Processing Ingot in Slicer* and *Creating Blank Wafers*.

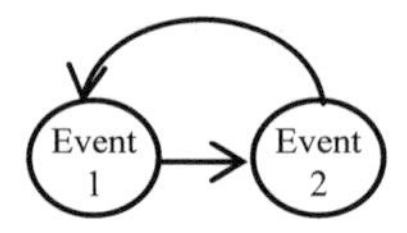

Fig. 8. The chronology of the two events *Processing Ingot* and *Creating Blank Wafers*.

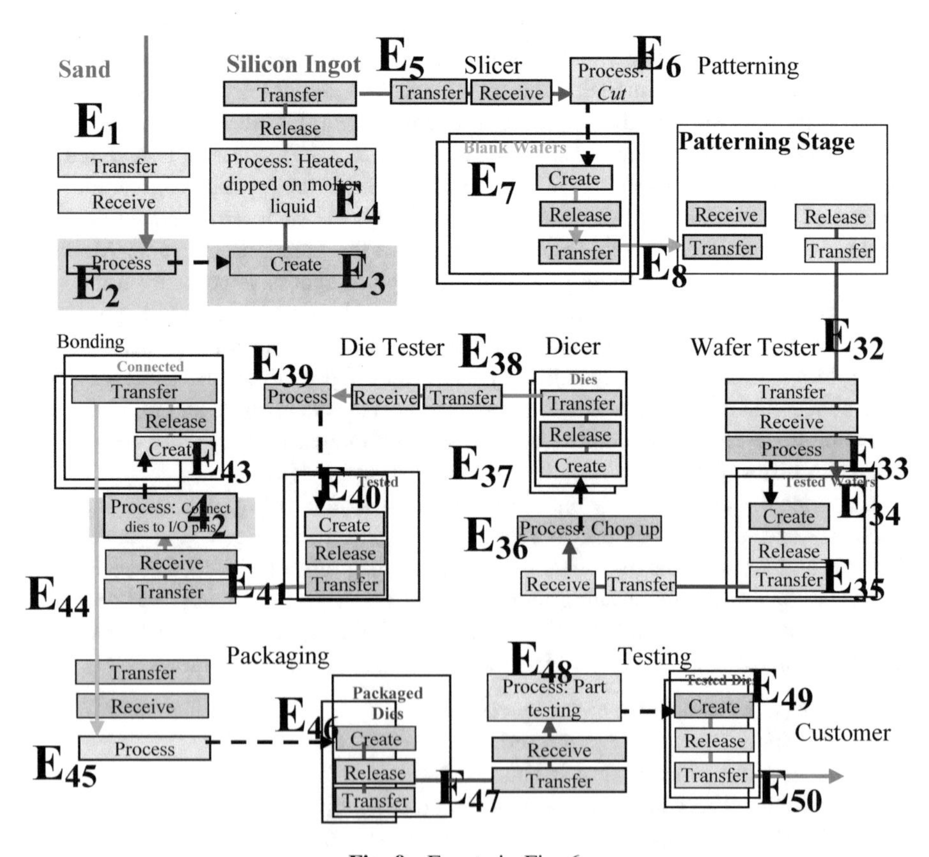

Fig. 9. Events in Fig. 6.

- Event 6 (E_6): The slicer cuts the silicon ingot.
- Event 7 (E_7): Blank wafers are created.
- Event 8 (E_8): Blank wafers are sent to the patterning stage.
- ....
- Event 50 (E_{50}): Tested packaged dies are sent to customer.

The space limitations for this paper do not permit all 50 events shown in Fig. 9 to be listed. These events are diagrammed in Fig. 10. Note that the events overlay the static description of Fig. 6.

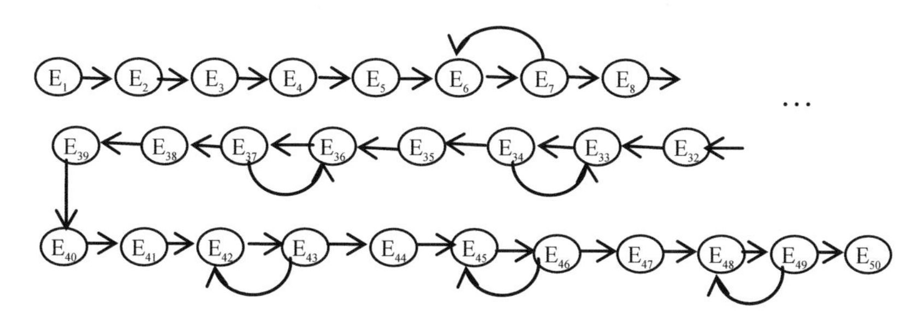

Fig. 10. The chronology of events in the manufacturing process for an integrated circuit.

5.3 Control

The manufacturing system described in the previous TM diagrams is a viable general solution for control and management by itself. Examples of its utilization include for a customized system dedicated to monitoring operations and facilitating the management and logistical operations of public transportation. Figure 11 shows an example of applying monitoring and control to the TM diagram.

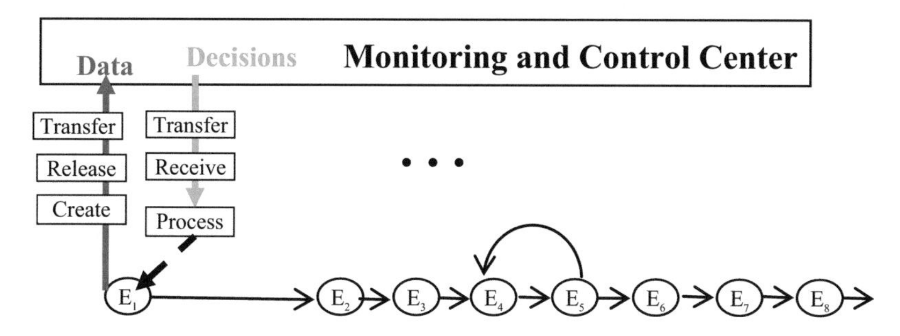

Fig. 11. TM representation as a basis for monitoring and control.

References

1. OMG. OMG Systems Modeling Language, September 2013
2. Finance, G.: SysML Modelling Language Explained (2010). http://www.omgsysml.org/SysML_Modelling_Language_explained-finance.pdf
3. Lu, S., Parsons, J.: Enforcing ontological rules in UML-based conceptual modeling: principles and implementation. In: Proceedings of 10th Workshop on Evaluating Modeling Methods for Systems Analysis and Design, held in conjunction with the 17th Conference on Advanced Information Systems, Porto, Portugal, pp. 451–462 (2005)
4. Dori, D.: Object-Process Methodology and Its Application to the Visual Semantic Web (2003). www.er.byu.edu/er2003/slides/ER2003PT1Dori.pdf
5. Kishore, K.L., Prabhakar, V.S.V.: VLSI design: VLSI design flow. In: Proceeding of the I. K. International Pvt. Ltd., pp. 79–107 (2009)
6. Patterson, D.A., Hennessy, J.L.: Computer Organization and Design, 5th edn. Elsevier Inc., Waltham (2014)
7. Hodges, S., Rounce, P.: A VLSI design management environment. Design Management Environments in CAD (INSPEC Accession No. 3876802). IEE Colloquium, p. 31 (1991). http://ieeexplore.ieee.org/stamp/stamp.jsp?tp=&arnumber=180960
8. Marinescu, D.C.: Internet-Based Workflow Management: Towards a Semantic Web. Wiley, New York (2002)
9. Shepelev, V.A.: Automatic workflow generation. In: Proceeding of the Design Automation Conference, Geneva, 16–20 September 1996, pp. 104–109 (1996). https://doi.org/10.1109/eurdac.1996.558191
10. Jullien, C., Leblond, A., Lecourvoisier, J.: A database interface for an integrated CAD system. In: Proceedings of the ACMIIEEE, 23rd Design Automation Conference, pp. 760–767 (1986)
11. Eppinger, S.D., Nukala, M.V., Whitney, D.E.: Generalized models of design iteration using signal flow graph. Res. Eng. Des. **9**, 112–123 (1997)
12. Magott, J., Skudlarski, K.: Combining generalized stochastic Petri nets and PERT networks for the performance evaluation of concurrent processes. In: International Workshop on Petri Nets and Performance Models, pp. 229–256 (1989)
13. Xiu, L.: VLSI Circuit Design Methodology Demystified: A Conceptual Taxonomy. Wiley, New Jersey (2008)
14. Wellstead, P.E.: Introduction to Physical System Modelling, Electronic edn. Academic Press Ltd., Chennai (2005)
15. Wikibooks: Chip Design Made Easy (2011). http://en.wikibooks.org/wiki/Chip_Design_Made_Easy
16. Sherwani, N.: Algorithms for VLSI Physical Design Automation, 3rd edn. Kluwer Academic, New York (2002)
17. Al-Fedaghi, S.S., AlQallaf, A.: Modelling and control of engineering plant processes. Int. J. Appl. Syst. Stud. **8**, 255–277 (2018)
18. Al-Fedaghi, S.S., Al-Dwaisan, E.: Framework for managing the very large scale integration design process. Am. J. Appl. Sci. **9**, 213–222 (2012)
19. Carreira, J.: Philosophy is not a luxury [blog], 2 March 2011. https://philosophyisnotaluxury.com/2011/03/02/to-thing-a-new-verb/
20. Heidegger, M.: The thing. In: Hofstadter, A. (Trans.) Poetry, Language, Thought, pp. 161–184. Harper & Row, New York (1975)
21. Perinbanayagam, R.S.: Identity's Moments: The Self in Action and Interaction. Lexington Books, Lanham (2012)

22. Williams, D.: Physics and flux: comment on professor Capek's essay. In: Boston Studies in the Philosophy of Science, vol. II, pp. 465–466. Humanities Press (1965)
23. Tang, C.-C.: Towards a real theory of event: a dialogue between Luhmann and Sewell. Paper presented at Niklas Luhmann's Systems Theory in the World Society: A Resonance from Taiwan. Institute of Sociology, Academia Sinica, Taipei (2010)

A Simple and Cost-Effective Anomaly Detection Paradigm on the Basis of Computational Intelligence for Mobile Ad-Hoc Networks from a Security Viewpoint

K. Pradeep Kumar[1(✉)] and B. R. Prasad Babu[2]

[1] Jawaharlal Nehru Technological University, Kakinada, India
pradeepkumarkrisnappa@gmail.com
[2] Department of Computer Science and Engineering, APSCE, Bengaluru, India

Abstract. A Mobile Adhoc Network (MANET) comprises of wide range of applications, since past one decade it has offered a tremendous breakthrough in the field of wireless communication. Therefore, its potential cost-effective routing and resource scheduling operations have not only enhanced the communicational benefits but also made it a hot topic of interest among different scientific and research communities. However, despite having so many diverse potential aspects, it lacks efficiency in terms security owing to the dynamic and non-centralized way of topology formation while performing communication. It also often found undergoing through an increasingly diverse application requirement which made it resource constrained. Thereby, in this type of network formation, the probability of threat occurrence for the purpose of disrupting communication process is more which also affects the data reliability to a huge extent. This study inclines towards addressing the security issue which is still an open research problem in the domain of MANET and further introduces a cost-effective computational intelligence driven mechanism which can enhance the resistivity of network model by discretizing all form of intrusions. Identifying network intrusion within a specified time increases the network reliability factor which is quite essential in MANET data computation and analysis. The proposed formulated analytical design is a computational process and experimented in a numerical computing platform to validate the performance in terms of throughput and packet delivery ratio.

Keywords: Network impactful anomaly detection · Security · Routing · Computational intelligence · Mobile Ad-hoc Networks (MANET)

1 Introduction

The conceptualization of MANET design aspects has led the wireless communication realities to a different aspect where the cost-effective, optimized communication performances are currently being comprehended [1, 2]. Since many years MANET has offered a wide range of advanced communication capabilities, in order to simplify the problem which has arisen due to resource-constrained network operations. It accompanies various communicational benefits with respect to optimizing the cost of

R. Silhavy (Ed.): CSOC 2019, AISC 986, pp. 78–86, 2019.
https://doi.org/10.1007/978-3-030-19813-8_9

computation and also reduces the network traffic load. MANET operations can be distinguished with certain characteristics which include (i) dynamic topology formation by self-organizing nodes followed by (ii) intermittent-link breakage paradigm in every instance of communication round. Although there exist various research archives, which talk about diverse applications of MANET but in reality, very few commercial applications are witnessed, and most of them become theoretically alive. Most of the conventional studies also lack in-depth analysis where different issues such as security, energy efficiency, and resource utilization are considered. One of the most significant issues in the domain of MANET is security loop-holes and its vulnerability to different susceptible attacks such as Sybil, falsification, wormhole, etc. [3–5]. The prime factor which makes MANET communication exposed to different attack level threats is its decentralized pattern of communication where the dynamicity and the frequent intermittent communication link breakage leads to an inconsistent situation. Thereby, in this situation, the possibility of threat level cannot be distinguished in a comprehensive manner. Thereby, it is yet to be resolved and remain an open research problem. From this viewpoint, an extensive study also revealed a fact that implementing a security protocol with higher resiliency in a challenging environment like MANET is not an easier task when all forms of network intrusion are concerned [6, 7].

The existing solution approaches have emphasized only towards resisting a specific form of attack scenario, but in MANET there can be any form of attack at any instant of time [8, 9]. It is also observed during a comprehensive critical analysis which is explored in the consecutive section.

Each attack pattern also has their unique way of strategized technique which can affect the network performance for some malicious intentions. To overcome the existing limitations of currently explored studies which have also addressed the similar problem, the proposed study has come up with a novel approach which mechanizes a light-weight and cost-effective network anomaly detection technique from a security viewpoint. The study analytically models the entire system and simulate it on the top of a numerical computing platform. It also attempts to balance the trade-off between both communication performance and security aspect associated with MANET routing operations which is more often found, overlooked in most of the existing studies. The study further performed a comparative analysis where the performance of the proposed model is compared with the existing SEND and SRP protocols. The extensive outcome in terms of anomaly detection accuracy along with network throughput shows that the proposed system accomplishes a higher percentage of intrusion detection accuracy without compromising the Quality-of-Service (QoS) aspect, i.e. the throughput factor. The overall pattern of organization of the paper is as follows- Sect. 2 discusses the existing research work followed by problem identification in Sect. 3. Section 4 discusses proposed conceptual modelling followed by an elaborated discussion of design analysis in Sect. 5. Further, Sect. 6 includes anticipated outcomes of the proposed design followed by the conclusion in Sect. 7.

2 Related Work

This section highlights the conventional research work which has exclusively addressed the security issue in MANET. Also, the thorough critical analysis of this existing approaches has assisted towards finding their limitations which implies their potential drawbacks confine a full-proof security solution in MANET. The study of [10] has worked on the similar line of research but their study subjected towards designing an efficient security modeling for mesh networks where they have specified the applicability of the presented framework into dynamic ad-hoc platforms. In the study of [11–13] a novel key management based crypto-system has been introduced. Whereas, in the study of [14] an efficient schema supported with the transmission time of data packets got introduced with some unique features which have later found to accomplish maximum possible threat detection in the Adhoc network. A thorough and critical analysis of ad-hoc network attacks, presented in [15] and [16] where key-based mechanisms are introduced to resist maximum possible security attacks in an ad-hoc network.

3 Problem Description

A thorough analysis of the related work has provided a significant research gap which clearly shows that most of the conventional security solutions adopted encryption based and key-management based approaches for securing routing operations and defining resistivity against different forms of potential attacks in MANET. As MANET is accompanied with highly dynamic topology formation paradigm, thereby regardless of centralized infrastructure support, implementing heavier cryptosystem on the top of it prolong the network latency, and it is long-term affects the performance efficiency. Thereby, implementing heavier cryptosystem may achieve security design goal but at the same time do not assure equivalent QoS performance efficiency. There is various limitation associated with the existing systems which include,

- Most of the cryptographic security solutions are resistive towards a particular form of attack and fails to achieve a security goal when the attack pattern gets altered.
- Majority of the conventional cryptosystems brings computational overhead to the system which also affects the network performance by slowing it down.
- Very few studies found to incorporate cognitive intelligence for identifying different levels of threats in the ad-hoc environment.
- The existing system does not offer optimal performance when both aspects pertaining to communication efficiency and security operations are considered.

There, the problem statement of the proposed research work is framed as – *"To design a cost-effective security framework assisted with computational intelligence which can assure a higher degree of resiliency against maximum possible attacks in a MANET environment."*

4 Proposed Methodology

The prime purpose of the proposed system is to accomplish the design goal associated with the proposed anomaly detection mechanism from a security viewpoint. For this purpose, the study has introduced a novel analytical approach which assists in the detection of different forms of routing attacks and threat behavioral aspects in a large-scale MANET. The formulation of the proposed solution is multi-fold and also it apples a cognitive computational intelligence for the detection of malicious network events. The cognitive pattern of communication here introduced to assist a node in making a situational decision if any node capture or other forms of attacks take place. At that time, it should have the capability to potentially become defensive to safeguard the entire network operation from being captured or compromised.

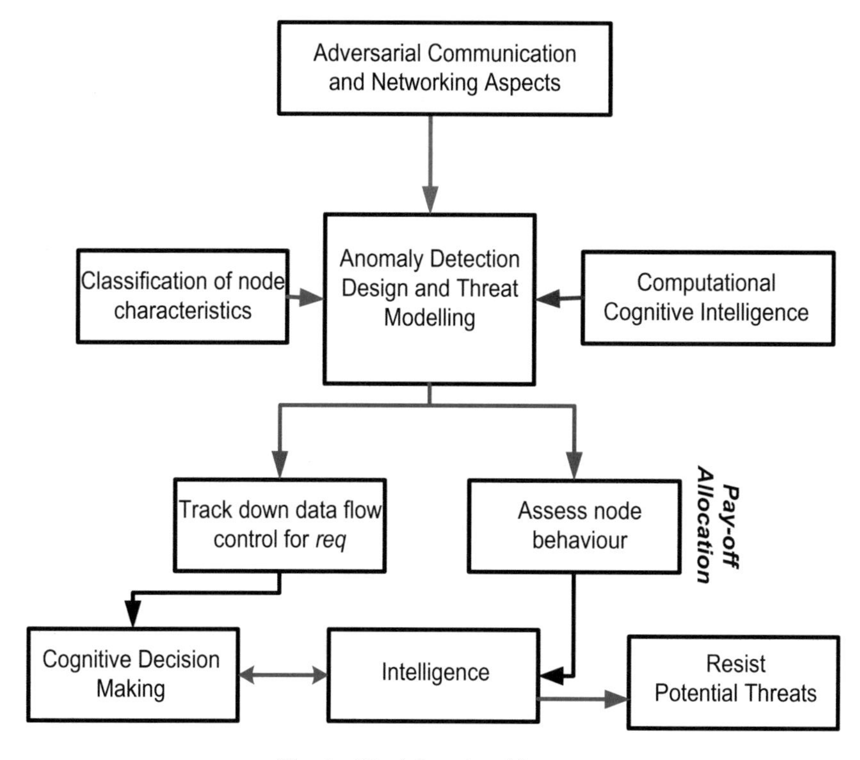

Fig. 1. Block-based architecture

The above Fig. 1 introduces the conceptual design methodology which is the prime theme of this network anomaly detection modeling. The intelligence-based security modeling in this aspect targets discretizing the unpredictable vulnerability factors which induce potential threats to the networks in a MANET. The novel cognitive factors are more inclined towards enhancing the decision-making process on the basis of intelligence which assists in detecting maximum possible threats within a specified time range and preventing the intrusion and the network anomaly with timeliness—the threat modeling in this context, considered in a way where the study assumed all forms of adversarial aspects and their common communication and routing patterns. Therefore, it is quite clear that the proposed system design forms an analytical base which could include the possibility to resist all forms of attacks with its potential intelligent capabilities. It learns continuously numerous attack patterns and update its system about their operational factors. It assists the system model to apply its cognitive skill to determine whether a captured node can disrupt the communication process or only meant for data to be dropped. The next section further discusses the design analysis of the presented framework.

5 Comprehensive Design Analysis

The multi-fold design pattern of the proposed system comprises four different segments of modeling which are (i) Dynamic Group-based modeling, (ii) Classification of node characteristics, (iii) Cognitive computational intelligence and finally (iv) Defensive threat modeling and pay-off allocation. The conceptualization of the proposed model intends to employ a computational intelligence which is also accompanied by a cognitive pattern of communication. The higher degree of precision and accuracy of detection is achieved through this cognitive communication where pay-off allocation policy plays a crucial role. The design principle of the proposed solution involves different types of routing configurations as different nodes are having different forms of characteristic designs.

- *Dynamic Group-based modeling:* As MANET is accompanied and characterized by dynamic topology, thereby predicting node behavior with respect to time is critical. It is also observed that the movement vector associated with a mobile node is also not specific which makes the routing phenomena quite inconsistent in every aspect. However, this type of phenomena is quite common in the domain of MANET. In order to address the device heterogeneity problem, the conceptualization comes up with distinct modeling which is group based. Here, a heterogeneous group formulation of mobile nodes performed with respect to their specific transmission characteristics. The study also considered that the number of transmission zones could not be higher than the number of nodes being deployed within a simulation environment. The novelty associated with this part is it narrows down the node assessing task which significantly saves enough time which is quite essential to attain the design goal of the proposed modeling.
- *Classification of node characteristics:* The study also considered different types of node behavioral aspects formulation which completes the task of clustering transmission zones. Usually, conventional literature has mostly considered two different types of nodes which are normal and adversary respectively, but to enhance the security operation, the proposed framework has discretized the grouping of nodes with respect to four different sets—the 1st set of nodes belonging to a distinct set of entities which are having normal behavioral aspects. The 2nd set of nodes are characterized in a way where it belongs to the adversarial entities, and in the 3rd set,

all forms of exploiter nodes are considered. Finally, in the 4th set nodes are characterized by a distinct pattern of faulty node. The characterization of nodes is performed on the basis of the best possible features associated with a node.

- *Defensive threat modeling and pay-off allocation:* The cognitively based pattern of communication has introduced a set of communication among the nodes which are interacting based on certain routing principles. It has also introduced a pay-off allocation policy to both the normal and adversarial nodes based on their set of actions. The following Fig. 2 illustrates the computational steps involved in pay-off allocation strategy which has strengthened up the proposed anomaly detection and threat modeling from both computational and security viewpoint.

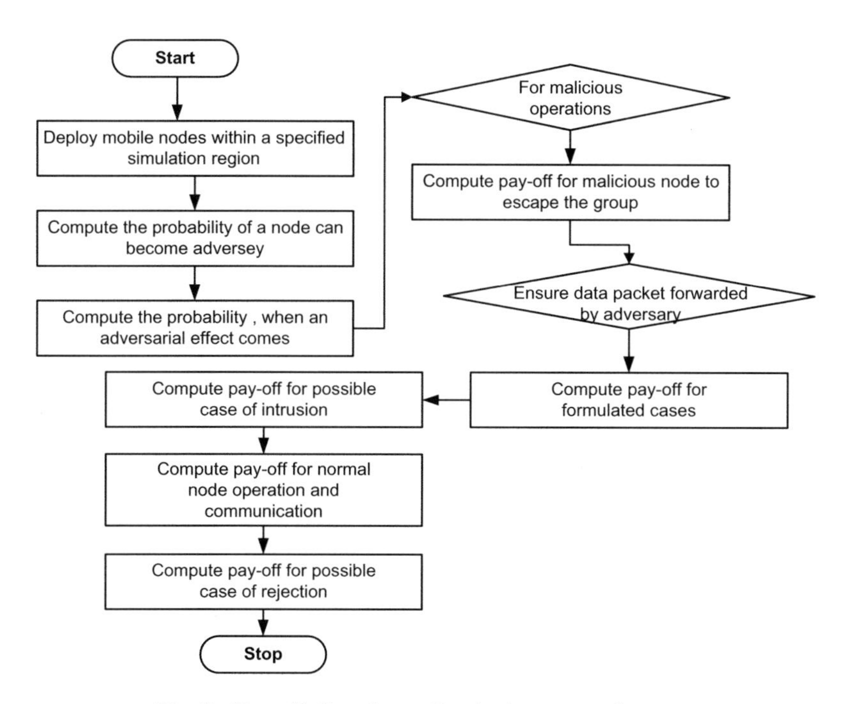

Fig. 2. Pay-off allocation policy in the proposed system

- *Cognitive computational intelligence*: The study employs analytical modeling to assess the intelligence metric associated with each node which is participating in the communication process. In this phase, the computation of structured resilient intelligence factors is computed. It has also formulated a design where two sets of attributes ε_1 and ε_2 represents, identifying successful communication event and identifying intrusion level respectively. The design associated with this module is highly robust and capable of extracting the confusive behavioral aspects associated with an adversarial node.

6 Results Discussion

The extensive simulation of the proposed security model has been performed in a numerical computing platform where the design process does not require a sophisticated simulation. The design of the model is done with respect to different possible communication aspects among nodes. The simulation has been carried out for 1800 number of iterations, and the performance of the system has been validated in terms of throughput and packet delivery ratio shows in Fig. 3.

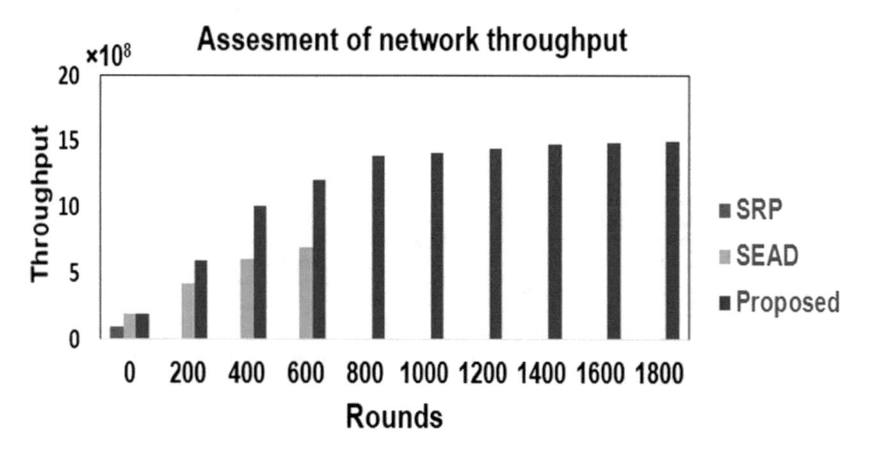

Fig. 3. Comparative analysis of throughput

The extensive performance of the proposed system is evaluated by comparing its outcome with the most cited work which is SEAD protocol [17] and SRP protocol [18] respectively. The outcome accomplished shows that the proposed system achieves very less packet drop by employing cognitive intelligence-based security mechanism and at the same time also achieves equivalent communication performance.

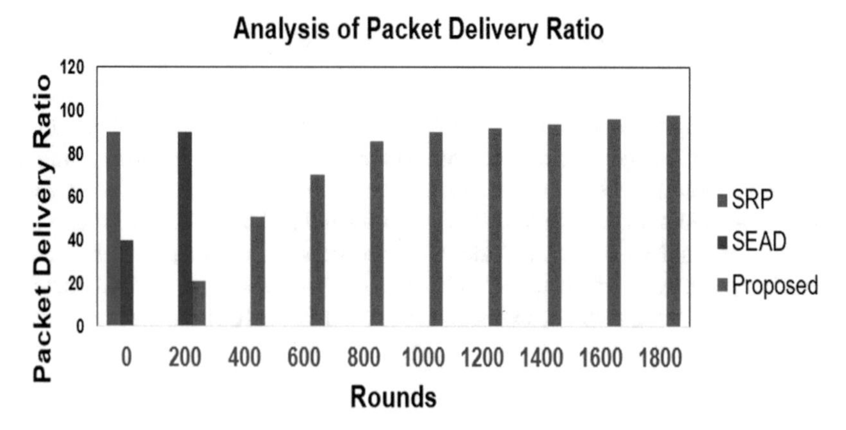

Fig. 4. Comparative analysis of packet delivery ratio

The outcome associated with the packet delivery ratio factor also shows that the proposed system significantly improves the network and communication performance while resisting maximum possible communication threats by reducing the probability of data getting dropped (in Fig. 4).

7 Conclusion

The study presented a very light-weight cognitive intelligence-based anomaly detection system in order to address the limitations of the existing security protocols in MANET. A thorough analysis of the existing security policies clearly shows that most of the techniques are based on cryptographic approaches and irrespective of resisting maximum attacks it suffers due to the degradation of network performance from the communicational aspects. The proposed design methodology is intended to resist maximum possible routing attacks by assessing the node behavior and characteristics whereas it also introduces a novel pay-off mechanism to boost up the positive factors which influences a regular node to detect another malicious node within a MANET. The extensive simulation outcome shows that the proposed system attains better performance in terms of both throughput and packet delivery ratio in comparison with the conventional baselines.

References

1. Jamalipour, A., Ma, Y.: Intermittently Connected Mobile Ad Hoc Networks: From Routing to Content Distribution. Springer, New York (2011)
2. Hinze, A.M.: Principles and Applications of Distributed Event-Based Systems. IGI Global, Hershey (2010)
3. Lakhtaria, K.I.: Technological Advancements and Applications in Mobile Ad-Hoc Networks: Research Trends: Research Trends. IGI Global, Hershey (2012)
4. Awad, W.S.: Improving Information Security Practices through Computational Intelligence. IGI Global, Hershey (2015)
5. Jambulingam, V.: Secure and Anonymous Routing for Mobile Adhoc Networks: Secure and Anonymous On-Demand Routing for Mobile Adhoc Networks. Lap Lambert Academic Publishing GmbH KG, Saarbrücken (2014)
6. Javaid, A.: Securing Mobile Ad Hoc Networks: Resource-Aware Self-Adaptive Security. Syngress, Rockland (2014)
7. Dinesh, Kumar, A., Singh, J.: A literature survey of secure routing protocols for MANET. Int. J. Appl. Sci. Eng. Res. **3**(6) (2014)
8. Ranaut, D., Lal, M.: A review on security issues and encryption algorithms in mobile ad-hoc network. Int. J. Sci. Res. **3**(6), 146–148 (2014). ISSN 2319-7064
9. Kukreja, D., Singh, U., Reddy, B.V.R.: A survey of trust based routing protocols in MANETs. J. Adv. Comput. Netw. **1**(4), 280–285 (2013)
10. Meganathan, N.T., Palanichamy, Y.: Privacy preserved and secured reliable routing protocol for wireless mesh networks (2015). Hindawi Publishing Corporation, Article ID 636590
11. Madhusudhanan, B., Chitra, S., Rajan, C.: Mobility based key management technique for multicast security in mobile ad hoc networks (2015). Hindawi Publishing Corporation, Article ID 801632

12. Alwan, H., Agarwal, A.: A multipath routing approach for secure and reliable data delivery in wireless sensor networks (2013). Hindawi Publishing Corporation, Article ID 232798
13. Guo, Y., Ma, J., Wang, C., Yang, K.: Incentive-based optimal nodes selection mechanism for threshold key management in MANETs with selfish nodes. Int. J. Distrib. Sens. Netw. (2013). Hindawi Publishing Corporation, Article ID 416983
14. Kim, D.-U., Kim, H.-W., Kim, G., Kim, S.: A counterattack-detection scheme in transmission time-based wormhole detection methods. Int. J. Distrib. Sens. Netw. (2013). Hindawi Publishing Corporation, Article ID 184931
15. Askoxylakis, I.G., Tryfonas, T., May, J., Siris, V., Traganitis, A.: A family of key agreement mechanisms for mission critical communications for secure mobile ad hoc and wireless mesh internetworking. EURASIP J. Wirel. Commun. Netw. (2011). Hindawi Publishing Corporation, Article ID 807684
16. Caballero-Gil, P., Hernández-Goya, C.: Efficient public key certificate management for mobile ad hoc networks. EURASIP J. Wirel. Commun. Netw. (2011). Hindawi Publishing Corporation, Article ID 935457
17. Hu, Y.-C., Johnson, D.B., Perrig, A.: SEAD: secure, efficient distance vector routing for mobile wireless ad hoc networks. Ad Hoc Netw. **1**, 175–192 (2003)
18. Papadimitratos, P., Haas, Z.J.: Secure routing for mobile ad hoc networks. In: Proceedings of the SCS Communication Networks and Distributed Systems Modeling and Simulation Conference (2002)

A Technique of Adaptation of the Workload Distribution Problem Model for the Fog-Computing Environment

I. Kalyaev[1], E. Melnik[2], and A. Klimenko[3](✉)

[1] Southern Federal University, Rostov-on-Don, Russia
[2] Southern Scientific Center of the Russian Academy of Science, Rostov-on-Don, Russia
[3] Scientific Research Institute of Multiprocessor Computer Systems of Southern Federal University, Taganrog, Russia
anna_klimenko@mail.ru

Abstract. In this paper the technique of workload distribution problem model adaptation to the fog-computing environment is presented. Workload distribution problem is solved for a wide range of systems, including reconfigurable ones, but its generic formalization doesn't pay attention to the peculiarities of the fog-computing environment. If the system is migrated to the mentioned environment, all algorithms of tasks re-distribution (if some are there) should be revised. We propose a technique for the workload distribution problem adaptation to the fog-computing environment. It includes additional parameters, the variety of additional objective functions which have to be chosen according to the computational process model, and the additional constraints. These components are injected into the problem formal model and allow to get the solution related to the fog-computing environment.

Keywords: Fog computing · Workload distribution · Optimization problem · Model adaptation

1 Introduction

The workload distribution problem is topical for a wide range of applications, and for the reconfigurable information and control systems, in particular. Such systems, described in [1–3], operate in the following manner: when the failure of the computational units occurs, the tasks from the faulted nodes are launched on the operational ones. Those nodes, where tasks can be launched are determined on the design stage of the system, or in on-line mode, but the problem is common for both cases: how to distribute the workload through the operational nodes of the system with predetermined constraints, optimization criteria and bordering conditions?

The problem mentioned above is solved usually as a kind of scheduling problem [4–6], or as a strip-packing problem [7, 8]. In this paper we will consider this optimization problem as a scheduling one with some additional variable parameters, which are considered in this paper.

R. Silhavy (Ed.): CSOC 2019, AISC 986, pp. 87–96, 2019.
https://doi.org/10.1007/978-3-030-19813-8_10

A considerable part of contemporary information and control systems is built in the fog-computing environment.

Fog-computing is rather new, but extensively growing technological field. Announce in 2012 for the first time, fog-computing aims to support the Internet of Things concept, to deliver facilities of big data processing with communicational environment offloading and the decrease of system latency [9–11].

In this paper the workload distribution problem model, which considers the fog computing environment peculiarities, is studied. Most problem models have been created for the heterogeneous computational system, and don't pay appropriate attention to the peculiarities of the systems, based on fog-computing concept. Hence, the solving of such optimization problems don't bring the results of a good quality in terms of fog-computing environment. The current paper proposes a technique of workload distribution problem model adaptation to the fog-computing environment, which allows to modify the previously developed problem model with fog-computing systems peculiarities consideration.

The following sections of the paper contain:

- a brief review of the workload distribution problem formalization approaches;
- an analysis of the fog-computing concept features, in other words, which features distinguish the fog-computing-based system from the heterogeneous one;
- the technique of workload distribution problem model adaptation;
- some selected experimental results;
- conclusion.

2 The Workload Distribution Problem Models

As was mentioned earlier, usually the workload distribution problem can be considered as a 2D strip-packing problem or as a scheduling one. Some good examples of a workload distribution modelled via strip-packing problem are given in [12], other examples of scheduling optimization, including structural optimization of the computational system, are presented in [13–15]. The way of problem modelling is determined by the general aim of the simulation: usually the strip-packing problem is used to model problems in which tasks require a continuous subset of identical resources that are arranged in a linear topology [12], including the strip-packing with precedence constraint, while the models of "classical" scheduling theory are much more comprehensive. The extended scheduling models are used to describe problems related to parallel computations, in particular, the problem of task scheduling on the set of connected (or stand-alone) computational units. Tasks can be described as a directed graph with vertexes weighed by computational complexities and ribs weighed by volumes of data to be transferred from task to task. Yet, it must be mentioned that the most known models of task distribution among the computational resources use the constraint "one task per node at the moment" [4–6], while the computational resources may vary.

So, to consider the problem of workload distribution we assume that:

More than one task can be performed by one computational unit (CU);

To carry out a task, a certain amount of resources is allocated, and this amount is a constant for all time of task completion.

The amount of resources allocated is a variable and is a parameter of the scheduling optimization problem.

To illustrate our assumption in a simplified manner we turn to the strip-packing problem.

Usually, the strip $W * T$ is given, and a set of, say, rectangles $w_i * t_i$, where w_i is a resource needed, t_i, accordingly, the time of task processing. The problem is to place rectangles into the strip so that to minimize, e.g., timespan.

Assume that just $c = w_i * t_i$ is fixed, so w_i can be varied. It allows to vary the $w_i < W$ and to decrease t_i.

Relating to the scheduling problem, assume the new parameter u_{ij} which describes the amount of the computational resource is given by CU j to task i. Parameter u_{ij} is a variable, so, to form a scheduling for the system, now we must consider not only task binding to the CUs and their times of beginning of calculations, but the amounts of resources allocated to process the task. Obviously, for identical CUs the bigger u_{ij} gives the shorter t_{ij}, where t_{ij} is the time of task processing.

So, the general example of scheduling model of the workload distribution among the set of heterogeneous computational units is as follows:

Assume $G = (X, Y)$ to be the acyclic weighed graph, where $X = \{x_i\}$ determines the computational complexity of task i, $i = 1,\ldots N$, $Y = \{y_i\}$ is the amount of information to be passed to the computational environment for the next task to be performed.

Consider $P = (M, S)$ is a weighed graph, where $M = \{m_j\}$ determines the set of CUs with performances m_j, $j = 1, \ldots, K$, $S = \{<j, l, b_{jl}>\}$ determine the ribs of the resource graph, connecting the vertexes j, l with the velocity of information transfer b_{jl}.

T is the maximum completion time for tasks described in G.

The tasks distribution is described by the following two matrixes:

$$R = \begin{vmatrix} t_{11} & 0 & \ldots & 0 \\ \ldots & \ldots & \ldots & \ldots \\ \ldots & \ldots & \ldots & t_{N-1,K-1} \\ t_{N1} & 0 & \ldots & \ldots \end{vmatrix}, \quad (1)$$

where t_{ij} is the time, when the CU j begins to process task i.

$$U = \begin{vmatrix} u_{11} & \ldots & \ldots & \ldots \\ u_{21} & \ldots & \ldots & \ldots \\ \ldots & \ldots & \ldots & u_{N-1,K-1} \\ \ldots & \ldots & \ldots & u_{NK} \end{vmatrix}, \quad (2)$$

where u_{ij} is the amount of a computational resource of the $\mathrm{CU_j}$ given to process task i.

Here, with the time constraint, the following important constraint must be considered:

$$\sum_{i=1}^{N} u_{ij} \leq 1, \forall 0 < j < K, \tag{3}$$

and the arbitrary objective function might be as follows:

$$F = \underset{U,R}{MIN}(Smth). \tag{4}$$

One can see that in such scheduling problem model not only processing periods of time vary, but the resource shares, which affect the task completion time.

Nowadays such problems are solved by multiple heuristic and metaheuristic methods, which are out of the scope of this paper. The scheduling problem model formalized above, is rather generic and relevant for multiple heterogeneous environments. Yet, the fog-computing environments will be considered in the following section and some conclusions will be made regarding the face that in general formalization the scheduling problem for the heterogeneous environment do not suit properly the fog-computing one and do not take into account some fog computing peculiarities.

3 An Analysis of the Fog-Computing Concept Features

Fog-computing is relatively new computational concept and is an extension of the well-known cloud computing (though there are works where the fog-computing is considered as a stand-alone concept). Fog-computing aims at providing of big data processing in an acceptable time with appropriate quality of services. To gain this, the key idea was generated: to preprocess the data as near to their sources as possible.

In contemporary concept descriptions the network is considered as a three-layer structure:

- edge of the network, which consists of end-point devices, sensors, computers, notepads, etc.;
- fog layer, which consists of communication facilities, routers, gateways, etc.;
- cloud, which consists of servers located in datacenter and, as usual, connected by fast network.

There is a significant difference between edge-computing and fog-computing: the first concept presupposes that the logic of tasks distribution at the edge of the network is implemented at the edge devices, while the fog-computing presupposes that the decision how and where to process the data is made on the fog-layer node. Besides, fog-computing does not exist without a cloud, and the part of the computations has to be done there.

We leave the detailed review of the fog-computing concept out of this paper's focus because of multiple works written in this field, e.g., [10, 11].

We concentrate on the formal distinctions between the fog-computing environment and the heterogeneous environment.

As was mentioned in [16], there are three main computational models in the fog-computing:

- The offloading model, when the data generated from edge devices are offloaded to the nearest fog node and then at the Cloud (i.e., up-offloading) and in the reverse order from the Cloud to edge devices (i.e., down-offloading).
- The aggregation model, when data streams generated by multiple edge devices are aggregated and possibly processed at the nearest Fog node before being uploaded to Cloud datacenter.
- The peer-to-peer (P2P) model, Fog nodes, which are at the proximity of edge devices, share their computing and storage capabilities and cooperate in order to offer an abstraction storage and computing layer to edge users.

It is obvious, that in case of the "offloading" model the nodes selecting priority should be given to those nodes, which are close to the remainder of the workload. In other words, if there is a need to offload the cloud nodes, it is expedient to distribute the workload through the fog-nodes situated not far from the cloud, and vice versa, if we have to offload some user devices. It is pointless to distribute the workload through the cloud nodes.

For the aggregation model the place to put the workload depends on the variety of the factors, including the capacity of the fog-nodes, the distance between the data sources and the nodes-preprocessors of the information, and so on.

The Peer-to-peer model presupposes some nodes cooperation which are not far from each other.

Besides, there is an important feature of the fog-computing: "cloud" devices are the integral participant of the computational process: some procedures are must be processed in the cloud.

The peculiarities listed above allow to form the major features of the workload distribution problem:

- The nodes of the fog-computing system are unequal, i.e. for different computational models the different nodes are in priority, but not prohibited. It is the first distinction we assume between the fog environment and the "classical" heterogeneous system.
- As the cloud must be involved in the computational process, some special constraints must be used to determine this.
- Some new parameters must be involved in the problem model.

4 A Technique of Workload Distribution Problem Model Adaptation to the Fog-Computing Environment

Taking into account the key features of the fog-computing environment and its distinction from the heterogeneous environment in terms of the workload distribution, the technique of workload distribution problem model adaptation was developed. It is presented in the picture below (Fig. 1).

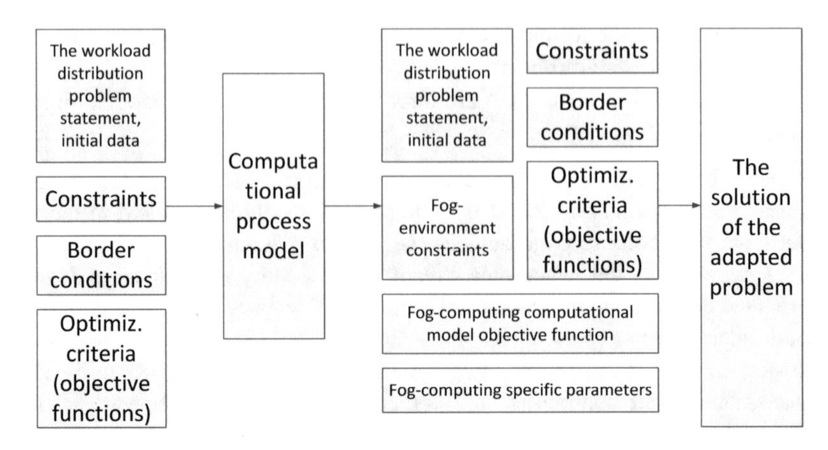

Fig. 1. A technique of workload distribution problem model adaptation to the fog-computing environment

The general stages of the technique are as follows:

- to get a problem, oriented to the heterogeneous computational environment;
- to choose the appropriate computational model in the fog-computing environment;
- to add the appropriate objective functions and constraints;
- to solve the adapted optimization problem of the workload distribution.

Let us consider some new variables, which have to be added to the problem model.

As the nodes in the system are not equal in view of priority of usage, the set of nodes is split to the subsets as is presented below:

$$P_{overall} = P_{fog} \cup P_{cloud} \tag{5}$$

$$P_{fog} \cap P_{cloud} = \varnothing \tag{6}$$

Also the set D of tuples $\{<j, d_1, d_2>\}$ is considered, where j is the node identifier, d_1 is the distance to the edge of the ne1twork, d_2 is the distance from the node j to the cloud layer, counted in network hops. This variables must be added to make the distribution with priorities possible.

The following important constraint is because of the cloud involvement to the computational process. Describe the functional which is able to determine if the device is in the fog-layer or not:

$$rel(j) = \begin{matrix} 1, & \textit{if the node in the cloud}, \\ 0, & \textit{in other cases}. \end{matrix} \tag{7}$$

Then the constraint, which determines the requirement to process the information in the P_{fog} and P_{cloud} is as follows:

$$\forall i \sum_{j} rel(j) > 0 \tag{8}$$

Then, consider the objective functions which are specific to the fog-computing environment and must be added to the problem model.

For the "offloading" model we have the following objective function cases:

$$F_1 = \sum_{i=1,j=1}^{N,M} d_{1ij} \rightarrow \min \tag{9}$$

when the edge devices are offloaded;

$$F_2 = \sum_{i=1,j=1}^{N,M} d_{2ij} \rightarrow \min \tag{10}$$

where the cloud devices are offloaded.

In case of F_1 the criterion is the distance between the edge of the network and the fog-node candidates;

In case of F_2 the criterion is the distance between the cloud and the node candidates.

For the "aggregation" model of computational process the additional objective function might be as follows.

The question is how to distribute the workload and where to put the data processing tasks to minimize the network workload. It must be mentioned that more than one processing subtasks can have the intensive input data streams. To simplify the problem in the scope of this paper we assume that the volumes of data transferred from the user devices are equal and so the communication channels capacities are. Then the rule of the data processing tasks placement is as follows: if for the task i of graph G the number of input ribs is more than the number of output ones, the task i must be placed so as to minimize the time of data receiving.

Determine the function of comparison of the input and output ribs number:

$$comp(i) = \begin{cases} 1, \ \textit{if the input ribs number is more } l \arg e \\ 0, \ \textit{in other cases.} \end{cases} \tag{11}$$

Then the objective function of aggregation tasks placement is as follows:

$$\forall i,j, \ comp(i) > 0, \ F_3 = \sum \frac{y_{ki}}{b_{lj}} \rightarrow \min \tag{12}$$

The third computational strategy model, P2P, presupposes the resource allocation on the set of fog-nodes to provide a special resource type, for example, in case of data storage organizing. So, the additional criterion for this strategy might be the strength of nodes connection and the velocity of the data transmission process, e.g. as is shown in expression (13):

$$F_4 = \sum \frac{v_{jk}}{K} \to \min, \ j, k \in P_{fog}, \tag{13}$$

where K is the number of nodes gathered into the cluster.

So, to adapt the workload distribution problem model, developed for the heterogeneous network, to the fog environment, it is expedient to add some constraints and objective functions. This makes the decision to be more relative to the fog-computing concept than simple workload distribution problem solving.

5 The Experimental Results

To check the results of model adaptation we generated two arbitrary graphs, *G* and *P* with set of 150 tasks with random computational complexity and the set of computational units (K = 16). For every unit the tuple <j, d_1, d_2> is set, which determines the distances between the node, edge of the network and the cloud. We presuppose that there can be maximum 4 hops distance between the fog nodes and the cloud. The objective function of the initial model is the workload dispersion minimization, and the aim of the experiment is to compare the resulting workload distribution within the system.

The results are presented in Fig. 2.

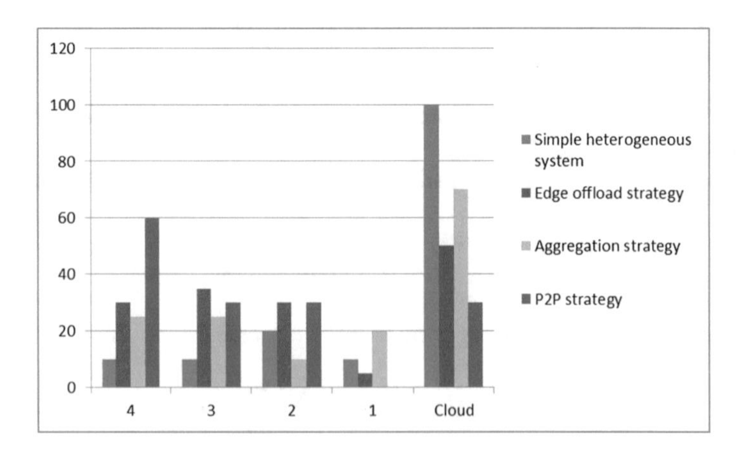

Fig. 2. The workload distribution through the system for the initial model and the adapted ones.

As is presented in Fig. 2, if the problem of the workload distribution is solved via distribution through the heterogeneous system, the most of the computational complexity is mapped on the cloud devices due to their high computational performance, and in general more capable devices take more workload. Such distribution is not acceptable in case of intensive data stream from the edge of the network and proves our assumption about the heterogeneous system model insufficiency in case of fog-computing infrastructure.

Then, the objective functions specific for to the fog-computing offload structure are added to the general model. The results of the workload distribution are shown with legend "Edge offload strategy". The workload has been distributed such as to increase the workload closer to the edge of the network. In case of "Aggregation strategy" the workload is distributed so as to meet the requirement to process the data closer to its sources. And, finally, P2P strategy additional objective function gathers the computational workload into the clusters of tightly connected devices.

So, the adaptation of the workload distribution problem model to the peculiarities of the fog-computing environment allows to get task distribution through the system, which is different from the one for the heterogeneous systems. The workload distributions, got in the boundaries of general computational strategies of the fog-computing concept, are more preferable for usage in the systems, based on the fog-computing infrastructure due to its orientation to the fog-computing features.

6 Conclusion

In this paper the technique of workload distribution problem model adaptation to the fog-computing environment is presented. The question is quite topical: the workload distribution problem is solved for a wide range of systems, but its generic formalization doesn't pay attention to the peculiarities of the fog-computing environment. At the same time, fog-computing is applied ubiquitously, and, for example, if the system is migrated to the mentioned environment, all algorithms of tasks re-distribution (is there some) should be revised.

We propose the generic technique for the workload distribution problem adaptation to the fog-computing environment. It includes additional parameters, the variety of additional objective functions which have to be chosen according to the computational process model, and the additional constraints. These components are injected into the problem formal model and allow to get the solution related to the fog-computing environment.

Acknowledgement. The paper has been prepared within the RFBR project 18-29-03229.

References

1. Avizienis, A., Laprie, J.C., Randell, B.: Fundamental concepts of dependability. Technical report, Seriesuniversity of Newcastle Upon Tyne Computing Science, vol. 1145, pp. 7–12 (2001). https://pld.ttu.ee/IAF0530/16/avi1.pdf
2. Melnik, E.V., Klimenko, A.B., Korobkin, V.V.: The method providing fault-tolerance for information and control systems of the industrial mechatronic objects. In: IOP Conference Series: Materials Science and Engineering (2017). https://doi.org/10.1088/1757-899x/177/1/012004

3. Melnik, E., Korovin, I., Klimenko, A.: Improving dependability of reconfigurable robotic control system. In: Lecture Notes in Computer Science (including subseries Lecture Notes in Artificial Intelligence and Lecture Notes in Bioinformatics), pp. 144–152 (2017). https://doi.org/10.1007/978-3-319-66471-2_16
4. Pinedo, M.L.: Scheduling: Theory, Algorithms, and Systems. Springer, New York (2008). https://doi.org/10.1007/978-0-387-78935-4
5. Barskiy, A.B.: Parallelniye process v vychislitelnih sistemah. Planirovaniye i organizatsya. «Radio i svyaz», Moskva (1990)
6. Gonchar, D.R., Furugyan, M.G.: Effectivniye algoritmy planirovaniya vychisleniy v mnogoprocessornyh sistemah realnogo vremeny. In: UBS 2014, №. 49, 19 November 2018. https://cyberleninka.ru/article/n/effektivnye-algoritmy-planirovaniya-vychisleniy-v-mnogoprotsessornyh-sistemah-realnogo-vremeni
7. Dell'Amico, M., Díaz, J.C.D., Iori, M.: The bin packing problem with precedence constraints. Oper. Res. (2012). https://doi.org/10.1287/opre.1120.1109
8. Ciscal-Terry, W., Dell'Amico, M., Iori, M.: Bin packing problem with general precedence constraints. IFAC-PapersOnLine (2015). https://doi.org/10.1016/j.ifacol.2015.06.386
9. Moysiadis, V., Sarigiannidis, P., Moscholios, I.: Towards distributed data management in fog computing. Wirel. Commun. Mob. Comput. (2018). https://doi.org/10.1155/2018/7597686
10. Wang, Y., Uehara, T., Sasaki, R.: Fog computing: issues and challenges in security and forensics. In: Proceedings - International Computer Software and Applications Conference, pp. 53–59 (2015). https://doi.org/10.1109/compsac.2015.173
11. Cisco and/or its Affiliates: Fog Computing and the Internet of Things: Extend the Cloud to Where the Things Are (2015). https://www.cisco.com/c/dam/en_us/solutions/trends/iot/docs/computing-overview.pdf
12. Augustine, J., Banerjee, S., Irani, S.: Strip packing with precedence constraints and strip packing with release times. Theor. Comput. Sci. (2009). https://doi.org/10.1016/j.tcs.2009.05.024
13. Proto, D., Mottola, F., Carpinelli, G.: Optimal scheduling of a microgrid with demand response resources. IET Gener. Transm. Distrib. (2014). https://doi.org/10.1049/iet-gtd.2013.0758
14. Zhang, M., Chen, J.: Islanding and scheduling of power distribution systems with distributed generation. IEEE Trans. Power Syst. (2015). https://doi.org/10.1109/tpwrs.2014.2382564
15. Chan, K.W., Luo, X.: Real-time scheduling of electric vehicles charging in low-voltage residential distribution systems to minimise power losses and improve voltage profile. IET Gener. Transm. Distrib. (2014). https://doi.org/10.1049/iet-gtd.2013.0256
16. Moysiadis, V., Sarigiannidis, P., Moscholios, I.: Towards distributed data management in fog computing. Wirel. Commun. Mob. Comput. **2018** (2018). Article ID 7597686. https://doi.org/10.1155/2018/7597686

Approach of Process Modeling Applied in Particular Pedagogical Research

Marek Vaclavik[1(✉)], Zuzana Sikorova[1], and Tomas Barot[2]

[1] Department of Education and Adult Education, Faculty of Education, University of Ostrava, Fr. Sramka 3, 709 00 Ostrava, Czech Republic
{Marek.Vaclavik, Zuzana.Sikorova}@osu.cz

[2] Department of Mathematics with Didactics, Faculty of Education, University of Ostrava, Fr. Sramka 3, 709 00 Ostrava, Czech Republic
Tomas.Barot@osu.cz

Abstract. Possibilities of process modeling have been generally appeared in the control system theory, which belongs to the technical cybernetics. Process modeling of real systems are essential in a synthesis of control systems. Synthesis methods are based on an identification of a considered controlled process. The theoretical approaches of the technical cybernetics belong to the general cybernetics. However, in case of the other cybernetic disciplines, a majority of abstract layer can be seen e.g. in pedagogical cybernetics. This cybernetic discipline is a priori based on a feedback strategy of a process control. Abstract objects have not been widely considered with any mathematical background. A connection of the mathematical description for purposes of the pedagogical cybernetics using process models has been proposed yet. In this paper, an implementation of the proposed mathematical process models is demonstrated in case of applied pedagogical research. Practical results are then verified using statistical methods for hypotheses testing.

Keywords: Process modeling · Technical cybernetics · Pedagogical cybernetics · Statistical methods · Testing hypothesis · Applied research

1 Introduction

Utilization of the mathematical modeling has been dominantly appeared in the control system theory [1], which has a significant position in the industrial practical applications, e.g. [2–6]. Various innovative methods for a process identification and for a synthesis of a control has been published yet. In other particular control strategies, e.g. in Model Predictive Control [7], the including of a process model in a controller can be seen with applied strategy of receding horizon. Many various strategies have the similar aim focused on optimal control of the various types of mathematical models [1].

The control system theory can be considered as a part of a general cybernetics known as a technical cybernetics. A feedback control strategy [8] appears in this specific technical and in the general form of cybernetics. However, in non-technical disciplines, which belong to cybernetics, the feedback control loop can be seen also e.g.

R. Silhavy (Ed.): CSOC 2019, AISC 986, pp. 97–106, 2019.
https://doi.org/10.1007/978-3-030-19813-8_11

in the pedagogical cybernetics [9]. As a disadvantage, the absence of mathematical background can be considered in the non-technical fields of the cybernetics.

However, the possibilities of a process modeling can be also applied on the abstract models in context of general cybernetic laws. Even strategies can be transferred e.g. the adaptive control strategy [10, 11]. Concretely, in the particular discipline of the pedagogical cybernetics, a consideration of the mathematical models from the field of the technical cybernetics, can be inspirational, as can be seen in [12]. In [12], the process modeling, known in field of the technical cybernetics, was demonstrated on simulation of the statistical analysis. For this purposes, the particular form of a mathematical description was corresponded with the descriptive statistical characteristics of the measured variables. Proposed structure of models was based on the transfer function of process of the first order. Parameters a gain of a process [12] and a time-constant [12] were substituted by the new attributes arithmetical average and variation bounded on measured variables.

For the described extended proposed connection [12] between the technical and pedagogical cybernetics, verification of this modeling approach should be confirmed. The confirmation of proposal [12] were built on statistical evaluation of performed model in form of step functions. Comparing step functions for each item of a categorical variable, which were defined as an alternative form for modeling purposes, was successfully proved using statistical methods for testing hypothesis.

For purposes of modeling phenomenon in frequently appeared research in pedagogy, e.g. [13–16], the proposed mathematical description could be considered as advantageous in the sense of the cybernetic approaches. Generally known principles of the quantitative research methods, including e.g. descriptive statistical methods [17, 18] or methods for testing hypotheses [17, 19], are essential; however, they can be suitable complemented to the cybernetic principle, as the proposed approach [12] shows. In this paper, a particular pedagogical research, which was analysed only using statistical methods in [20], is actually extended by a new analysis using proposed models [12] of the pedagogical cybernetics. Confirmation of obtained results is provided using testing hypotheses.

2 Structure of Proposed Process Models

The technical cybernetics [1–8] belongs to the general cybernetics with appearance of various strategical principles. In the non-technical cybernetic disciplines, theories have not been widely focused on any mathematical background. It could be suitable to extend the previous theories by some new aspects or approaches.

As a typical example of the non-technical cybernetic discipline, the pedagogical cybernetics can be considered. There have been proposed many theoretical strategies in the context of the general cybernetic principles e.g. the feedback control loop [9] applied in the environment of the education. In [10, 11], the adaptive control strategy, known from the technical cybernetics, was applied in the education. In paper [10], possibilities of improving the voluntary tutoring of students were proposed and realized in Maths Support Centre at Tomas Bata University in Zlin [21]. Other application of the adaptive control strategy can be seen in [11], where this strategy can be suitably applied in education of math courses for foreign students.

In this paper, the more detailed approach for modeling of abstract models is implemented in the sense of the cybernetic laws in the particular example of the pedagogical research. In utilized proposal [12], only simulation experiments were provided. In this paper, the practically realized research is presented.

Concrete aims of the proposal [12] are consisted of transforming the statistical properties of measured data, which belongs to the concrete abstract object (e.g. students), into the parameters of the proposed structured mathematical descriptions (1), (2) [12].

$$G_r(s) = \frac{K}{Ts+1} \equiv \frac{\bar{x}_r}{\sigma_r^2 s + 1};\ r = 1, \ldots, q \quad (1)$$

$$h_r = K\left(1 - e^{-\frac{t}{T}}\right) \equiv \bar{x}_r\left(1 - e^{-\frac{t}{\sigma_r^2}}\right);\ r = 1, \ldots, q \quad (2)$$

$$\left.\begin{aligned} \bar{x}_r &= n_r^{-1} \sum_{i=1}^{n_r} x_{ri} \\ \sigma_r^2 &= (n_r - 1)^{-1} \sum_{i=1}^{n_r} (x_{ri} - \bar{x}_r) \end{aligned}\right\} \quad (3)$$

In the contribution [12], the comparison of the models (2) for sets of data obtained the same conclusions as the quantitative research techniques [17]. However, the proposal [12] only presents the alternative possibility of modeling the statistical behaviour of some considered abstract models. In comparison with the quantitative research techniques, the main advantage is consideration, that proposal [12] can be classified as a potential extension of the cybernetic theory [12].

In Eq. (1), the proposed models from [12] of statistical behaviour of data sets can be seen in form of transfer functions of the first order with complex variable s. Technical parameters a gain K and a time-constant T, known in the technical cybernetics, were equalled to descriptive characteristics average and variation of data set, as can be seen in (3) [12].

In Eqs. (1) and (2), a division into r-th data set from q data sets is considered for reason of existence of r items in the categorical statistical variable [17]. In each r-th item of this categorical variable, the numerical data x_{ri} are included. Range of r-th data set is n_r [17].

3 Statistical Methods Used for Analysis and Verification

In a frame of a quantitative research [17], statistical analyses are focused usually on processing the descriptive statistical characteristic [17, 18] or on testing hypothesis [17]. Testing hypothesis can be divided into parametrical and non-parametrical statistical tests [17]. The first category of the parametrical methods can be applied only on data, which fulfills the condition of normality of data, e.g. ANOVA (for q items in a categorical variable fulfilling condition $q > 2$) or T-test ($q = 2$). In the second category, e.g. Kruskal-Wallis ($q > 2$) or Mann-Whitney test ($q = 2$). Chi-squared test with both categorical variables belongs also into the second category of the non-parametrical tests. Evaluation of normality of each data sets can be performed by normality tests e.g. Shapiro-Wilk or Anderson-Darling. Conclusion for testing normality and testing hypotheses are depended on the results in form of p-value, which can be obtained by statistical software [17].

In case of testing hypothesis on existence of statistical significant dependences between numerical variable on the items of the categorical variable, ANOVA (respectively Kruskal-Wallis test) or T-test (respectively Mann-Whitney test) can be applied respecting the number of items q [17].

In this testing hypotheses, the zero hypothesis has form "There are not statistical significant important dependences between *a numerical variable* and *a categorial variable*." The alternative hypothesis has a form: "There are statistical significant important dependences between *a numerical variable* and *a categorial variable*." In the research, definition of significance level α is an important assumption. If is $p < \alpha$, the zero hypothesis is rejected in favor of the alternative hypothesis. In opposite case, the zero hypothesis is failed to reject [17].

In this paper, u hypotheses are generally considered. For purposes of the quantitative research [17], which is a verification procedure for evaluation of applied proposed approach [12] of process modeling, each hypothesis j-th H has a zero hypothesis jH_0 and an alternative hypothesis jH_1.

In case of application of proposed model (1)–(3), u hypotheses influent the notation of each particular models (4)–(6) for each measured data in r-th item of q-th categorical variable. Where n_{jr} is number of measured data x_{jri} in r-th item of j-th and in i-th row in the categorical variable. In a confirmation analysis, models (5) are tested on testing dependences of modelled progresses (5) on the categorical variables. The same results should be demanded in comparison with the statistical analyses based on analysis of the hypotheses jH using the described method for testing hypotheses [17].

$$G_{jr}(s) = \frac{\bar{x}_{jr}}{\sigma_{jr}^2 s + 1}; \; r = 1, \ldots, q; \, j = 1, \ldots, u \tag{4}$$

$$h_{jr} = \bar{x}_{jr}\left(1 - e^{-\frac{t}{\sigma_{jr}^2}}\right); \; r = 1, \ldots, q; \, j = 1, \ldots, u \tag{5}$$

$$\left.\begin{aligned} \bar{x}_{jr} &= n_{jr}^{-1} \sum_{i=1}^{n_{jr}} x_{jri} \\ \sigma_{jr}^2 &= (n_{jr} - 1)^{-1} \sum_{i=1}^{n_{jr}} (x_{jri} - \bar{x}_{jr}) \end{aligned}\right\} \tag{6}$$

4 Results

Application of the proposed approach of process modeling was performed only on simulation data in [12]. In this paper, the practical realization of provided particular pedagogical research [20] is complemented using the introduced proposed rules [12]. In frame of the realized pedagogical research [20], aims were focused on the empiric analyses [17] of dependences of assessment of quality of teaching on a year of study in the particular pedagogical course at Faculty of education at University of Ostrava.

Respondents returned 84 questionnaires with 31 included items. In the quantitative analysis [20], results of questionnaires were based on a student evaluation of a teaching. In this paper, following 6 hypotheses (Table 1) were selected from a whole 9 hypotheses [20]. As a categorical variable, a year of study was considered. The numerical variable was based on an interval from 1 (not important) to 4 (most important) [20].

Table 1. Selected hypotheses in realized pedagogical research [20]

Hypothesis *j*-th *H*	*j*-th H_0 "There are not statistical significant important dependences between ..."	*j*-th H_1 "There are statistical significant important dependences between ..."
1*H*	"The positive expectation of student and the year of study"	
2*H*	"Student's assessment of formal attributes of teaching and the year of study"	
3*H*	"The student's assessment based on providing teacher's supported questions to students and the year of study"	
4*H*	"The student's assessment based on teacher's description of practical examples with applicable possibilities and the year of study"	
5*H*	"The student's assessment of the teacher as the professional and the year of study"	
6*H*	"The student's assessment based on methods utilization in frame of application in the future practice and the year of study"	

Descriptive statistical characteristics (6) arithmetical average and variation were obtained for each data set in [20], as can be seen in Tables 2 and 3.

Table 2. Arithmetical averages for categorized data sets

Hypothesis *j*-th *H*	Average $\bar{x}_{j1}$ Item: 1st year of bachelor study	Average $\bar{x}_{j2}$ Item: 2nd year of bachelor study	Average $\bar{x}_{j3}$ Item: 1st year of following master study	Average $\bar{x}_{j4}$ Item: 2nd year of following master study
1*H*	1.86	1.89	1.96	2.17
2*H*	1.71	1.26	1.00	1.00
3*H*	1.67	1.33	1.21	1.30
4*H*	1.31	1.59	1.38	1.50
5*H*	1.67	1.67	1.21	1.50
6*H*	1.81	1.41	1.33	2.15

Table 3. Variations for categorized data sets

Hypothesis j-th H	Variation σ_{j1}^2 Item: 1st year of bachelor study	Variation σ_{j2}^2 Item: 2nd year of bachelor study	Variation σ_{j3}^2 Item: 1st year of following master study	Variation σ_{j4}^2 Item: 2nd year of following master study
1H	0.41	0.32	0.370	0.470
2H	1.44	0.19	0.002	0.001
3H	0.41	0.22	0.160	0.200
4H	0.22	0.39	0.230	0.250
5H	1.17	0.30	0.250	0.250
6H	0.73	0.24	0.220	0.430

In Table 4, results of testing normality of measured data in each item of the categorical variable were obtained using Shapiro-Wilk and Anderson-Darling tests in [20]. In [20], conclusions about testing hypotheses were achieved using Kruskal-Wallis test, as can be also seen in Table 4. Both statistical testing was performed with consideration of the significance level α equal to 0.05. For statistical methods performed in [20] and in this paper, statistical software PAST Statistics [22] was used.

Table 4. Testing normality of measured data and testing hypotheses for measured data on sign. Level $\alpha = 0.05$ using Kruskal-Wallis test

Hypotheses	Testing Normality of Measured Data	Testing Hypotheses (p-value)	Conclusion of Testing Hypotheses
$1H_0, 1H_1$	Unconfirmed on Sign. Level α	$0.5604 > 0.05$	Fail to Reject $1H_0$ on Sign. Level α
$2H_0, 2H_1$	Unconfirmed on Sign. Level α	$0.0007 < 0.05$	Reject $2H_0$ in Favor of $2H_1$ on Sign. Level α
$3H_0, 3H_1$	Unconfirmed on Sign. Level α	$0.05764 > 0.05$	Fail to Reject $3H_0$ on Sign. Level α
$4H_0, 4H_1$	Unconfirmed on Sign. Level α	$0.4388 > 0.05$	Fail to Reject $4H_0$ on Sign. Level α
$5H_0, 5H_1$	Unconfirmed on Sign. Level α	$0.0172 < 0.05$	Reject $5H_0$ in Favor of $5H_1$ on Sign. Level α
$6H_0, 6H_1$	Unconfirmed on Sign. Level α	$0.0034 < 0.05$	Reject $6H_0$ in Favor of $6H_1$ on Sign. Level α

In this paper, the realized quantitative pedagogical research [20] is extended by proposed models (4)–(6). Numbered measured data, which belong to particular items of categorical variables, were approximate using models in form of the transfer functions (Table 5) and in form of the step functions (Table 6).

Table 5. Modeled transfer functions for categorized data sets

Hypothesis *j*-th *H*	Transfer function G_{j1} (s) Item: 1st year of bachelor study	Transfer function G_{j2} (s) Item: 2nd year of bachelor study	Transfer function G_{j3} (s) Item: 1st year of following master study	Transfer function G_{j4} (s) Item: 2nd year of following master study
1*H*	$\frac{1.86}{0.41\,s+1}$	$\frac{1.89}{0.32\,s+1}$	$\frac{1.96}{0.37\,s+1}$	$\frac{2.17}{0.47\,s+1}$
2*H*	$\frac{1.71}{1.44\,s+1}$	$\frac{1.26}{0.19\,s+1}$	$\frac{1}{0.002\,s+1}$	$\frac{1}{0.001\,s+1}$
3*H*	$\frac{1.67}{0.41\,s+1}$	$\frac{1.33}{0.22\,s+1}$	$\frac{1.21}{0.16\,s+1}$	$\frac{1.3}{0.2\,s+1}$
4*H*	$\frac{1.31}{0.22\,s+1}$	$\frac{1.59}{0.39\,s+1}$	$\frac{1.38}{0.23\,s+1}$	$\frac{1.5}{0.25\,s+1}$
5*H*	$\frac{1.67}{1.17\,s+1}$	$\frac{1.67}{0.3\,s+1}$	$\frac{1.21}{0.25\,s+1}$	$\frac{1.5}{0.25\,s+1}$
6*H*	$\frac{1.81}{0.73\,s+1}$	$\frac{1.41}{0.24\,s+1}$	$\frac{1.33}{0.22\,s+1}$	$\frac{2.15}{0.43\,s+1}$

Table 6. Modeled step functions for categorized data sets

Hypothesis *j*-th *H*	Step function h_{j1} Item: 1st year of bachelor study	Step function h_{j2} Item: 2nd year of bachelor study	Step function h_{j3} Item: 1st year of following master study	Step function h_{j4} Item: 2nd year of following master study
1*H*	$1.86(1-e^{-\frac{t}{0.41}})$	$1.89(1-e^{-\frac{t}{0.32}})$	$1.96(1-e^{-\frac{t}{0.37}})$	$2.17(1-e^{-\frac{t}{0.47}})$
2*H*	$1.71(1-e^{-\frac{t}{1.44}})$	$1.26(1-e^{-\frac{t}{0.19}})$	$1\,(1-e^{-\frac{t}{0.002}})$	$1\,(1-e^{-\frac{t}{0.001}})$
3*H*	$1.67(1-e^{-\frac{t}{0.41}})$	$1.33(1-e^{-\frac{t}{0.22}})$	$1.21(1-e^{-\frac{t}{0.16}})$	$1.3(1-e^{-\frac{t}{0.2}})$
4*H*	$1.31(1-e^{-\frac{t}{0.22}})$	$1.59(1-e^{-\frac{t}{0.39}})$	$1.38(1-e^{-\frac{t}{0.23}})$	$1.5(1-e^{-\frac{t}{0.25}})$
5*H*	$1.67(1-e^{-\frac{t}{1.17}})$	$1.67(1-e^{-\frac{t}{0.3}})$	$1.21(1-e^{-\frac{t}{0.25}})$	$1.5(1-e^{-\frac{t}{0.25}})$
6*H*	$1.81(1-e^{-\frac{t}{0.73}})$	$1.41(1-e^{-\frac{t}{0.24}})$	$1.33(1-e^{-\frac{t}{0.22}})$	$2.15(1-e^{-\frac{t}{0.43}})$

In Table 7, the successful verification of application can be seen. Cybernetic models (4)–(6), included in hypotheses on existence of statistical significant dependences of progresses of step functions in frame of *jH*, achieved the same conclusions as testing hypotheses (Table 4) performed by quantitative methods [17]. Utilized proposed approach was verified using Kruskal-Wallis test applied on models (5) in software PAST Statistics.

Table 7. Modeled step functions - tested on normality and tested by hypotheses on sign. Level $\alpha = 0.05$ using Kruskal-Wallis test

Hypothesis j-th H	$h_{j1}, \ldots, h_{j4}$ Step Functions for Data of j-th H	Testing Normality of Modeled Data	Hypotheses Testing (p-value)	Conclusion of Hypotheses Testing on Modeled Data
$1H$	$h_{11}, \ldots, h_{14}$	Unconfirmed on Sign. Level α	0.4794>0.05	Fail to Reject $1H_0$ on Sign. Level α
$2H$	$h_{21}, \ldots, h_{24}$	Unconfirmed on Sign. Level α	1.495E-8<0.05	Reject $2H_0$ in Favor of $2H_1$ on Sign. Level α
$3H$	$h_{31}, \ldots, h_{34}$	Unconfirmed on Sign. Level α	0.1445>0.05	Fail to Reject $3H_0$ on Sign. Level α
$4H$	$h_{41}, \ldots, h_{44}$	Unconfirmed on Sign. Level α	0.2015>0.05	Fail to Reject $4H_0$ on Sign. Level α
$5H$	$h_{51}, \ldots, h_{54}$	Unconfirmed on Sign. Level α	2.167E-6<0.05	Reject $5H_0$ in Favor of $5H_1$ on Sign. Level α
$6H$	$h_{61}, \ldots, h_{64}$	Unconfirmed on Sign. Level α	0.01434<0.05	Reject $6H_0$ in Favor of $6H_1$ on Sign. Level α

5 Conclusion

In frame of a field of the pedagogical cybernetics, mathematical process models were applied on particular pedagogical research. In comparison with statistical analysis of testing hypotheses on existence of a statistical significant dependences, used models provided same conclusions about behaviour of measured variables. Where measured variables were bound with the particular realized pedagogical research. Abstract objects were concretely considered as models, which form can be obviously seen in the technical cybernetics. It was proved that both disciplines, the technical and pedagogical cybernetics, can use analogue modeling possibilities in case of fitting the original properties of modeled process. In form of transfer function with variable s in complex plane, mathematical model was able to appropriate approximate the statistical character of variables. Therefore, practical realizations of quantitative research can have also other complementing style of expression of behaviour of measured processes.

References

1. Corriou, J.P.: Process Control: Theory and Applications. Springer, Berlin (2004). ISBN 1-85233-776-1
2. Lech, P., Wlodarski, P.: Analysis of the IoT WiFi mesh network. In: 6th Computer Science On-line Conference: Cybernetics and Mathematics Applications in Intelligent Systems. Advances in Intelligent Systems and Computing, vol. 574, pp. 272–280. Springer (2017). https://doi.org/10.1007/978-3-319-57264-2_28. ISBN 978-3-319-57263-5

3. Matusu, R., Prokop, R.: Control of time-delay systems with parametric uncertainty via two feedback controllers. In: 6th Computer Science On-line Conference: Cybernetics and Mathematics Applications in Intelligent Systems. Advances in Intelligent Systems and Computing, vol. 574, pp. 197–205. Springer (2017). ISBN 978-3-319-57263-5
4. Achuthan, K., Murali, S.S.: Virtual Lab: an adequate multi-modality learning channel for enhancing students' perception in chemistry. In: 6th Computer Science On-line Conference: Cybernetics and Mathematics Applications in Intelligent Systems. Advances in Intelligent Systems and Computing, vol. 574, pp. 419–433. Springer (2017). https://doi.org/10.1007/978-3-319-57264-2_42. ISBN 978-3-319-57263-5
5. Navratil, P., Pekar, L.: Possible approach to control of multi-variable control loop by using tools for determining optimal control pairs. Int. J. Circ. Syst. Sig. Process. **7**(6), 335–349 (2013). ISSN 1998-4464
6. Spacek, L., Bobal, V., Vojtesek, J.: LQ digital control of ball & plate system. In: 31st European Conference on Modelling and Simulation, pp. 403–408. European Council for Modelling and Simulation (2017). ISBN 978-0-9932440-4-9
7. Ingole, D., Holaza, J., Takacs, B., Kvasnica, M.: FPGA-based explicit model predictive control for closed loop control of intravenous anesthesia. In: 20th International Conference on Process Control (PC), pp. 42–47. IEEE (2015). ISBN 978-1-4673-6627-4
8. Kucera, V.: Analysis and Design of Discrete Linear Control Systems. Nakladatelstvi Ceskoslovenske akademie ved, Praha (1991). ISBN 80-200-0252-9
9. Cevik, Y.D., Haslaman, T., Celik, S.: The effect of peer assessment on problem solving skills of prospective teachers supported by online learning activities. Stud. Educ. Eval. **44**, 23–35 (2015). https://doi.org/10.1016/j.stueduc.2014.12.002. ISSN 0191-491X
10. Barot, T.: Complemented adaptive control strategy with application in pedagogical cybernetics. In: 7th Computer Science On-line Conference: Cybernetics and Algorithms in Intelligent Systems. Advances in Intelligent Systems and Computing, vol. 765, pp. 53–62. Springer (2018). https://doi.org/10.1007/978-3-319-91192-2_6. ISBN 978-3-319-91191-5
11. Barot, T.: Adaptive control strategy in context with pedagogical cybernetics. Int. J. Inf. Commun. Technol. Educ. **6**(2), 5–11 (2017). University of Ostrava. ISSN 1805-3726
12. Barot, T.: Possibilities of process modeling in pedagogical cybernetics based on control-system-theory approaches. In: 6th Computer Science On-line Conference: Cybernetics and Mathematics Applications in Intelligent Systems. Advances in Intelligent Systems and Computing, vol. 574, pp. 110–119. Springer (2017). https://doi.org/10.1007/978-3-319-57264-2_11. ISBN 978-3-319-57263-5
13. Korenova, L.: GeoGebra in teaching of primary school mathematics. Int. J. Technol. Math. Educ. **24**(3), 155–160 (2017). Research Information. ISSN 17442710
14. Kostolanyova, K.: Adaptation of personalized education in e-learning environment. In: 1st International Symposium on Emerging Technologies for Education, pp. 433–442. Springer (2017). https://doi.org/10.1007/978-3-319-52836-6_46. ISBN 978-3-319-52835-9
15. Schoftner, T., Traxler, P., Prieschl, W., Atzwanger, M.: E-learning introduction for students of the first semester in the form of an online seminar. In: Pre-conference Workshop of the 14th E-Learning Conference for Computer Science, pp. 125–129. CEUR-WS (2016). ISSN 1613-0073
16. Sikorova, Z., Cervenkova, I.: Styles of textbook use. New Educ. Rev. **35**(1), 112–122 (2014). Adam Marszalek Publishing House. ISSN 17326729
17. Kitchenham, B., Madeyski, L., Budgen, D., et al.: Robust statistical methods for empirical software engineering. Empirical Softw. Eng. 1–52 (2016). Springer. ISSN 1573-7616
18. Krivy, I., Tvrdik, J., Krpec, R.: Stochastic algorithms in nonlinear regression. Comput. Stat. Data Anal. **33**(12), 277–290 (2000). https://doi.org/10.1016/S0167-(99)00059-6. Elsevier

19. Sulovska, K., Belaskova, S., Adamek, M.: Gait patterns for crime fighting: Statistical evaluation. In: Proceedings of SPIE - The International Society for Optical Engineering, vol. 8901. SPIE (2013). https://doi.org/10.1117/12.2033323. ISBN 978-081949770-3
20. Vaclavik, M., Cervenkova, I.: Analysis of dependences between assessment of quality of teaching and year of study. In: Otazky evaluace vyuky na vysokych skolach, pp. 63–76. University of Ostrava (2018). ISBN 978-80-7599-024-2
21. Patikova, Z.: Podpurna centra pro vyuku matematiky na vysokych skolach, Setkani ucitelu matematiky vsech typu a stupnu skol 2016, pp. 97–100. JCMF (2016). (in Czech)
22. Hammer, O., Harper, D.A.T., Ryan, P.D.: PAST: paleontological statistics software package for education and data analysis. Palaeontologia Electronica **4**(1) (2001). http://palaeo-electronica.org/2001_1/past/issue1_01.htm. Palaeontological Association

Architectural Framework for Industry 4.0 Compliance Supply Chain System for Automotive Industry

Kiran Kumar Chandriah[1(✉)] and N. V. Raghavendra[2]

[1] Mercedes Benz Research and Development India, Bengaluru, India
kiran.chandriah@daimler.com
[2] Department of Mechanical Engineering, NIE, Mysuru, India
nvr@nie.ac.in

Abstract. The vision of Industry 4.0 impacts the mechanism as well as its benefits to the supply chain system. The biggest challenge to achieve a robust, stable and efficient eco-system for Supply chain system is to have an architecture that initiates the framework for both the eco-system and predictive analysis. This paper proposes an architecture that can be realized on the cloud infrastructure considering a 360° view of the requirements. The three-core component of the architecture includes cloud, analytics and Internet of Things. The synchronization of these core components aims to achieve reliability in optimal latency. The proposed novel architecture of SCM exploits the potential of the cyber-physical system, big-data and predictive methods to minimize the demand -supply gap irrespective of uncertainty and unpredictable events. The model validation is done by Delphi method of validation and case studies of automotive sector. It was found to be acceptable and useful for adopting the architecture for synchronized supply chain system to Industry 4.0 as well as provisions many disruptive innovations, which is quite useful for both social and economic view point.

Keywords: Supply chain · Industry 4.0 · Automotive industry · Cyber-physical · Cloud computing architecture

1 Introduction

The phrase that "*necessity is the mother of inventions*", is found quite true if the core reason of Industry 4.0 vision is analyzed. The fast-human population growth and limitations of resources brings an ever-widening gap between the demand of the product/services and the supply. This ever-growing gap laid down the foundation of the fourth industrial revolution vision namely Industry-4.0 (I-4.0). The collaborative platform of cyber-physical system for smart and intelligent as well as fully automated process of the industry is the core goal of I-4.0 [1–3], that impact greatly the synchronous supply chain system. The enabler quadrants of the I-4.0 vision. The system emerges out by a non-linear correlation of the key constituents, viz. (1) socializing (2) transparency, (3) network collaboration and (4) autonomy. The emerged technological collaborations for the supply chain system (SCS) provisions opportunity for the

R. Silhavy (Ed.): CSOC 2019, AISC 986, pp. 107–116, 2019.
https://doi.org/10.1007/978-3-030-19813-8_12

disruptive innovations because of these interlinked technical granular pillars. A clear and simplified architecture understanding of the framework correlated technology will facilitate the strategy makers to explore more insights for the utilization of data analysis that has impact on the supply chains. This paper introduces a framework that facilitates the automotive industry to follow the norms of Supply Chain Management (SCM) compliance of upcoming Industry 4.0 standards. The organization of the manuscript is as follow: Existing automation-based approaches are discussed in Sect. 2 followed by problem identification in Sect. 3. Proposed solution is discussed in Sect. 4 followed by result analysis in Sect. 5, while Sect. 6 briefs of the paper contribution.

2 Brief Overview of Literature

This section extends the review discussion of our prior work [4]. At present, there are various work carried out considering implementation of an Internet-of-Things (IoT) and Industry 4.0 e.g. usage of fog computing (Aazam et al. [5]), supervisory control system (Gonzalez et al. [6]), intelligent manufacturing system (Wan et al. [7]), data processing approach in Industry 4.0 (Patel et al. [8]), usage of graphic-based operation (Posada et al. [9]), usage of biometric in securing IoT (Condry and Nelson, [10]). Exclusive study towards automating an automotive sector is carried out by many researchers e.g. Fernandes et al. [11]. Various approaches implemented towards automotive industry are decision making approach (Sun and Zhu [12]), evaluation-based approach (Zhang et al. [13]), security (Ray et al. [14]), forecasting (Yildiz et al. [15]), simulation-based approach (Fahhama et al. [16]), case study specific (El-Farouk and Fouad [17]), analysis of existing methodology (Faycal [18], Liu et al. [19], Shi [20]). Apart from this, there are also various work carried out for improving SCM over cloud environment. The work carried out by Choi [21] has presented a discussion of SCM with system of systems approach associated with big data. Optimization-based approach is discussed by Yue et al. [22] where fuzzy logic was utilized considering some of few uncertainty factors in SCM. Evolutionary approach was advocated by Wang et al. [23] towards improving control system on SCM over highly distributed environment.

3 Problem Identification

Irrespective of diversified work in IoT in reference to automotive industry, still there are bigger set of open issues viz. (i) none of the existing approaches have been proven compliance of industry 4.0 standard, (ii) the actual concept of data fusion from vehicles followed by management and analysis is never discussed, (iii) there is no real-time synchronization between the demand and supply with respect to vehicular component as demands and relaying of services by suppliers in automotive industry, (iv) most importantly, a validated framework for automotive industry using IoT and industry 4.0 standard is missing. Therefore, statement of problem is "*To develop a cost-effective and scalable framework that offers capability of automating the process bridging the gap between demand and supply.*"

4 Proposed Architecture

The prime purpose of the proposed architecture is to develop a very novel framework of Industry 4.0 explicitly considering the case study of automotive industry. The architecture harnesses the potential communication system in cyber-physical system which focuses on enhancing the cost-effective productivity very smartly as well as enhancing the consumer experience. The planned and evolved architecture is shown in Fig. 1.

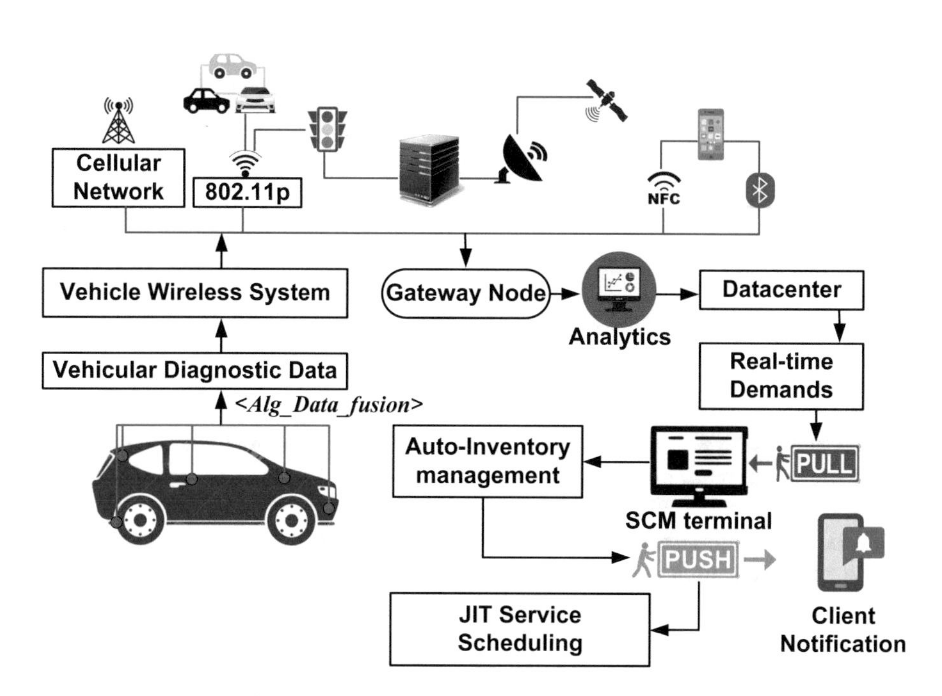

Fig. 1. Proposed architecture as per Industry 4.0

The illustration of the essential components of the architecture is as follows:

(i) Aggregating Fused Vehicular Component Data: In the present time, vehicles are already reported to use diversified sensors e.g. wheel speed sensor, steering angle sensor, collision angle sensor, blind spot detector, slide curtain sensor, tire pressure sensor, drowsiness sensor, nighttime pedestrian warning sensor, front airbag sensor, oxygen sensor, fuel temperature sensor, mass air flow sensor, etc. Figure 1 highlights the process of communication within all the sensing units and monitoring panels, which is usually on the dashboard of a vehicle. But it is not necessary that all the monitoring units to be on the dashboard. The model considers that there exists heterogeneous sensor on different place that gathers the information on the basis of occurrences of events. It is also assumed that there is also actuator that undertakes necessary actions when specific event information is acquired beyond its threshold limit. However, the study ignores this actuator involvement much in modeling as it is focused processing the fused sensor information and forward where an intelligence-based application could be executed for performing analytical operations on it.

Algorithm for Data Fusion
Input: M_i (Monitoring units), S_j (Sensing units), *e* (physical information of event)
Output: data (fused sensory data)
Start:
1. *init* M_i and S_j, $i << j$, $i \in Z$, and $j \in R$
2. **For** $e = e_1 : e_2$
3. $S_j \rightarrow (data)_j$
4. **If** $((data)_j == pr)$
5. $(data)_p = (data)_j$
6. **Else**
7. apply $f_1((data)_j)$
8. **End**
9. **End**
10. $(data)_p \rightarrow (M_i)_p$ && $(data)_j \rightarrow (M_i)_{p-1}$
End

According to this algorithm, the sensors (S_1, S_2, …. Sj) captures data (data(S_1) data(S_2) …. ≠ data (S_j)). The algorithm will need to define the number of sensors installed along with number of components running over monitoring panels (M_1, M_2, …., M_i). The study assumes that monitoring units (*i*) is always less than number of sensing units (*j*), which also show a representation of smart environment where there are good possibilities that information exhibited by one monitoring unit could be amalgamated from multiple number of sensing units (Line-1). For a complete set of event *e*, the complete observation is carried out for start and end event points i.e. e_1 and e_2 respectively. In order to retain better practical situation, the proposed system also considers that some of the sensory data are quite prioritized with prioritization value fixed. It is assumed that a matrix *pr* is considered to be repository of programmed threat level (Line-4). Hence, the system forwards two types of data viz. normal sensed data $(data)_j$ and prioritized data $(data)_p$ (Line-5 and Line-7). The prioritized data are forwarded to monitoring units *M* followed by applying statistical function $f_1(x)$ on the normal data (Line-10).

(ii) Transmitting Extracted Knowledge to Cloud Ecosystem: This step of operation is carried out by aggregating the fused data and storing it in cloud datacenter via gateway node. Figure 1 shows the process of applying analytics and extracting knowledge to be stored in datacenter. The description of the above-mentioned flow is illustrated in terms of the algorithm that takes the input of data (fused data) which after processing yields datastore (data stored in cloud). The steps of the algorithm are as follows:

Algorithm for repositing analytical data in data center
Input: data (fused data)
Output: $data_{store}$ (data stored in cloud)
Start
1. Apply f_2(data)
2. **For** all e_i
3. **If** $e_i > e_{th}$
4. flag $e_i \rightarrow$ *critical*
5. **If** $e_{lw} < e_i < e_{th}$
6. flag $e_i \rightarrow$ *medium*
7. **If** $e_i < e_{lw}$
8. flag $e_i \rightarrow$ *normal*
9. **End**
10. Perform data forwarding, $e_i \rightarrow g_{node} \rightarrow$ available $(select(med)_k)$
11. Store ei as datastore in data center.
End

After obtaining the fused data, the algorithm to carry out analytical operation is applied to the fused data in order to obtain the scheduled data with respect to criticality (Line-1). The proposed system considers all the eventual information *e* which is actually a form of real-time data (Line-2). With respect to the threshold event information eth (Line-3), the proposed system makes a decision of criticality of the eventual information i.e. critical, medium, or normal (Line-5 to Line-7). This generates a discrete set of information which is pushed forward into the mobile wireless gateway system gnode that further forwards to the respective mobile network using different sets of communication system i.e. satellite, wireless networks, cellular network (Line-10). There is a possibility of k number of antenna system which the gateway node has to select on the basis of priority level of the data packet pre-configured. The information is forwarded finally to the datacenters.

(iii) Interfacing Mined Data with Supply Chain Management Terminal: This is the final operation that completely automates the process of demands and supply. Based on the analytical data of the vehicular diagnostic, the algorithm could easily find out the necessarily of service/maintenance, called as a *demand*. This information of demand is something that is pulled autonomously from the terminal of client SCM machine. Based on demand, the service provider schedules it service that saves production cost as well as offers faster response time. The schema of its implementation is showcased as follows;

The implementation of the proposed scheme is discussed in following step of algorithm implementation as below:

Algorithm for demand-supply conformation using pull-push
Input: $data_{store}$(stored data in datacenter)
Output: *c* (automated control of demand-supply)
Start
1. Extract $data_{store}$ ('critical', 'medium');
2. Forward datastore→f_2(x)
3. **If** demand(*critical*)=high
4. q_1→*pull_msg*(*critical*)
5. **Else**
6. q_2→*pull_msg*(*medium*)
7. **End**
8. Forward (q_1|q_2)→SCM_interface
9. c→generate *pull_msg*(ρ_1 && ρ_2)
End

The prime intention of this algorithm is to ensure that real-time information of the demands are generated and is forwarded precisely to the service provider autonomously, which can assists in offering well-planned services to clients in automotive industry. The complete algorithm actually works in two phases in parallel. In the first phase, the algorithm just forwards only the data-store elements matching with *critical* and *medium* order as they could only be precise representative of demands (Line-1). The shortlisted information is then forwarded to an explicit function f_2(x) which is responsible for auto-identification and evaluating the newly generated demands (Line-2). Upon finding the criticality of the demand (Line-3 and Line-5), the function generates a pull message q_1 and q_2, which is majorly meant for classifying the form of the demands newly generated (Line-4 and Line-6). Upon successful confirmation, the algorithm further generates a push message (ρ_1 and ρ_2). The first push message ρ_1 acts as a customized and automated message that can be only accessible for clients about the information related to services associated with the upcoming generated demands. On the other hand, the second push message ρ_2 is meant for service scheduling for the service provider. The interesting point is the production or planning of service delivery can be started well-in advanced on the basis of the analytical information of the demand and service delivery can be carried out on or before time.

5 Result Analysis

As the proposed system is basically a typical architecture focusing on implementing 4.0 standard with respect to 4.0 service industry, therefore, there is a very big scenario of implementation of such scheme. The proposed system implements a Delphi technique in order to perform validation of the model choosing 175 participants associated with SCM process in automotive industry since last 7–10 years of profession experience. The complete assessment is carried out with respect to 5 performance metrics of 4.0 industry standard viz. (i) demand analysis, (ii) precision in decision making, (iii) quality in service delivery, (iv) cost optimization, (v) new product innovation. The consolidated outcome was shown in Figs. 2, 3, 4, 5 and 6, where it can be seen that proposed system offers better visualization and interpretation of demand particulars very distinctively (Fig. 2). The model also ensures that it can cater up the exact demands of both supply and demands and is also capable of meeting the demands in shortest span of time (Fig. 3). The quality of the service delivery is dominantly proven to be enhanced owing to the proposed function of performing analytical operations towards extracting information of real-time demands (Fig. 4). Apart from this, there is no inter-dependencies towards adopting any complex infrastructure or sophisticated engineering process in order to execute the norms of the proposed architecture. Moreover, it doesn't require much controlling from the operators too (Fig. 5). Finally, the proposed system is also in favor of disruptive innovation as working on new products/services as per dynamic demands will also generate more line of automotive applications as well as it will introduce a new arena of business (Fig. 6). The proposed system will also ensure higher degree of simplicity in order to formulate any planning towards different forms and types of demands and incorporate a true edge to competitive market. Apart from this, there is all round of monitoring level in order to ensure that all the delivered services/product meets the actual quality that is demanded by the client in real-time basis. Moreover, the proposed system offers enhanced computational efficient.

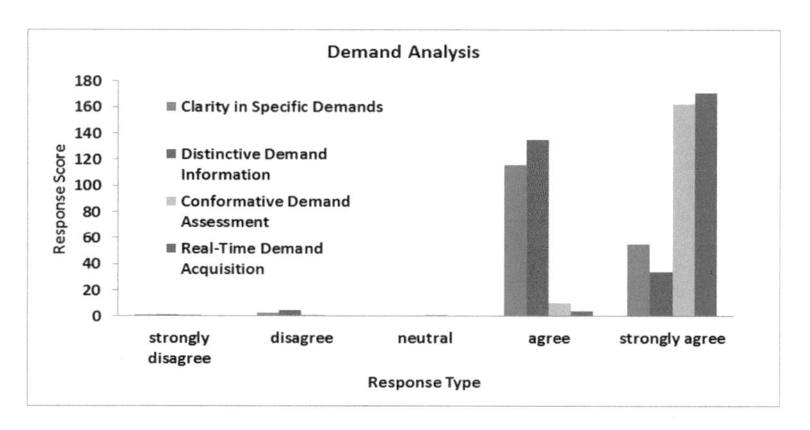

Fig. 2. Evaluation of demand analysis

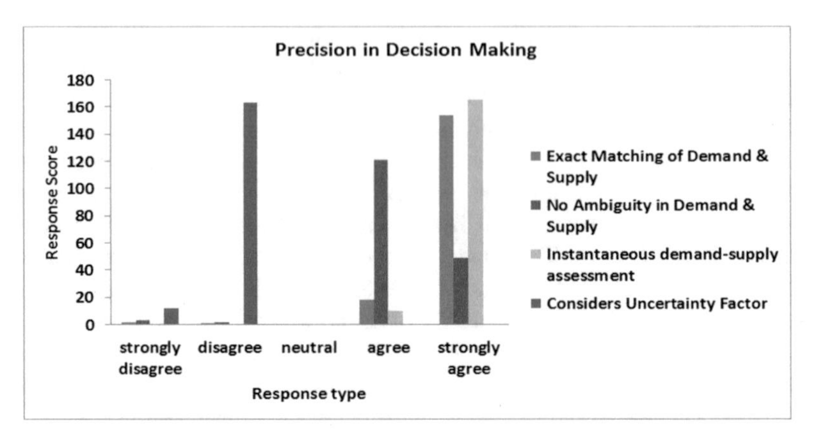

Fig. 3. Evaluation of precision in decision making

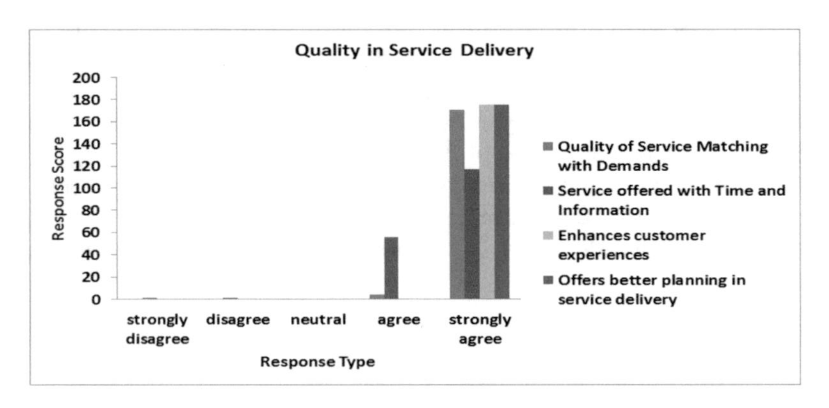

Fig. 4. Analysis of quality in service delivery

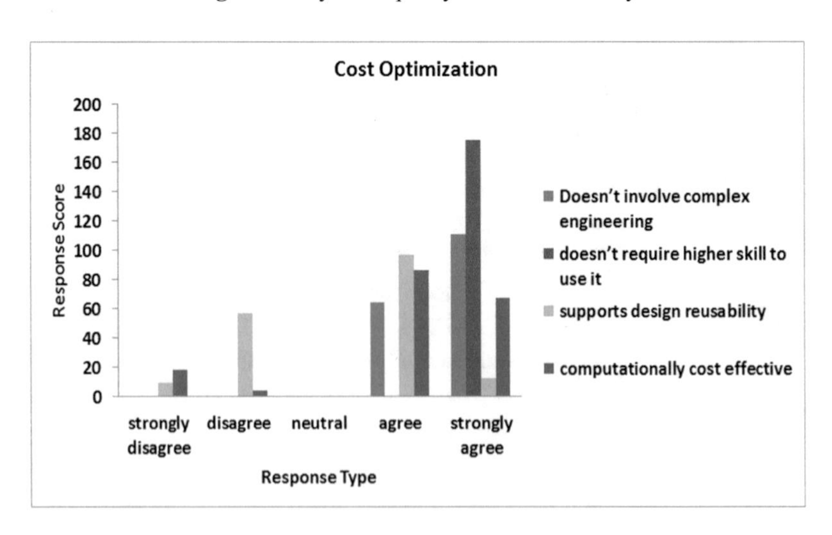

Fig. 5. Analysis of cost optimization

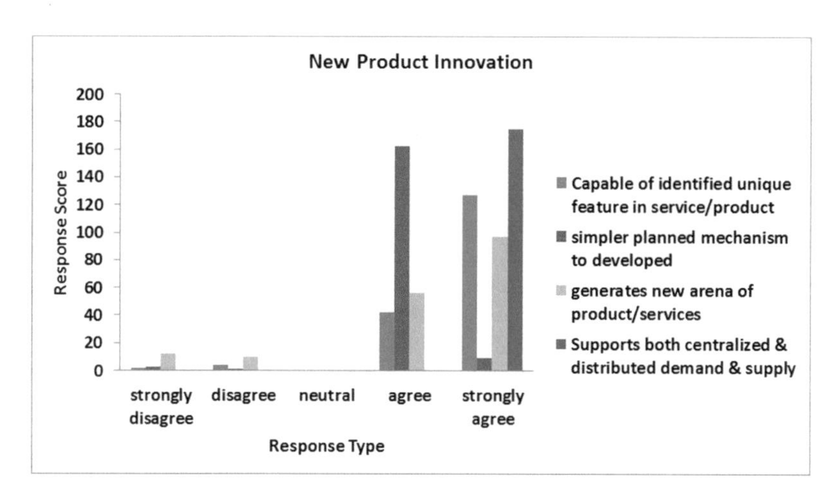

Fig. 6. Analysis of new product innovation

6 Conclusion and Future Work

This paper has presented a framework that is capable of automating the process of supply chain management in order to close down the gap between real-time demands and supply issues considering Industry 4.0 standard over IoT ecosystem. The best part of the framework is that it offers better cost effectiveness in implementation with better response rate that directly influences the quality of service delivery in SCM. The future work could be carried out in the direction of addressing the problems of inclusion of uncertainty while performing prediction using machine learning approach. Such design will offer more clarity and preciseness over the actual need of the demands more specifically which can further reduce the cost of production and offer more benefits in the form of automating the decision making in SCM.

References

1. Haverkort, B.R., Zimmermann, A.: Smart industry: how ICT will change the game! IEEE Internet Comput. **21**(1), 8–10 (2017)
2. Gilchrist, A.: Industry 4.0: The Industrial Internet of Things. Apress (2016)
3. Velda, A.M.E., Dhiba, Y.: Supply chain management performance in automotive sector: case of Morocco. In: 2017 International Colloquium on Logistics and Supply Chain Management (LOGISTIQUA), Rabat, pp. 106–111 (2017)
4. Chandriah, K.K., Raghavendra, N.V.: Scaling research efficacy in supply chain management towards industry 4.0 automation. Communications **7**, 32–40 (2018)
5. Aazam, M., Zeadally, S., Harras, K.A.: Deploying fog computing in industrial internet of things and industry 4.0. IEEE Trans. Ind. Inform. **14**(10), 4674–4682 (2018)
6. Gonzalez, A.G.C., Alves, M.V.S., Viana, G.S., Carvalho, L.K., Basilio, J.C.: Supervisory control-based navigation architecture: a new framework for autonomous robots in industry 4.0 environments. IEEE Trans. Ind. Inform. **14**(4), 1732–1743 (2018)

7. Wan, J., Yi, M., Li, D., Zhang, C., Wang, S., Zhou, K.: Mobile services for customization manufacturing systems: an example of industry 4.0. IEEE Access **4**, 8977–8986 (2016)
8. Patel, P., Ali, M.I., Sheth, A.: From raw data to smart manufacturing: AI and semantic web of things for industry 4.0. IEEE Intell. Syst. **33**(4), 79–86 (2018)
9. Posada, J., et al.: Graphics and media technologies for operators in industry 4.0. IEEE Comput. Graph. Appl. **38**(5), 119–132 (2018)
10. Condry, M.W., Nelson, C.B.: Using smart edge IoT devices for safer, rapid response with industry IoT control operations. Proc. IEEE **104**(5), 938–946 (2016)
11. Fernandes, A.C., et al.: Towards an approach to assess Supply Chain Quality Management maturity. In: 2017 IEEE International Conference on Industrial Engineering and Engineering Management (IEEM), Singapore, pp. 1445–1449 (2017)
12. Sun, J., Zhu, Q.: Organizational green supply chain management capability assessment: a hybrid group decision making model application. IEEE Eng. Manag. Rev. **46**(1), 117–127 (2018)
13. Zhang, J., Guo, Z., Chen, X.: Evaluation of automotive supply chain risks: an empirical research. In: 2016 13th International Conference on Service Systems and Service Management (ICSSSM), Kunming, pp. 1–6 (2016)
14. Ray, S., Chen, W., Cammarota, R.: Invited: protecting the supply chain for automotives and IoTs. In: 2018 55th ACM/ESDA/IEEE Design Automation Conference (DAC), San Francisco, CA, pp. 1–4 (2018)
15. Yildiz, H., DuHadway, S., Narasimhan, R., Narayanan, S.: Production planning using evolving demand forecasts in the automotive industry. IEEE Trans. Eng. Manag. **63**(3), 296–304 (2016)
16. Fahhama, L., Zamma, A., Mansouri, K., Elmajid, Z.: Towards a mixed method model and simulation of the automotive supply chain network connectivity. In: 2017 International Colloquium on Logistics and Supply Chain Management (LOGISTIQUA), Rabat, pp. 13–18 (2017)
17. El Farouk Imane, I., Fouad, J.: Synchronous flow in automotive industry: case study of RENAULT. In: 2017 International Colloquium on Logistics and Supply Chain Management (LOGISTIQUA), Rabat, pp. 199–203 (2017)
18. Faycal, M., Abouabdellah, P.A.: Proposition of a methodology for evaluating the maintenance's performance in a global supply chain integrating reverse logistics: case study: automotive wiring. In: 2016 3rd International Conference on Logistics Operations Management (GOL), Fez, pp. 1–7 (2016)
19. Liu, Y., Zhang, B., Wang, Y.: Automotive supply chain management research. In: 2010 International Conference on Optoelectronics and Image Processing, Haikou, pp. 306–308 (2010)
20. Shi, X.: Research on logistics mode in automobile enterprises based on industry conformity. In: 2011 2nd International Conference on Artificial Intelligence, Management Science and Electronic Commerce (AIMSEC), Dengleng, pp. 1402–1405 (2011)
21. Choi, T.: A system of systems approach for global supply chain management in the big data era. IEEE Eng. Manag. Rev. **46**(1), 91–97 (2018)
22. Yue, X., Chen, Y.: Strategy optimization of supply chain enterprises based on fuzzy decision making model in internet of things. IEEE Access **6**, 70378–70387 (2018)
23. Wang, Y., Geng, X., Zhang, F., Ruan, J.: An immune genetic algorithm for multi-echelon inventory cost control of IOT based supply chains. IEEE Access **6**, 8547–8555 (2018)

Dynamic Routing Protocol Convergence in Simulated and Real IPv4 and IPv6 Networks

Tomas Sochor(✉) and Hana Sochorova

University of Ostrava, 30. dubna 22, 70103 Ostrava, Czech Republic
tomas.sochor@osu.cz
http://prf.osu.eu/kip

Abstract. The paper focuses primarily on the research of IGP routing protocols, namely RIP, EIGRP and OSPF from the point of view of the convergence time in small and medium-sized networks. Experiments using GNS3 and in real routed networks are shown for both IPv4 and IPv6 protocols and their results are described in detail. The results demonstrate EIGRP as the fastest converging protocol closely followed by OSPF while RIP convergence is much slower. Also, the data traffic generated by all three protocols were compared. Here, RIP is the best while EIGRP and OSPF generated 5–6 times more data in the beginning and 2–3 times more later. The comparison of simulated results shows that routing protocol behavior in GNS3 differ significantly from real routers. Corresponding results were obtained for both IPv4 and IPv6 and both are presented. The difference between IPv4 and IPv6 are not significant. On the other hand, significant difference has been observed between the corresponding experiments in GNS3-simulated and real networks.

Keywords: RIP · OSPF · EIGRP · IPv4 · IPv6 · GNS3 · Small network · Medium network · Routing convergence time · Routing update size · Cisco 2811 routers

1 Introduction

Routing is one of key functions for keeping interconnected network in a state when each network is able to communicate to others. Routing informs every router where to forward packets destined to a specific IP address. Sometimes, primarily in small networks or networks with no or rare redundant links, static routing based on manual setting of forwarding paths (next hops) is sufficient. Nevertheless, dynamic routing is required in the majority of networks because only a dynamic routing can respond to changes in the network interconnections promptly and in an automated way.

There are several dynamic routing approaches reflected in various routing protocols. Here, the focus is on IGP protocols performing dynamic routing in isolated (autonomous) systems composed of several networks. Almost

R. Silhavy (Ed.): CSOC 2019, AISC 986, pp. 117–126, 2019.
https://doi.org/10.1007/978-3-030-19813-8_13

every dynamic routing approach applied in IP networks rely solely on routers as active elements in the routing process, thus leaving the end-nodes out of the routing process completely. In currently prevailing IP-based networks, two main approaches to dynamic routing problem are used, namely distance-vector routing protocols based on Bellman-Ford algorithm [1–3] and link-state routing protocols based primarily on Djikstra algorithm [4,5].

Many significant differences between the vector-distance and link-state approach exist. Among them, the most fundamental difference consists in the way how routes are updated among the involved routers. The distance-vector routing represented primarily by still highly popular RIP protocols [3,6] relies primarily on distribution of the whole routing table among neighboring routers. On the other hand, the operation of link-state protocols (e.g. OSPF described in [7]) is based on the topological database (in OSPF, it is called Link-state database) generated independently by each of the routers using reception and transmission of link-state packets.

This fundamental difference implies a lot of consequences. While distance-vector approach is easy to be implemented and operated even on a low-cost router with limited hardware resources, it is prone to errors (like routing loops) due to the lack of information about the network topology in the routers. In addition, it can be characterized by slightly higher consumption of bandwidth for routing update distribution. On the other hand, link-state routing approach is resilient to causing routing loops and other errors due to the existence of the link-state database in every router. It is compensated with the higher consumption of router hardware resources (more CPU capacity and memory is needed for more complex calculations) and of the bandwidth, primarily at the beginning of routing process operation.

In all networks where dynamic routing is applied, it is always desirable that the response of routers to a change in the network is as quick as possible. The reason is obvious: the longer the response time of a router is, the higher the likelihood of routing error is. Once the first router (usually the router that is directly connected to the network where a change, e.g. a loss of connectivity between two routers, happened) responses to the network change, it causes the routing tables of all other routers became incorrect (at least partially), and the network as a whole is in inconsistent state. This can result in an error like routing loop or others. Therefore, the time necessary for changing the network from an inconsistent state to the new consistent state (this is usually called convergence time) should be as short as possible. Thus, the convergence time is one of the key factors when deciding on a dynamic routing protocol to apply in a network.

2 Previous Research

Despite the fact that other aspects of routing, namely routing in MANETs (Mobile Ad-hoc NETworks), have become much more popular in recent years as a research topic focusing on routing efficiency (e.g. [8–10]), as well as routing protocol security studies (e.g. [11]) and QoS (e.g. [12]), there are several studies published in last ten years focusing on various aspects of routing protocols in IP networks (e.g. [13–17]).

For example, the paper [15] focuses on both popular RIP, OSPF and EIGRP as well as the retired Cisco's IGRP protocols. While most conclusions seem to be generally acceptable (e.g. higher link utilization and router overhead by OSPF and EIGRP in comparison to RIP), it should be noted that the presentation of results is a bit too brief as well as the discussion on them. Other conclusions need to be verified and could be true only under specific circumstances (e.g. the claimed more difficult configuration of OSPF in comparison to EIGRP).

On the other hand, the paper [13] compared EIGRP, OSPF, and, despite their title, only partly RIP. The main conclusion of the paper is that EIGRP overperforms OSPF in almost all measured aspects, namely network convergence time, routing protocol traffic, CPU utilization and network bandwidth. Such a conclusion is far from being surprising because EIGRP is proprietary protocol that is known to be better fitted on the specific hardware that open-source OSPF can be. Slightly later, the paper [18] has published similar conclusions, too.

The more recent paper [16] focuses solely on comparison of RIP and OSPF. The paper presents similar results to the previously discussed one obtained from the small network consisting of 6 routers. The most notable results are the differences measured in convergence time (approx. 4 times higher for RIP than OSPF) and packet low (cca 2 times higher for RIP) while latency and throughput did not show significant differences.

The paper [19] demonstrates a comparison among RIPv2, EIGRP and OSPF again in a small network consisting of 8 routers. Unfortunately, the paper does not present any quantitative results and focuses rather on the configuration of route redistribution.

Finally, the most recent paper [14] demonstrates much more comprehensive view on the comparison of RIP, EIRGP and OSPF on the network consisting of 14 routers including routing update authentication. The paper demonstrates the comparison concluding that from the point of view of convergence time, packet loss and throughput, RIP produced the worst results. While in convergence time, RIP results are more than 20 times worse comparing the competitors while, in packet loss, RIP demonstrated hundreds of packets lost while just units of them were observer in OSPF and EIGRP. The differences in throughput is not so significant but here RIP was the worst, while EIGRP dominated over OSPF.

Regarding routing protocols for IPv6, there are few publications devoted to IPv6 routing performance in networks. For example, [20] shows certain comparison but it focuses on MANET. Similarly, [21] published a performance evaluation in IPv4 and IPv6 networks but their focus lies on packet loss and comparison among various types of networks from the point of view of the transmission parameters vs. network technology and they do not deal with the convergence. Moreover, their results can be hardly verified because commercial simulation tool was used. On the other hand, the paper [22] compared EIRGP, OSPFv3 and RIPng for routing in IPv6 network. However, their comparison focuses to the traffic parameters like delay, packet loss and throughput. Again, EIGR dominated over OSPF and RIPng. Nevertheless, no relevant publication focusing on convergence time comparison in IPv6 networks has been found so far.

Altogether, almost all of the previous research regarding the convergence time of various IGP protocols, including [23], concluded that the convergence times for RIP is much worse than the ones for link-state protocols like EIGRP and OSPF. On the other hand, the more precise quantification (like how many times the link-state protocols are better than link-state ones) varies significantly across the existing publications. Therefore, this study focuses to the convergence time comparison among existing IGP protocols.

3 Measurement Methodology

The main aim of the presented study was to verify and possibly precise the findings mentioned above. Because of the fact that it could be complicated to build a real network with a sufficient number of routers (primarily due to the limited laboratory hardware availability), a decision was made to build the network in GNS3 network virtualization software. For the purpose of verification, the measurements in the small network were performed using a real network made of Cisco 2811 routers as well.

The limitations on GNS3 measurements are quite acceptable: the GNS3 does not perform the simulation of the routers (that is always an abstraction of many features). On the other hand, GNS3 runs the real router software (e.g. Cisco IOS) in the virtualization environment. Therefore, the network behavior in GNS3 is quite close to the real network behavior.

First, the model network (as described later in the Subsect. 4.1) was set up in GNS3. As a next step, the IPv4 addresses from the selected IPv4 ranges as shown in Table 1 were assigned to all the network devices. This network was used as a basis for configuring a selected routing protocol (among RIP, EIGRP, OSPF) and subsequent measurements. In parallel, the copy of the model network was configured with IPv6 addresses (see Table 1). Subsequently, the selected IPv6 routing protocol (among RIPng, EIGRP and OSPFv3) was configured and routing convergence was measured. The measurement of routing convergence was performed via monitoring the change of routing tables on all routers after defined change in the network (manual shutting down a router interface).

When a specific routing protocol convergence was measured, its specific features were taken into account. For RIP, it is known than for historical reasons, because of its prone to form routing loops, its timer default values are quite high (i.e. Update: 30 s, Invalid: 180 s, Hold Down: 180 s, Flush: 240 s). Therefore, another measurement with significantly shortened timers (Update: 1 s, Invalid: 5 s, Hold Down: 5 s, Flush: 10 s) was done, too. For EIGRP where Feasible Successor (FS, that is a backup route) plays an important role in finding a new path after a change, another measurement was added where FS was set and finding of an alternate route is much faster.

In the measurement of routing protocol convergence, it should be taken into account that there could be a significantly stronger traffic just after the change in the network than later. Therefore, the first minute traffic was measured separately. Anyway, despite the fact that numerous factors can play their role in routing protocol selection, the convergence time is the most important because the longer the network it in incoherent state, the higher the likelihood of a routing/forwarding error occurrence is.

4 Results

The measurements were performed in two networks, namely small and medium-sized ones. While the measurements in the small model network were performed both in GNS3 and in the real network made of Cisco 2811 routers, the results from the medium-sized network were obtained solely using GNS3. Both networks are described in the following subsection.

4.1 Model Network Description

The small model network consisted of 4 routers and 4 networks. Among them, 2 networks were LANs and 2 networks were interconnecting. The detailed layout of the network is shown in Fig. 1. In addition, another larger network was used for additional measurements of IPv6 routing convergence. The network consisted of 11 routers with multiple redundant connections as shown in Fig. 2.

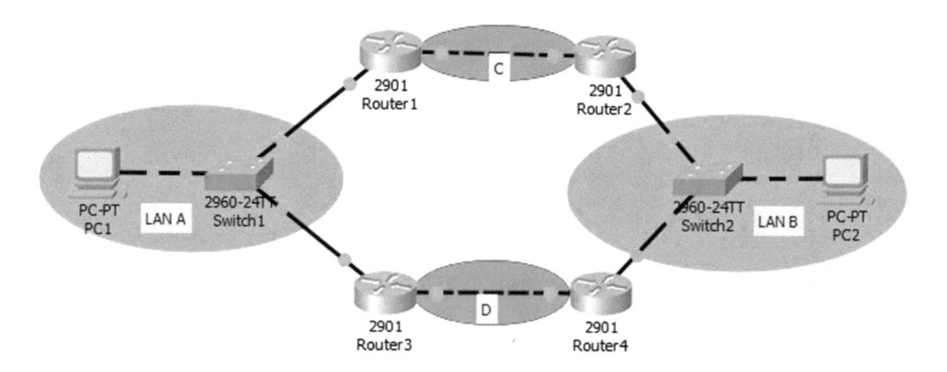

Fig. 1. Diagram of the small model network used for IPv4 and IPv6. Orange ellipses indicate LANs (LAN A, LAN B) while violet ellipses indicate the interconnecting networks (C and D)

IP addresses for experiments with IP version 4 and IP version 6 are summarized in Table 1. In all routers, the selected routing protocol was run in default configuration. No redistribution and/or static routes were applied.

Table 1. IP address ranges assigned to the small model network

IP address ranges assigned		
Network name	IPv4 range	IPv6 range
LAN A	192.168.1.0/24	1003:2018::/64
LAN B	192.168.2.0/24	2004:2018::/64
Network C	10.10.10.0/30	1002:2018::/64
Network D	20.20.20.0/30	3004:2018::/64

4.2 Results for IPv4

As mentioned earlier, results in the small network were obtained both from GNS3 simulation and from real network based on Cisco 2811 routers. As it can be seen from Table 2, there are significant differences among routing protocol convergence in GNS3 and in the real networks.

Table 2. Convergence times [in seconds] in the small network using IPv4

Routing protocol	GNS3	Real network
RIP default timers	69.4	17.1
RIP shortened timers	8.0	2.7
EIGRP	15.9	12.6
EIGRP with FS	12.3	3.4
OSPF	4.9	7.1

It is obvious that RIP with default timers provides the worst results that can be hardly acceptable in a real network. On the other hand, when RIP timers were set to shorter values, its convergence time competes with other protocols (namely EIGRP with FS and OSPF). It should be noted nevertheless that the timers were shorten in an aggressive way that can lead into network oscillations etc. This result confirms that RIP is a real competitive option for small network where it offers reasonable convergence times and low resource consumption along with the ease of use.

4.3 Results for IPv6

Similarly to the results for IPv4, the results for IPv6 in the small network were obtained both from GNS3 simulation and from real network based on Cisco 2811 routers. The results are summarized in Table 3. Again, significant differences between GNS3 and real network were observed, in certain cases (RIPng with default timers) the difference was even more significant.

Table 3. Convergence times in the small network using IPv6 [in seconds]

Routing protocol	GNS3	Real network
RIPng default timers	113.3	18.3
RIPng shortened timers	49.3	12.9
EIGRP	12.9	13.9
EIGRP with FS	4.9	3.1
OSPFv3	4.9	6.9

For IPv6, similarly to the former case of IPv4, RIPng in default setting produced the worst convergence times. What is different, is the fact that even with shortened timers, RIPng could hardly compete to link-state protocols EIGRP and OSPFv3. There is another difference in comparison EIGRP and OSPFv3. Here, EIGRPv3 with FS and OSPF produced comparable results in GNS3 and EIGRPv3 with FS even excelled on real devices.

Results from GNS3 Differ Significantly from Real Network Measurements. As one can see in Tables 2 and 3, the difference between convergence time in GSN3 and real network devices seems to be partly systematic in the following sense: For RIP, GNS3 convergence time in GNS3 in much longer than measured on real devices. For EIGRP in IPv4, a small difference exists without FS while more significant with FS. For EIRGP in IPv6, differences are quite negligible. For OSPF, GNS3 provides more favorable times than real network devices (Table 4). However, no explanation of these phenomena could be found.

Table 4. GNS3 Convergence times in the medium–sized network using IPv6 [seconds]

Routing protocol	Convergence delay
RIP default timers	52.4
EIGRP with FS	2.3
OSPF	7.4

4.4 Routing Protocol Traffic

There is another important aspect of routing protocols that should be taken into account in routing protocol selection for a real network that is the amount of traffic generated by the protocol itself. It is widely known that link-.state protocols perform so-called “initial flood” of packets when their operation started. Therefore, a small comparison was made to quantify its size. The results are shown in Table 5.

As one can see, the initial flood caused higher traffic in the first minute for both EIGRP and OSPF in IPv4 and IPv6. However, the increase was not too

Table 5. Traffic generated by routing protocols in the small network

Protocols	Bytes 1st min.	Bytes next mins
RIP IPv4	436	436
EIGRP IPv4	2400	962
OSPF IPv4	3864	1316
RIPng IPv6	868	504
EIGRP IPv6	4324	1598
OSPFv3 IPv6	1312	1128

high, and in certain cases it was even negligible (namely for EIGRP in IPv6). On the other hand, the difference in subsequent traffic amount is much smaller but still significant.

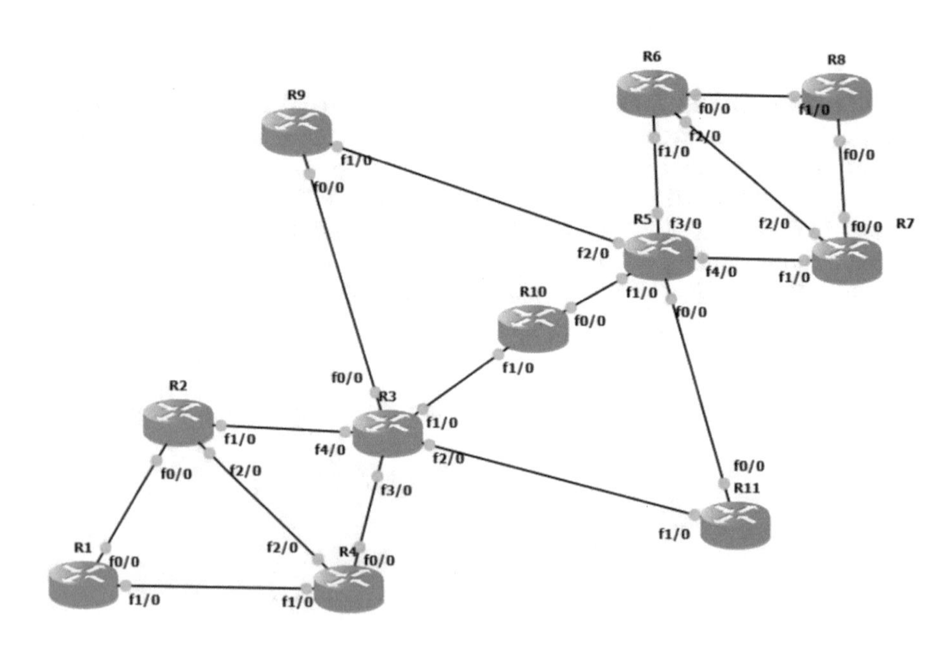

Fig. 2. The medium-sized network used for GNS3 measurements.

5 Conclusions

The results described here partially confirmed previously published findings claiming that link-state protocols over-perform their distance–vector competitors. This was confirmed in the sense that RIP in default setting provides much slower convergence that EIGRP and OSPF for both IPv4 and IPv6 regardless of the network. On the other hand, when RIP timers are configured for faster convergence, RIP can become a real competitor to link-state protocols, primarily in small networks using IPv4.

The difference between GNS3 modeled network and real network convergence times observed in this study will require a deeper investigation to reveal the reasons.

Regarding the traffic generated by the routing protocol, the study demonstrated that this is not a decisive factor for small and medium–sized networks. The difference exist but they are quite small.

Acknowledgment. The paper was supported by the project *Network Infrastructure Security* of the Student Grant Competition of the University of Ostrava.

References

1. Awerbuch, B., Bar-Noy, A., Gopal, M.: Approximate distributed Bellman-Ford algorithms. IEEE Trans. Commun. **42**(8), 2515–2517 (1994)
2. Hedrick, C.: Routing Information Protocol. IETF (1988). https://www.rfc-editor.org/rfc/rfc1058.txt
3. Malkin, G.: RIP version 2-carrying additional information. IETF (1994). https://www.rfc-editor.org/rfc/rfc1723.txt
4. Noto, M., Sato, H.: A method for the shortest path search by extended Dijkstra algorithm. In: IEEE International Conference on Systems, Man, and Cybernetics, pp. 2316–2320. IEEE (2000)
5. Jianya, Y.Y.G.: An efficient implementation of shortest path algorithm based on Dijkstra algorithm. J. Wuhan Tech. Univ. Surv. Mapp. (Wtusm) **3**(004) (1999)
6. Malkin, G., Minnear, R.: RIPng for IPv6. IETF (1997). https://www.rfc-editor.org/rfc/rfc2080.txt
7. Moy, J.: OSPF Version 2. RFC 2328. IETF (1998). https://www.rfc-editor.org/rfc/rfc2328.txt
8. Pandey, K., Swaroop, A.: A comprehensive performance analysis of proactive, reactive and hybrid MANETs routing protocols. arXiv preprint arXiv:1112.5703 (2011)
9. Sochor, T., Klimes, C.: Overview of web anonymization. Eur. J. Bus. Sci. Technol. **3**(2), 96–105 (2017)
10. Gupta, A.K., Sadawarti, H., Verma, A.K.: Performance analysis of AODV, DSR & TORA routing protocols. Int. J. Eng. Technol. **2**(2), 226 (2010)
11. Lysenko, S., Savenko, O., Bobrovnikova, K., Kryshchuk, A., Savenko, B.: Information technology for botnets detection based on their behaviour in the corporate area network. In: Communications in Computer and Information Science, vol. 718, pp. 167–181 (2017)
12. Zach, P., Pokorny, M., Balej, J.: Quality of experience of voice services in corporate network. Procedia Econ. Finan. **12**, 771–779 (2014)
13. Thorenoor, S.G.: Dynamic routing protocol implementation decision between EIGRP, OSPF and RIP based on technical background using OPNET modeler. In: 2010 Second International Conference on Computer and Network Technology (ICCNT), pp. 191–195. IEEE (2010)
14. Athira, M., Abrahami, L., Sangeetha, R.G.: Study on network performance of interior gateway protocols–RIP, EIGRP and OSPF. In: International Conference on Nextgen Electronic Technologies: Silicon to Software (ICNETS2), pp. 344–348. IEEE (2017). https://doi.org/10.1109/ICNETS2.2017.8067958

15. Vetriselvan, V., Patil, P.R., Mahendran, M.: Survey on the RIP, OSPF, EIGRP routing protocols. Int. J. Comput. Sci. Inf. Technol. **5**(2), 1058–1065 (2014)
16. Jayakumar, M., Rekha, N.R.S., Bharathi, B.: A comparative study on RIP and OSPF protocols. In: International Conference on Innovations in Information, Embedded and Communication Systems (ICIIECS), pp. 1–5. IEEE (2015). https://doi.org/10.1109/ICIIECS.2015.7193275
17. Circiumarescu, L.D., Predusca, G., Angelescu, N., Puchianu, D.: Comparative analysis of protocol RIP, OSPF, RIGRP and IGRP for service video conferencing, E-mail, FTP, HTTP. In: 2015 20th International Conference on Control Systems and Computer Science (CSCS), pp. 584–589. IEEE (2015). https://doi.org/10.1109/CSCS.2015.17
18. Dwyer, J., Jasani, H.: An analysis of convergence delay caused by link failures in autonomous systems. In: Proceedings of IEEE Southeastcon, pp. 1–6 (2012). ISSN 1558-058X
19. Dey, G.K., Ahmed, M.M., Ahmmed, K.T.: Performance analysis and redistribution among RIPv2, EIGRP & OSPF routing protocol. In: 2015 1st International Conference on Computer and Information Engineering (ICCIE), pp. 21–24. IEEE (2015)
20. Chauhan, D., Sharma, S.: Performance evaluation of different routing protocols in IPv4 and IPv6 networks on the basis of packet sizes. Procedia Comput. Sci. **46**, 1072–1078 (2015)
21. Chandel, S.T., Sharma, S.: Performance evaluation of IPv4 and IPv6 routing protocols on wired, wireless and hybrid networks. Int. J. Comput. Netw. Appl. (IJCNA) **3**(3), 57–62 (2016)
22. Al Farizky, R.F., et al.: Routing protocol RIPng, OSPFv3, and EIGRP on IPv6 for video streaming services. In: 5th International Conference on Cyber and IT Service Management (CITSM), pp. 1–6. IEEE (2017). https://doi.org/10.1109/CITSM.2017.8089250
23. Jerabek, D.: Routing Protocols Efficiency. Bachelor thesis. University of Ostrava (2017). (in Czech)

Survey on Messaging in the Internet

Tomas Sochor(✉) and Nadezda Chalupova

Mendel University in Brno, Zemedelska 1, 613 00 Brno, Czech Republic
tomas.sochor@mendelu.cz

Abstract. The prevailing messaging service in the internet has always been the electronic mail. Despite multiple potential competitors have emerged since it started in early 1970's, e-mail still has the highest share among users. The properties of e-mail and their major competitors are briefly reviewed in order to understand current challenges for e-mail. There are unconfirmed expectations for e-mail to weaken its position primarily among young generations. Therefore, prospects for a future for messaging services (not exclusively e-mail) are presented by results of a small-scale poll among more than 200 messaging users in universities.

Keywords: Electronic mail · Messaging service · Messaging reliability · Spam elimination · IRC · ICQ · Skype · SMTP · Facebook Messenger · Google Hangouts · Viber · WhatsApp

1 Introduction

Messaging in general is usually understood as a service for sending (primary text–based) messages to the recipient (and delivering them). In the internet, messaging can be understood in numerous ways because the interpersonal communication plays an important role from its beginning in 1960's. According to [1], the history of messaging between users (extremely primitive from the present perspective) in the ARPANET (as the internet predecessor) started even in early 1960's on the CTSS system where about hundreds of users (but maximum 30 users simultaneously logged-in) used to send messages using shared files named according to the recipient's nick (e.g. TO_JOHN) in stored in commonly accessible directories. On CTSS, the first mail program was written and made available in 1965. In 1971, first attempts to develop a standardized messaging protocol were made (e.g. RFC196, RFC221 and RFC 278). These protocols were predominantly based on the older FTP protocol. Since then, the electronic mail gradually became the first widely used electronic messaging service.

Much later, in 1981, the SMTP protocol was standardized that became widely used for transferring electronic mail messages. The SMTP protocol has established the basic principles for transferring e-mail messages for quite a long period and it is still used (in a significantly updated version defined in 2008 in RFC5321 (see [2])) as a main standardized messaging service in the internet.

R. Silhavy (Ed.): CSOC 2019, AISC 986, pp. 127–136, 2019.
https://doi.org/10.1007/978-3-030-19813-8_14

1.1 Electronic Mail Still Prevails in Internet Messaging

Even before SMTP standardization, the usage of the electronic mail started growing and the user increase have been going still. For instance, the 2014's study [3] had estimated the number of e-mail users to 2.5 billion (in the 1st quarter of 2014)[1] and expected increase to 2.8 billion of users in 2018. Later, in 2017, the total number was reported as 3.7 billion e-mail users (see [4]) that exceeded the previous estimate significantly. Moreover, in 2017, the total averaged e-mail traffic was estimated to 269 billion messages per day (again according to [4]) while other resources estimate differ quite a lot. For instance, recent figures from [5] indicate the total daily number of legitimate messages to almost 71 billion that gives (combined with daily spam average 406 billion) the gross total of 477 billion e-mail messages sent every day.

There are numerous reasons for such a massive e-mail usage. Among them, the following worth mentioning as the most significant for users:

- E-mail messages are used as an alternative authentication tool and bot discriminator in various internet services like social networks and others.
- E-mail service is available on almost any operating system and hardware unlike other more resource–demanding messengers.

Despite the figures, there have been recurring expectations that e-mail service will become obsolete, e.g. in [6]. More recently, other publications declare that e-mail usage is expected to decrease (primarily among young people), e.g. in [7] where millenians are claimed to hate e-mail, or [8] who estimates the e-mail to become obsolete by 2020. However, such claims usually suffer from the lack of real findings and solid basis. On the other hand, the more sophisticated and thorough study [9] tries to prove that e-mail service is necessary and still among the most useful. For the above reasons, this paper tries to study the user expectations regarding messaging future usage, primarily among young generation.

1.2 Differences in E-mail Service in the Past and Present

The features and conclusively the usage of the electronic mail was much different in its beginning in 1970's than nowadays. This was not only due to the fact that almost no systems with GUI were available and only so-called 7-bit characters were acceptable in the message body; enclosures could not be sent at all (see RFC 821 and RFC 822) not only because of usual strict limitations of acceptable message size. Nevertheless, the basic behavior of the e-mail service and the application protocol for sending messages (namely SMTP) have been kept almost unchanged (see [2]).

Because of the fact that in present, the great majority of e-mail messages in still sent via SMTP-based electronic mail service, the terms "e-mail" and "electronic mail" will be used here as the synonyms of "SMTP-based electronic mail". One should bear in mind that other e-mail messaging system using

[1] At that time, the total estimated number of active mailboxes was 4.1 billion.

different protocol exist and are used in specific environments (primarily in corporate and other large-scale networks), usually having an interface to SMTP-based e-mail service in the internet. Their features and behavior can significantly differ from the standard e-mail, however.

1.3 Current Electronic Mail Challenges and Issues

E-mail service, being used by a huge number of internet users worldwide, suffers from certain issues still. Quite soon after e-mail became available and used by more that dozens of people, its democratic nature consisting in its availability for free and virtually for anyone interested started causing certain troubles. Even in early 1970's, some people tried to misuse the e-mail for sending messages with contents beyond the original intent for that they joined the service. On one hand, it is quite obvious that once users can communicate quickly and for free (unlike using non-IT messaging like post, telephone etc.), they started to use the e-mail services for various purposes. On the other hand, some recipient felt unhappy for receiving unwanted messages e.g. of religious or other highly private nature from people not familiar to them. Such messages started to be called "spam" later.

The current rate of spam in e-mail traffic is very high, it exceeds 80% of all e-mail traffic as a rule (see e.g. [5]) despite numerous measures eliminating significant part hereof (as described e.g. in [10,11,15]). There is an adversary side-effect of certain anti-spam techniques affecting the e-mail message delivery. It is obvious that anti-spam measures greylisting (see [14]) and blacklisting (described in [12]) that rely on message delivery refusal (either temporary or permanent) can have an impact to either delayed or completely impossible delivery. The effect of greylisting on the message delivery has been investigated in detail in [13] where significant delay in the message delivery was observed for certain messages. On the other hand, techniques for sender authentication (e.g. SPF, DKIM and DMARC) can work well only if they are deployed among the majority of servers. Moreover, the limited message size that is exerted as a "natural" limit (unfortunately, seldom explicitly expressed for users) can also complicate the use of e-mail for some users (e.g. by the inability to send a message when connected to an "unknown" network). Most of the above drawbacks are much more important for e-mail than for other messaging services. Therefore, the above factors could motivate some users to prefer other messaging services than e-mail (despite they hardly can avoid their usage completely). This was one of the motives why to focus the poll described below to this aspect of messaging as well.

2 Alternative Messaging Services

In parallel with the growing user adoption of electronic mail service, other messaging services were emerging (and sometimes disappearing) as well. This paper does not have an ambition to cover all historical and presently existing messengers but some of them will be mentioned focusing on the most popular ones.

One of the first command-line text messengers was `continuum` on Multics (later renamed to `forum`) in 1970's (according to [1]). Some of such tools (usually limited to a logical group of servers with similar operating system) kept its existence up to now (e.g. wall or write) but their application is limited.

2.1 Commercial Messengers

On the other hand, certain vital alternative (primarily "instant") messaging services exist. In 1988, IRC emerged and started growing, first in Finland (see [16]), later around the globe. A bit later, several commercial instant messengers appeared. Among them, ICQ was first popular one (developed in 1996 in Israel). It started as a text messenger but gradually voice and video call services were added. ICQ still exists and its development continues but it was rather overshadowed by newer competitors. Subsequently, Skype became popular because it was among first messengers to offer voice calls.

At present, messaging services by great software and internet services seem to prevail. In addition to Skype (now owned by Microsoft), Facebook Messenger, Google Hangouts, Viber, and Whatsapp attracted significant number of users (see e.g. [17]). In this paper, all of them together are called "*commercial messengers*" so that the difference from open internet standardized solutions like SMTP is emphasized.

The number of users of any alternative messaging service is still significantly lower than the one for electronic mail. The most popular among others, Facebook Messenger, is estimated to have about 1.3 billion of users in mid 2017 (see [18]). From the point of view, Viber has almost same figures while other lag significantly behind according to [17]. As one can compare to figures above, the number of users of the most popular commercial messengers is about 3 times less than the current number of e-mail users, still. Nevertheless, the increase of users of e-mail is far less steep that the one for commercial messengers. Therefore it could be useful to investigate reasons whether e-mail users tend to switch to other messengers and why.

2.2 Inter–Process Messaging

It should be noted that for certain users, e.g. software developers, the term "messaging" could bear a slightly different meaning than mentioned here. Due to the fact that sending messages is one of fundamental forms of interprocess communication (even increasingly important due to the increase of running processes communicating beyond the scope of a single operating system), this messaging worth mentioning here as well. Interprocess messaging is usually facilitated using a sort of Message–oriented Midleware (MOM) as described in [19]. Among technologies and tools used for implementing the interprocess messaging, primarily Java Message Service and IBM MQ should be mentioned. Most of MOM supports both active (unicast or multicast) message sending and passive messaging (e.g. publish–subscribe systems). Because of different nature and purpose of such messaging, it is not covered in this research. This is true also for other emerging messaging methods whose real impact is still quite limited.

3 Research of Messaging Habits and Expectations

In order to identify the trends, expectations and potential obstacles in using various messaging among users, a poll was organized among messaging users. For example, [20] give some reasons for text messaging preference by (not only) young generation over the phone or video calls: the direct communication is socially more demanding or difficult. So the poll aim was to identify possible trends in messaging usage.

3.1 Poll Methodology

The primary target group was an academic community of two medium–sized Czech universities where the majority of respondents were from departments of economics and IT. Among them, the significant majority of respondents were students. This focusing on students was chosen due to the fact that students were supposed several times as a group where new IT trends can be identified first. The selection of poll respondents was semi–random. It means that only users belonging to the target group (part of Czech academicians and students in the filed of economics and IT) were addressed (via internal university e-mail) but no further limitation in responding the poll was applied. Moreover, because the access to the poll was not restricted[2] also other participant from universities outside the Czech Republic were invited. Therefore, some academicians from France, Poland, Turkey and other countries took part, too.

The poll consists in filling in a short anonymous questionnaire where there are 3 meritorious questions (with multiple options offered and choice allowed). The first question identified messaging services used (including corporate and private e-mail, Viber, Facebook Messenger, Hangouts, Skype and others). The second question asks about the trends in the number of sent/received messages in the specific services in the past while the third question asks for the equivalent data as expected in the near future. The age of respondents was not collected but from other signs it is estimated that more than 90% of respondents were below 40 years.

The collected responses were normalized and processed using descriptive statistic means. The inconsistent responses were excluded from the analysis, e. g. when the respondent declared decrease/increase expectation in using of services than he/she is not currently using. The results then should be more valid. The final number of consistent responses was 119 (from 224 originally).

[2] The poll is still available at https://goo.gl/forms/32R8x9yFVuBL4zyX2.

3.2 Poll Results

It should be noted first that more about 60% of the respondents use messaging intensely (send and/or receive more than 25 messages per day). Therefore, the results from the poll could be generalized only to the similar group of intense messaging users. Among the respondents, almost everybody declared to use a private e-mail. The usage of other messaging services (corporate e-mail, Facebook Messenger, WhatsApp, Google Hangouts, Viber, Skype and other were the poll options) was much variable.

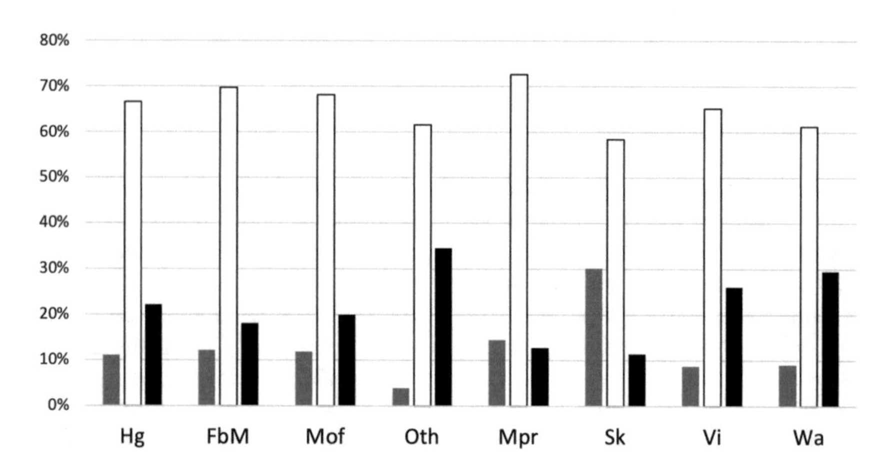

Fig. 1. Expected usage increase (left column, grey), no change (middle column, white) and increase (right column, black) for individual messaging services (from the left: Hangouts, Facebook Messenger, corporate e-mail, other, private e-mail, Skype, Viber, WhatsApp).

The main findings from the poll regarding the present usage were as follows:

1. The number of messaging services used does not correlate to the messaging intensity (the avg. number of messages send per day). The average is slightly more than 3 messaging services used per user.
2. Among the messaging services, e-mails (both office and private identically) were the most popular (96% of intense users and 92% in total) as well as Facebook messenger seems to be very popular (it is used by 91% of intense users and 83% in total). The popularity of other messengers was much lower (Skype 44%, WhatsApp 37%, even less for others)
3. Except Facebook Messenger (FM), the popularity of the specific messenger does not correlate to the messaging frequency. For FM, the strong positive correlation between the messaging frequency and the FM popularity.

Moreover, the poll include some questions regarding the future expectations in messaging usage. The overall results regarding the future expectation can be seen in Fig. 1. From the responses, the following conclusions were derived:

- Intense messaging users have higher expectation in the future increase of messaging services used. But the increase expectation is very small, not either one service in average (median is 1). On the other hand, the responses indicate that no decrease is expected in e-mail usage. It means that certain consolidation in the messaging market can be expected in the future.
- Slight increase in usage is expected in average in Viber and WhatsApp (however, the conclusion for WhatsApp should be considered as weaker because of significantly smaller share of users in the respondents).
- The expectation of the decrease noes not prevail for any messaging service in the poll.

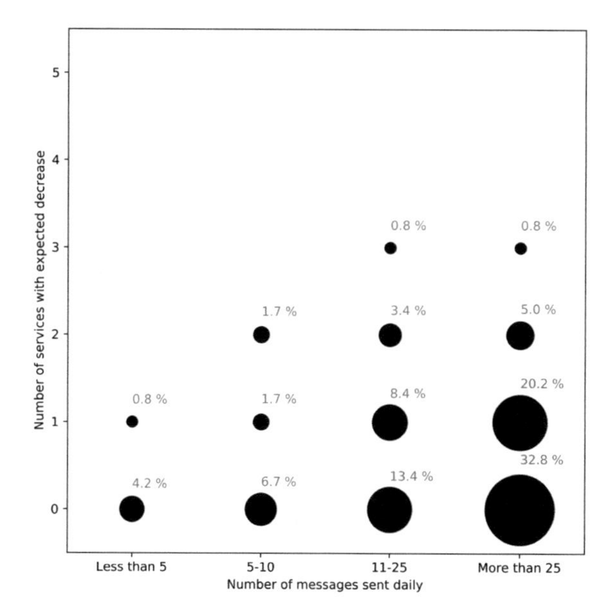

Fig. 2. Respondents with expected decrease in accordance to message intensity.

Despite the high variability of the poll responses, the most popular commercial messenger can be identified as Facebook Messenger leaving all their competitors far behind. This is roughly in accordance to the global estimates cited in Sects. 1.1 and 2. On the other hand, it seems obvious that the e-mail service will continue in its dominance (both for office and private purposes) even over the most popular commercial messenger, primarily among less experienced and less IT-skilled users. One of the main focuses of the poll was to estimate the trends in messaging use from the past to the future. The expectations for future use of messaging services are summarized in Fig. 2 where "pessimistic" users are summarized, and Fig. 3 where users with "optimistic" expectations are shown.

In spite of the fact that the number of respondents was not very high, some generalizable conclusion can be drawn still. This is because of the fact that the

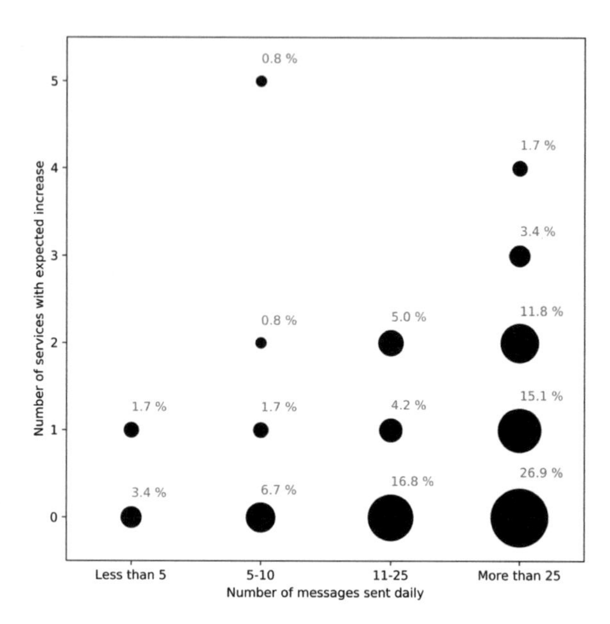

Fig. 3. Respondents with expected increase in accordance to message intensity.

similarity among respondents was quite high so that they represent the group of students in IT and economic programs reasonably.

Moreover, the reasons for potential decrease or ceasing in a specific messaging service usage were examined. Here, only about 64 respondents provided their response. Their results are summarized in Fig. 4. It is obvious that the most frequently presented reason for expectation that any (or more) of used services will the respondent use less is lowering number of respondent's contacts involved in service.

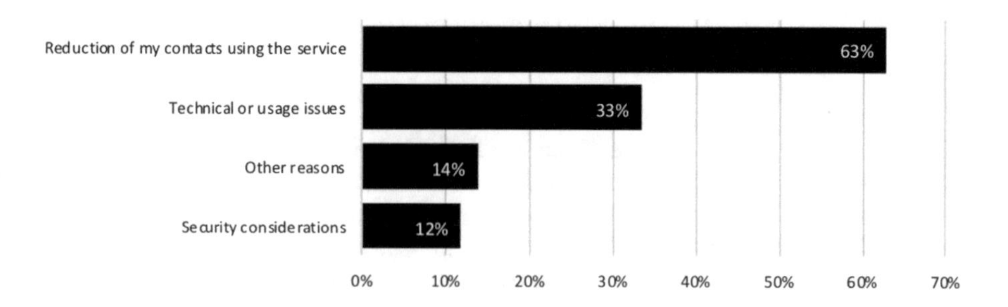

Fig. 4. Reasons for decrease in messaging service usage intensity.

4 Conclusions

The review and research indicate that at present, number one messaging service is still SMTP-based e-mail, primarily for private use. However, there are many issues in implementation of protective and security measures resulting in possible worsening of the service quality, primarily due to the weak sender authentication. As both global figures and results from the small–scale poll show, still no significant trend in users leaving e-mail in favor of commercial messengers can be observed. Nevertheless, because of the necessity of having an electronic mailbox for many reasons (e.g. for authentication in the case of the lost password for a web service including messaging), the decrease of e-mail users cannot be expected in the near future. Despite the fact that results cannot be generalized to the whole population, we can therefore conclude that the above messaging usage patterns and expectations are reasonably valid for the whole group of IT and economics students.

Acknowledgment. The paper was supported by the project *Advanced issues in network infrastructure security* of the Student Grant Competition of the University of Ostrava.

References

1. Vleck, T.V.: The history of electronic mail. http://www.multicians.org/thvv/mail-history.html. Accessed 07 Aug 2018
2. Klensin, J.: Simple Mail Transfer Protocol. IETF (2008). https://www.ietf.org/rfc/rfc5321.txt. Accessed 15 Jan 2018
3. Radicati Group: E-mail statistics Report: 2014–2018. Palo Alto: Radicati Group. http://www.radicati.com/wp/wp-content/uploads/2014/01/Email-Statistics-Report-2014-2018-Executive-Summary.pdf (2014). Accessed 15 Sept 2014
4. Radicati Group: Email Market, 2017–2021 – Executive Summary. Palo Alto: Radicati Group (2017). https://www.radicati.com/wp/wp-content/uploads/2017/06/Email-Market-2017-2021-Executive-Summary.pdf. Accessed 30 Dec 2017
5. Talos Intelligence: Email & Spam data. Talos Intelligence. https://www.talosintelligence.com/reputation_center/email_rep#tab=1. Accessed 15 Jan 2018
6. Sheehan, K.B.: E-mail survey response rates: a review. J. Comput.-Mediat. Commun. **6**(2), JCMC621 (2001)
7. Perez, L.: Millennials hate email (2015). https://medium.com/@lelper/millennials-hate-email-a283e84f4bf7. Accessed 25 Oct 2018
8. Brandon, J.: Why email will be obsolete by 2020. Inc.com (2015). https://www.inc.com/john-brandon/why-email-will-be-obsolete-by-2020.html. Accessed 25 Oct 2018
9. Lafrance, A.: The Triumph of Email. Why does one of the world's most reviled technologies keep winning? The Atlantic. https://www.theatlantic.com/technology/archive/2016/01/what-comes-after-email/422625/. Accessed 25 Oct 2018
10. Sochor, T., Farana, R.: Improving efficiency of e-mail communication via SPAM elimination using blacklisting. In: 2013 21st Telecommunications Forum Telfor, TELFOR 2013 - Proceedings of Papers. School of Electrical Engineering, University of Belgrade, Belgrade, pp. 924-927 (2013). ISBN 978-147991419-7

11. Volna, E., Sochor, T., Meli, C., Kominkova–Oplatkova, Z.: Soft computing-based information security. In: Multidisciplinary Perspectives in Cryptology and Information Security. IGI Global, Hershey, USA, pp. 29–60 (2014). ISBN 978-1-4666-5808-0
12. Levine, J.: DNS blacklists and whitelists. IETF (2010). http://tools.ietf.org/html/rfc5782. Accessed 15 Jan 2018
13. Sochor, T.: Greylisting method analysis in real SMTP server environment: case-study. In: Innovations and Advances in Computer Sciences and Engineering. Springer, Netherlands, pp. 423-427 (2010)
14. Harris, E.: The next step in the spam control war: greylisting (2003). http://projects.puremagic.com/greylisting/. Accessed 15 Jan 2018
15. Lysenko, S., Savenko, O., Bobrovnikova, K., Kryshchuk, A., Savenko, B.: Information technology for botnets detection based on their behaviour in the corporate area network. Commun. Comput. Inf. Sci. **718**, 167–181 (2017)
16. Oikarinen, J.: IRC History. IRC.org. http://www.irc.org/history_docs/jarkko.html. Accessed 15 Jan 2018
17. Statista.com: Most popular mobile messaging apps worldwide as of July 2018, based on number of monthly active users. https://www.statista.com/statistics/258749/most-popular-global-mobile-messenger-apps/. Accessed 07 Aug 2018
18. Statista.com: Number of monthly active Facebook Messenger users from April 2014 to April 2017. https://www.statista.com/statistics/417295/facebook-messenger-monthly-active-users/. Accessed 07 Aug 2018
19. Curry, E.: Message-oriented middleware. In: Middleware for Communications. Wiley, Blackwell, pp. 1-28 (2004). https://doi.org/10.1002/0470862084.ch1
20. Špok, D., Majer, J.: Nedostupní: umění nekomunikovat (Unavailable: the art of noncommunication, in Czech). In: Psychologie.cz: Seriál – Vrtání, Czechia, Prague, 20 September 2018 (2018). https://psychologie.bandcamp.com/track/nedostupn-um-n-nekomunikovat. Accessed 01 Sept 2018

Theatrical Notes in Perturbation Techniques Applied to the Fully Nonlinear Water Waves

Mustapha Mouhid(✉) and Mohamed Chagdali

Polymer Physics and Critical Phenomena Laboratory,
Faculty of Sciences Ben M'sik, University Hassan II,
P.O.BOX 7955, Casablanca, Morocco
Mouhidmustapha@gmail.com

Abstract. A perturbation techniques is presented in this paper to linearize the system of the fully nonlinear water waves, to become linear or weakly nonlinear. The present model, is a first applied to simulate the generation of monochromatic periodic nonlinear gravity waves, by applying a semi-analytical method to resolve the nonlinear water waves propagation have verified by different orders of linear problems.

Keywords: Nonlinear water waves · Gravity waves · Perturbation techniques · Free surface

1 Introduction

The study of the nonlinear water waves is in great importance to the coastal, offshore and ocean engineering, but is very difficult to deal with the nature of the system of equations governing the gravity waves problems.

The problem of the free surface water waves is well established since the works of Alembert, Euler, Bernoulli and Lagrange in the 18th century [1, 2]. However, only analytical or numerical approximations are accessible. Simple analytical approximations are interesting for physical insights. So, scientists and engineers have been proposed a several algorithms in the literature for the computation of the fully nonlinear water waves. Recently, Xiao et al. [2] proposed a rapid simulation of short waves in meshless numerical wave tank (NWT) based on the method of fundamental solutions, Loukili et al. [3], present a semi-analytical or semi-numerical method coupled with the method of fundamental solutions (MFS) to resolve the nonlinear gravity waves propagation, have verified by different orders of linear problems, and wave–structure interactions as an application. In the reference [3] have been done to illustrate the superior accuracy of the MQ and fast method when compared to the method of fundamental solutions. In this paper we present a perturbation techniques to linearize the system of the fully nonlinear water waves, to become linear or weakly nonlinear.

The linear problem solving procedure at the order 1 is generalized to the different orders, by constructing a recurring formulation in the general form: $L_n = L(\varphi_n(x, y, z)) = S(\varphi_{n-1}, \varphi_{n-2}, \ldots)$, which can be solved in a sequential manner.

R. Silhavy (Ed.): CSOC 2019, AISC 986, pp. 137–146, 2019.
https://doi.org/10.1007/978-3-030-19813-8_15

The procedure presented in this paper, is much more powerful than illustrated in the reference [4], which had been based on the projection of the free surface on the mean line by expressing the potential φ and its derivatives by a Taylor expansion.

2 Position of the Problem

We consider a monochromatic incident wave of small amplitude propagating in the presence of the flat bottom of a numerical wave tank (Fig. 1).

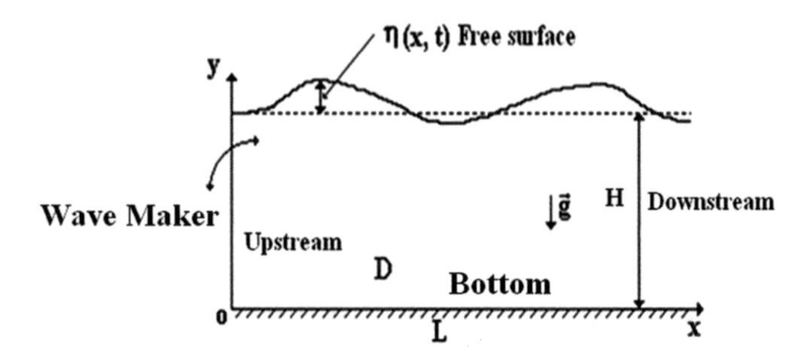

Fig. 1. Sketch of numerical wave tank (NWT)

As a part of the nonlinear wave theory, the movement is supposed to be plane and periodic in time, irrotational, incompressible and the fluid is assumed to be inviscid. The formulation can be made in terms of the velocity potential φ and the elevation of the free surface η.

The problem of propagation of the nonlinear water waves is reduced to resolve the system of equations:

$$\Delta\varphi = 0 \qquad \text{in the fluid domain} \tag{1}$$

$$\frac{\partial\varphi}{\partial y} = 0 \qquad \text{at the bottom} \tag{2}$$

$$\frac{\partial\varphi}{\partial t} + \frac{1}{2}\left\|\vec{\nabla}\varphi\right\|^2 + g\eta = 0 \qquad \text{at the free surface} \tag{3}$$

$$\frac{\partial\varphi}{\partial y} = \frac{\partial\eta}{\partial t} + \frac{\partial\eta}{\partial x}\frac{\partial\varphi}{\partial x} \qquad \text{at the free surface} \tag{4}$$

With: (x, y) the Cartesian coordinates, H the depth of the water; g the gravity and $\vec{\nabla}$ gradient operator. To these equations are added the upstream and downstream conditions as described in the references [5].

3 Theoretical Formulation of the Perturbation Techniques

In this section a perturbation techniques [6–8] it used for the fully nonlinearities of the water waves problem, we use a series expansion of perturbation with respect to a parameter ε assumed to be small, so the order of magnitude is proportional to the camber of the free surface as [5]:

$$\varepsilon = O\left(\frac{a}{\lambda}\right) \tag{5}$$

Where λ is the wavelength and a: is the amplitude of the wave, more precisely the velocity potential φ and the elevation of the free surface η are sought in the form:

$$\varphi = \varphi_0 + \varepsilon\varphi_1 + \varepsilon^2\varphi_2 + \ldots. \tag{6}$$

$$\eta = \eta_0 + \varepsilon\eta_1 + \varepsilon^2\eta_2 + \ldots. \tag{7}$$

The developments in series (6) and (7) are injected into the Eqs. (3) and (4), which are respectively represent the dynamic condition and the kinematic condition, we obtain:

- Dynamic Condition:

$$\left(\varepsilon\frac{\partial\varphi_1}{\partial t} + \varepsilon^2\frac{\partial\varphi_2}{\partial t} + \ldots + \varepsilon^p\frac{\partial\varphi_p}{\partial t} + \ldots\right) + 0.5\left(\begin{array}{c}\left(\varepsilon\frac{\partial\varphi_1}{\partial x} + \varepsilon^2\frac{\partial\varphi_2}{\partial x} + \ldots + \varepsilon^p\frac{\partial\varphi_p}{\partial x} + \ldots\right)^2\\ + \left(\varepsilon\frac{\partial\varphi_1}{\partial y} + \varepsilon^2\frac{\partial\varphi_2}{\partial y} + \ldots + \varepsilon^p\frac{\partial\varphi_p}{\partial y} + \ldots\right)^2\end{array}\right) + g\left(\varepsilon\eta_1 + \varepsilon^2\eta_2 + \ldots + \varepsilon^p\eta_p + \ldots\right) = 0 \tag{8}$$

- Kinematic Condition:

$$\left(\varepsilon\frac{\partial\eta_1}{\partial t} + \varepsilon^2\frac{\partial\eta_2}{\partial t} + \ldots + \varepsilon^p\frac{\partial\eta_p}{\partial t} + \ldots\right) - \left(\varepsilon\frac{\partial\varphi_1}{\partial y} + \varepsilon^2\frac{\partial\varphi_2}{\partial y} + \ldots + \varepsilon^p\frac{\partial\varphi_p}{\partial y} + \ldots\right) + \left(\varepsilon\frac{\partial\eta_1}{\partial x} + \varepsilon^2\frac{\partial\eta_2}{\partial x} + \ldots + \varepsilon^p\frac{\partial\eta_p}{\partial x} + \ldots\right)\left(\varepsilon\frac{\partial\varphi_1}{\partial x} + \varepsilon^2\frac{\partial\varphi_2}{\partial x} + \ldots + \varepsilon^p\frac{\partial\varphi_p}{\partial x} + \ldots\right) = 0 \tag{9}$$

We gather all terms according to the increasing powers of ε we obtain a following formulation form:

- Dynamic conditions

$$\varepsilon\left(\frac{\partial\varphi_1}{\partial t}+g\eta_1\right)+\varepsilon^2\left(\frac{\partial\varphi_2}{\partial t}+0.5\left(\left(\frac{\partial\varphi_1}{\partial x}\right)^2+\left(\frac{\partial\varphi_1}{\partial y}\right)^2\right)+g\eta_2\right)+$$
$$\varepsilon^3\left(\frac{\partial\varphi_3}{\partial t}+0.5\left(2\left(\frac{\partial\varphi_1}{\partial x}\frac{\partial\varphi_2}{\partial x}+\frac{\partial\varphi_1}{\partial y}\frac{\partial\varphi_2}{\partial y}\right)\right)+g\eta_3\right)+\varepsilon^4(\ldots)+\ldots+\varepsilon^p(\ldots)+\ldots \tag{10}$$

- Kinematic Conditions

$$\varepsilon\left(\frac{\partial\eta_1}{\partial t}-\frac{\partial\varphi_1}{\partial y}\right)+\varepsilon^2\left(\frac{\partial\eta_2}{\partial t}-\frac{\partial\varphi_2}{\partial y}+\frac{\partial\varphi_1}{\partial x}\frac{\partial\eta_1}{\partial x}\right)+\varepsilon^3\left(\frac{\partial\eta_3}{\partial t}-\frac{\partial\varphi_3}{\partial y}+\frac{\partial\varphi_1}{\partial x}\frac{\partial\eta_2}{\partial x}+\frac{\partial\varphi_2}{\partial x}\frac{\partial\eta_1}{\partial x}\right)+$$
$$\varepsilon^4(\ldots)+\ldots+\varepsilon^p(\ldots)+\ldots \tag{11}$$

In this part, we note that the problem at the order 1 corresponds exactly to the linear water wave problem. What can be deduced directly by neglecting the nonlinear feature of the dynamic and kinematic boundary condition of the free surface water waves problem, we show also that we can construct a recurring formulation in order P, we obtain a linear problems have the same linear operator, only the second members that changes what constitutes an advantage for the numerical resolution. We present problems in different orders written as the following form:

Problem at the order 1:

$$\Delta\varphi_1=0 \qquad \text{in the fluid domain} \tag{12}$$

$$\frac{\partial\varphi_1}{\partial y}=0 \qquad \text{at the bottom} \tag{13}$$

$$\frac{\partial\varphi_1}{\partial t}+g\eta_1=0 \qquad \text{at the free surface} \tag{14}$$

$$\frac{\partial\eta_1}{\partial t}-\frac{\partial\varphi_1}{\partial y}=0 \qquad \text{at the free surface} \tag{15}$$

Problem at the order 2:

$$\Delta\varphi_2=0 \qquad \text{in the fluid domain} \tag{16}$$

$$\frac{\partial\varphi_2}{\partial y}=0 \qquad \text{at the bottom} \tag{17}$$

$$\frac{\partial\varphi_2}{\partial t}+g\eta_2=SM_2^1 \qquad \text{at the free surface} \tag{18}$$

$$\frac{\partial \eta_2}{\partial t} - \frac{\partial \varphi_2}{\partial y} = SM_2^2 \qquad \text{at the free surface} \tag{19}$$

Where:

$$SM_2^1 = -0.5\left(\left(\frac{\partial \varphi_1}{\partial x}\right)^2 + \left(\frac{\partial \varphi_1}{\partial y}\right)^2\right) \tag{20}$$

$$SM_2^2 = -\frac{\partial \varphi_1}{\partial x}\frac{\partial \eta_1}{\partial x} \tag{21}$$

Problem at the order P:

$$\Delta \varphi_p = 0 \qquad \text{in the fluid domain} \tag{22}$$

$$\frac{\partial \varphi_p}{\partial y} = 0 \qquad \text{at the bottom} \tag{23}$$

$$\frac{\partial \varphi_p}{\partial t} + g\eta_p = SM_p^1 \qquad \text{at the free surface} \tag{24}$$

$$\frac{\partial \eta_p}{\partial t} - \frac{\partial \varphi_p}{\partial y} = SM_p^2 \qquad \text{at the free surface} \tag{25}$$

For p = 1 the second members SM_1^1 and SM_1^2 are null.
For p > 1

$$SM_p^2 = -\sum\nolimits_{k=1}^{p-1}\left(\frac{\partial \varphi_k}{\partial x}\frac{\partial \eta_{p-k}}{\partial x}\right) \tag{26}$$

If p is an odd number:

$$SM_p^1 = -0.5\sum\nolimits_{k=1}^{p-1}\left(\frac{\partial \varphi_k}{\partial x}\frac{\partial \varphi_{p-k}}{\partial x} + \frac{\partial \varphi_k}{\partial y}\frac{\partial \varphi_{p-k}}{\partial y}\right) \tag{27}$$

If p is an even number :

$$SM_p^1 = -0.5\sum\nolimits_{k=1}^{p-2}\left(2\left(\frac{\partial \varphi_k}{\partial x}\frac{\partial \varphi_{p-k}}{\partial x} + \frac{\partial \varphi_k}{\partial y}\frac{\partial \varphi_{p-k}}{\partial y}\right)\right) - \left(\frac{\partial \varphi_{p/2}}{\partial x}\right)^2 - \left(\frac{\partial \varphi_{p/2}}{\partial y}\right)^2 \tag{28}$$

4 Orthogonal Transformation Method

In this work, in order to facilitate the treatment of boundary conditions, we will use a system of orthogonal curvilinear coordinates. This coordinate system will be built numerically, since the problem considered is unsteady, the system of orthogonal curvilinear coordinates used will be evolutionary. At each time step, the n linear problems are solved successively by generating the mesh associated with the coordinate system for each problem and at each time step; the construction of a curvilinear coordinate system for curved border geometries is obtained using the methods of transforming the physical domain into a rectangular domain in the two-dimensional case.

We assume that the application T which transforms the physical region (D) into a rectangular region (D^t) (see Fig. 2) as [5]:

$$\boldsymbol{T} : \mathrm{M}(\mathrm{x}, \mathrm{y}) \in \mathrm{D} \rightarrow \mathrm{N}(\xi, \zeta) \in \mathrm{D}^{\mathrm{t}} \text{ is a bijection.}$$

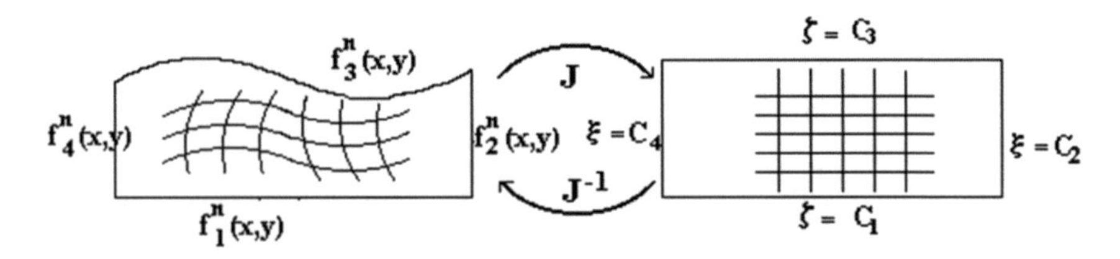

Fig. 2. (a) Physical domain, (b) Transformed rectangular domain

The differential operators involved in motion equations can be expressed in the orthogonal curvilinear coordinate system (ξ, ζ) as follows:

$$\Delta_{x,y}(F) = J^2 \Delta_{\xi,\zeta}(F) \tag{29}$$

$$\vec{\nabla}(F) = \sqrt{J}.\nabla^*(F^*) \tag{30}$$

The free surface is scalable and therefore the orthogonal curvilinear coordinate system (ξ, ζ) is explicitly depends on time, this temporal variation is introduced directly into the equations of the movement which makes appear additional terms in the unsteady parts as:

$$\left(\frac{\partial F}{\partial t}\right)_{x_i} = \left(\frac{\partial F}{\partial t}\right)_{\xi_i} + \vec{\nabla}^*(F).U_d^* \tag{31}$$

Where:

$$\vec{\nabla} = \left(\frac{\partial}{\partial t}, \frac{\partial}{\partial t}\right) \quad (32)$$

$$U_d^* = \left(\frac{\partial \xi}{\partial t}, \frac{\partial \zeta}{\partial t}\right) \quad (33)$$

The nonlinear water waves equations in the transformed rectangular domain are written as the following:

$$\frac{\partial^2 \varphi_p}{\partial \xi^2} + \frac{\partial^2 \varphi_p}{\partial \zeta^2} = 0 \qquad \text{in the transformed rectangular domain} \quad (34)$$

$$\frac{\partial \varphi_p}{\partial \zeta} = 0 \qquad \text{at the bottom} \quad (35)$$

$$\frac{\partial \varphi_p}{\partial t} + u_d \frac{\partial \varphi_p}{\partial \xi} + v_d \frac{\partial \varphi_p}{\partial \zeta} + g\eta_p = \mathrm{J}.S_p^1 \qquad \text{at the free surface} \quad (36)$$

$$\frac{\partial \eta_p}{\partial t} + u_d \frac{\partial \eta_p}{\partial \xi} - \frac{\partial \varphi_p}{\partial \zeta} = \mathrm{J}.S_p^2 \qquad \text{at the free surface} \quad (37)$$

Where J is assumed to be the transformation module written as:

$$\mathrm{J} = \frac{1}{x_\xi y_\zeta + x_\zeta y_\xi} \quad (38)$$

For p = 1 the second members SM_1^1 and SM_1^2 are null.
For p > 1

$$\mathrm{SM}_p^2 = -\sum_{k=1}^{p-1} \left(\frac{\partial \varphi_k}{\partial \xi} \frac{\partial \eta_{p-k}}{\partial \xi}\right) \quad (39)$$

If p is an odd number:

$$\mathrm{SM}_p^1 = -0.5 \sum_{k=1}^{p-1} \left(\frac{\partial \varphi_k}{\partial \xi} \frac{\partial \varphi_{p-k}}{\partial \xi} + \frac{\partial \varphi_k}{\partial \zeta} \frac{\partial \varphi_{p-k}}{\partial \zeta}\right) \quad (40)$$

If p is an even number:

$$\mathrm{SM}_p^1 = -0.5 \sum_{k=1}^{p-2} \left(2\left(\frac{\partial \varphi_k}{\partial \xi} \frac{\partial \varphi_{p-k}}{\partial \xi} + \frac{\partial \varphi_k}{\partial \zeta} \frac{\partial \varphi_{p-k}}{\partial \zeta}\right)\right) - \left(\frac{\partial \varphi_{p/2}}{\partial \zeta}\right)^2 - \left(\frac{\partial \varphi_{p/2}}{\partial \zeta}\right)^2 \quad (41)$$

5 Mesh Generation Method

The methods of generation of the orthogonal and evolutionary mesh are given in the reference [5]. This method makes it possible to define the curvilinear and orthogonal coordinate system (ξ, ζ) at each instant from the following orthogonality relation:

$$\vec{\nabla}\,\xi(x,y,t).\,\vec{\nabla}\,\zeta(x,y,t) = 0 \quad (42)$$

Which is valid at all points of the field including the boundaries.
In the two-dimensional case, the Eq. (40) has the following solution:

$$\frac{\partial \xi}{\partial x} = \frac{\partial \zeta}{\partial y} \quad (43)$$

$$\frac{\partial \xi}{\partial y} = -\frac{\partial \zeta}{\partial x} \quad (44)$$

Using the relationships between Cartesian coordinates (x, y) and curvilinear coordinates (ξ, ζ), we can write:

$$\frac{\partial \xi}{\partial x} = J\frac{\partial y}{\partial \zeta} \quad (45)$$

$$\frac{\partial \xi}{\partial x} = -J\frac{\partial x}{\partial \zeta} \quad (46)$$

$$\frac{\partial \zeta}{\partial x} = -J\frac{\partial y}{\partial \xi} \quad (47)$$

$$\frac{\partial \zeta}{\partial y} = J\frac{\partial x}{\partial \xi} \quad (48)$$

Where:

$$J = \left(\frac{\partial x}{\partial \xi}\frac{\partial y}{\partial \zeta} - \frac{\partial x}{\partial \zeta}\frac{\partial y}{\partial \xi}\right)^{-1} = \frac{\partial \xi}{\partial x}\frac{\partial \zeta}{\partial y} - \frac{\partial \zeta}{\partial x}\frac{\partial \xi}{\partial y} \quad (49)$$

6 Results and Discussions

In order to analyze the capacity of the present semi-analytical model presented above. The finite differences method (FDM) [9] is used to resolve the problem of the weakly nonlinear gravity waves propagation. In this work, our numerical experiment indicates that the numerical solutions coincide with the analytical solutions up to order 4. For that we present in the Fig. 3 a comparison of the numerical solution and the analytical solution at order 5 of stokes gravity waves theory along the numerical wave tank (NWT), taking as amplitude $a = 0.1$ m , length L = 14 m, height H = 2 m, and $T = 2.8434$ s. We note a good agreement overall with a slight dissymmetry upstream-downstream at the shape of each peak.

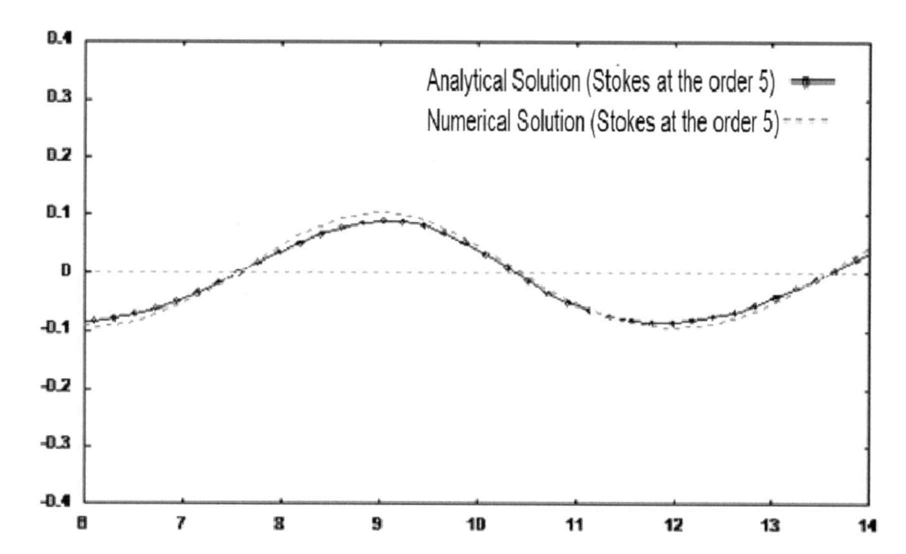

Fig. 3. Comparison of the numerical and the analytical solutions of the elevation of the free surface water waves at order 5

In the Fig. 4, we present the elevation of the free surface water waves at orders 5, 6 and 7 in comparison with the analytical solution of Stokes at order 5.

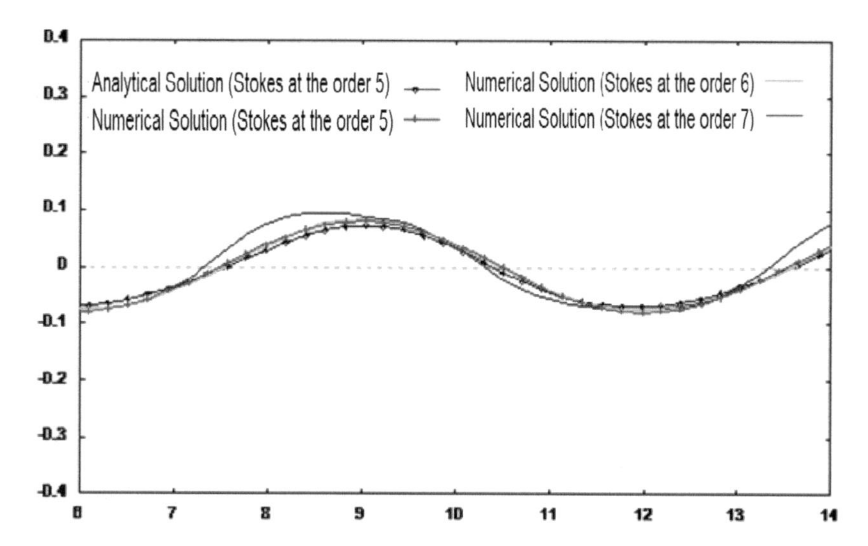

Fig. 4. Comparison of the numerical and the analytical solutions of the elevation of the free surface water waves for different orders

It can be seen clearly that for the small amplitudes there is a good concordance of the solutions up to the order 6. The solution at the order 7 is detached and there is a strong asymmetry upstream-downstream at each peak. The derivatives of higher orders that appear in the second members which make it difficult to approach the numerical solution at the order 7.

7 Conclusion

In this work, an analytical scheme was proposed by applying a perturbation technique to the nonlinear free surface water waves. The principle of this approach is to apply the perturbation method to the nonlinear problem which allows us to have linear problems checked for the different orders, we have presented solutions for higher orders (order 1, order 2, order3 …). We note that the problems of the different orders have the same form. Consequently, the problem solving procedure at order 1 is generalized to the order P. which can solved in a sequential manner. The evolution of the shape of the domain and the deformation of the boundaries as a function of time justify the use of an evolutionary mesh. The construction of this mesh at every moment requires the construction of the complete system of Cartesian coordinates (x, y) and orthogonal curvilinear coordinates (ξ, ζ) as well as the calculation of the transformation module J. This brings two difficulties. The first is the introduction of the time derivative on the Cartesian coordinate system and the second is the treatment of the boundary conditions on the boundaries of the coordinates of the physical domain and the transformed domain.

References

1. Craik, A.D.D.: The origins of water wave theory. Ann. Rev. Fluid Mech. **36**, 1–28 (2004)
2. Xiao, L.F., Yang, J.M., Peng, T., Tao, T.: A free surface interpolation approach for rapid simulation of short waves in meshless numerical wave tank based on the radial basis function. J. Comput. Phys. **307**, 203–224 (2016)
3. Loukili, M., Mordane, S.: New numerical investigation using meshless methods applied to the linear free surface water waves. In: Silhavy, R. (ed.) Cybernetics and Algorithms in Intelligent Systems. CSOC 2018. Advances in Intelligent Systems and Computing, vol. 765. Springer, Cham (2019). https://doi.org/10.1007/978-3-319-91192-2_33
4. Loukili, M., El Aarabi, L., Mordane, S.: Computation of nonlinear free-surface flows using the method of fundamental solutions. In: Silhavy, R. (ed.) Software Engineering and Algorithms in Intelligent Systems. CSOC 2018. Advances in Intelligent Systems and Computing, vol. 763. Springer, Cham (2019). https://doi.org/10.1007/978-3-319-91186-1_44
5. Loukili, M., Mordane, S., Chagdali, M.: Formulation semi analytique de la propagation non linéaire de la houle. XIV èmes Journées Nationales Génie Côtier – Génie Civil (2016)
6. Nayfey, A.: Perturbation Methods. Wiley, New York (1973)
7. Hinsh, E.J.: Perturbation Methods. Cambridge Texts in Applied Mathematics. Cambridge University Press, Cambridge (1991)
8. Fenton, J.D.: The numerical solution of steady water wave problems. Comput. Geosci. **14**(3), 357–368 (1988)
9. Loc, T.P., Daube, O.: Une méthode O(h2) et O(h4) combinée pour la résolution numérique des équations de Navier-Stokes. Comptes Rendus de l'Académie des Sciences, Série A Sciences Mathématiques, vol. 284, no. 19, pp. 1241–1243

The Optimisation of LDPC Decoding Algorithm Parameters for 5G Access Network Empirical Models

Tomas Knot(✉) and Karel Vlcek

Faculty of Applied Informatics, Tomas Bata University in Zlin, Nad Stranemi 4511, Zlín, Czech Republic
{knot, vlcek}@utb.cz

Abstract. This article is focused on the optimisation of LDPC codes in order to achieve high efficiency in encoding and decoding messages that also respect Transmission Channel Properties: this issue is related to the frequency band of the assumed transmission, and on models of - faults that affect individual symbols - or groups of symbols, according to their Transmission Environment Properties.

Keywords: Shannon inequality · Regular and irregular LDPC codes · Tanner Graph · MAP · LLR · Differential Evolution · Approximation of tanhx · FPGA + ARM · SoC · SystemC

1 Introduction

5G Mobile Access Networks form the core of the upcoming fifth-generation of telecommunication networks. The timetable for this global project was officially launched in 2015 - and is planned to be completed by 2020, when key telecom providers will not only be equipped with the appropriate technology – but, in the final year, standardization will have be completed as a precondition for their practical use in all parts of the 5G project. In view of the approval of the relevant recommendations, (standards) - the aim being to achieve so-called consensus – which involves a very ambitious commitment.

Achieving consensus in the offers of telecommunication services tends to be influenced by the specific interests of 5G manufacturers on the one hand, and users on the other all across the globe. The Standards Approval Process is significantly impacted not only by technical solutions possibilities - but also by the economic interests of all of the stakeholders in a worldwide context. The text describes physically often radically different technical principles – on which base to define new services – like, for instance, Organic Vapour Detection [7]. The aforedmentioned properties, due to their different principles of operation and design options, are suitable for integration, and consequently … for the creation of SoC Systems [21].

The proposed outlined procedure thus enables the conception, design and implementation of the process - from the application design stage to its final form - into the form of a component with a specific functionality targeted on fulfilling the target

R. Silhavy (Ed.): CSOC 2019, AISC 986, pp. 147–155, 2019.
https://doi.org/10.1007/978-3-030-19813-8_16

project requirements – in this way, a dedicated customer processor arises, with an immediate follow-up programme routine for adjusting the sensed quantity on an independently variable quantity. In this context, one can mention numerous MEMS applications - short for Micro-Electro-Mechanical Systems; or dimensional systems in which sensors and control electronics - including chip-integrated circuits, are manufactured using semiconductor manufacturing processes.

The method being analysed in the study [6] is used to further increase the computing power of embedded processors, allowing the possibility of extending the set of instructions. This has to do with the extension of a set of instructions for an optimised sequence of activities that is linked in a sophisticated manner to the output signals of the add-on instruction decoder. Freely programmable FPGAs are integrated on the same chip for this instruction decoder. The goal of creating additional hardware support - (an add-on instruction decoder), is the implementation of a decoder for such new instructions; which will make it easier to increase the computing performance of a given application task. Its utility is - for example, in the case of an Organic Vapour Detector [7]. In this case for example, for evaluating the relative amount of organic vapours in the surrounding environment.

The high efficiency of LDPC codes [1], has led - in recent years, to their widespread application in a number of technical recommendations - including the ITU Model [9]. The empirical models of Wireless Transmission Channels - subsequently described in the text below, are determined on the basis of Theoretical Relational Relationships while – at the same time, respecting the values obtained from the real environment of the transmission channel. According to ITU-R [10], empirical channel models are created, where the predominant attention is devoted to WiMAX - (Worldwide Interoperability for Microwave Access). The authors of the cited works base themselves on Real Environment Values that are obtained by measuring certain parameters in this environment. Simulations of Efficiency are based on Physical Environment Requirements, which lead to multi-path radio wave propagation - including moving transmitters and receivers.

2 Why Coded Modulation with LDPC Codes Exactly?

The Error Correction mechanism of the LDPC code - when using different decoding algorithms, causes different efficiencies with regard to the characteristic influence of the Channel Transmission Environment. Of these different decoding conditions - the greatest influence is that of word-length in the Block LDPC code, [3, 4]. The greatest difference between LDPC codes and Classic Block codes lies in their decoding. Classical Block codes are decoded according to the most likely likelihood of symbols in their ML - (Maximum Likelihood); thus, they are usually short - and are designed according to Algebraic Criteria so as to make the decoding calculation simple. In contrast to this, LDPC codes - if decoded iteratively using a graphical representation of their control matrix, then the decoding process is much longer and less structured [2].

It is generally valid that the number $N - K = {}_{rank2}H$, gives the sum of linearly independent rows in matrix H; above body $GF(2)$. If Control Matrix H contains – in each column and every line, the same number of non-zero elements; this has to do with a Regulation Code Control Matrix, [11]:

$$R = \frac{K}{N} = \frac{N-m}{N} \approx 1 - \frac{\sum_i v_i i}{\sum_i h_i i} = 1 - \frac{\sum_j \rho_j / j}{\sum_j \lambda_j / j} \quad (1)$$

The use of the Non-zero Code increases in line with the average number of non-zero elements - and thereby, Code Efficiency; if one compares it with a code with the same initial parameters, which were used for regular codes. This change is accompanied by changes in parameters - called the Code Ratio. In such cases, Non-linear LDPC code decryption can be resolved by using an identical decoding algorithm; just like decoding regular code). The symbol "$\approx$" - in relation to (1), indicates that - in spite of the difference between Regular and Non-regular codes, we are dealing with codes close to each other- in both cases. Based on this, the generator matrix is derived as an orthogonal matrix to the H matrix.

Using the Gaussian-Jordan Elimination method, we obtain from control matrix H a matrix that generates the following procedure: if one sets the control matrix as $H = [A\ I_{N-K}]$; where A is a binary matrix of dimensions $(N - K) \times$ K and second matrix I_{N-K} is the unit matrix of the order $N - K$. The generating matrix is $G = [I_k\ A^T]$. Thus, the code ratio here matches the orthogonality of the matrices $GH^T = 0$. The following paragraph may serve as an introduction to LDPC codes - and thus to their decoding. These findings can improve the idea of the implementation of both Coding and Iterative LDPC code decoders.

A deeper understanding of the role of LDPC Codes in Information Transmission mechanism is given as follows: How does a bug fix using Parity Sums work? The basic concept of Code Security is to deliberately increase the number of message symbols in such a way that these added symbols create redundant information in the transmitted message. These added symbols are therefore, sufficiently distinguishable from one other - and can serve in the Message Reception Point as an "aide" - in the event of a violation of the symbols during transmission through the channel environment. The simplest way to do this is by simply adding another symbol. For Binary Reports, this can be done by adding a parity bit; the value of which is calculated from the message bits.

In the case of an Even Parity Code, the parity bit will have such a value that the sum of the integers in each of the code words is even. Should inversion of a bit occur in the code-word, it would be recognizable that an error has occurred - but it would not be possible to determine which bit of the code-word was changed by the influence of the transmission channel environment.

If multiple bits of the coded word are changed, it may happen that the parity bit may indicate a faultless value – yet, despite this however, the dubious text message is transmitted. Nevertheless, it can be stated that multiple use of the Parity Security Principle increases the minimum code distance - which is defined as the number of different (binary) symbols between the block code vectors; but, on the other hand, it can greatly complicate the Decoding Algorithm.

3 Customisation of the Decoding Algorithm's Transmission Channel Properties

The physical properties of the transmission channel are - for part of the 5G frequency band, substantially dependent on the humidity in the air. This is the reason for studying and creating empirical models of Mobile Access Networks. The Decoding Algorithms used to decode LDPC codes are called "Message Passing Algorithms". The reason for this nomenclature is that it depicts the reality of progressive (iterative) calculation performed taking place during the evaluation of the received code message symbols when using the Tanner Graph, (TG). This has the same role as Control Matrices and Sparse Control Matrices.

The adjective "iterative" - as mentioned, issued. Due to the citation rate of the core work - increasingly commonly promoted, especially in translations into the Czech language; this name belonging to a whole group of codes where decoding is based on iterative algorithms. This has to do with codes with iterative error correction procedures - and therefore, these codes are used for code "iterative codes". The standardisation of these codes over a short time-frame, has contributed to their expansion, especially thanks to monographs, [11].

Iterative Codes are also used in a range of services provided by Mobile Communications Networks, including speech services, digital video broadcasting, and wireless LANs, (Local Area Networks). In all of these areas, so often in demand today; its application is an Iterative Correction of Error approach which represent the main ideas - needed for an understanding of these issues; to analyse, design and implement these exceptionally efficient codes. Depending on the algorithms and the arrangement of the code message - in the course of which the decoding process repeatedly combs through the message between the data and control nodes of the Tanner Graph, and also according to the type of message interference - it is possible to approach this with different algorithms.

The following paragraph defines LDPC (Low-Density Parity-Check) codes, the class of these code security measures - which were first introduced by Gallager in his dissertation in 1962, twelve years after Hamming published the first bug-fix code in 1950. Both Hamming and Gallager codes belong to block codes group: the message is divided into blocks so as to be able to be decoded prior to its transmission; in a similar way, upon receipt by the receiver, these blocks are decoded in the decoder. While Hamming Codes are short and can be distributed in a very structured manner - according to their ability to correct errors; LDPC codes on the other hand, are created pseudo-randomly with a hypothetical, theoretically justified probability of Error Correction Capability.

The Description process describes the addition of parity bits as a means of error detection and correction in digital messages. Code Security with Block Codes is described with the aid of equations representing Parity Totals, which correspond to the Control Matrix. The Tanner Graph is an example of a bipartite graph, which becomes a tool for resolving iterative decoding algorithms.

The Iterative Bit-flipping Algorithm - which is the basic solution to the problem, is used in the explanation - and the solution is then modified into a "Sum-Product

Decoding Algorithm", motivated by the aim of achieving more accurate for expressing message symbols.

4 Reasons for Exploiting LDPC Codes – High Error Correction Efficiency

The evaluation of LDPC code properties is performed from the perspective of its effectiveness under various properties defined by their control matrices. The variable called the "Logarithm Likelihood Ratio", (LLR) serves here as a random variable, and will have - according to the definition: $p(x = 0) > p(x = 1)$ will acquire a positive value; in the opposite case, the values will be negative. This is consistent with the definition of the relationship for the $L(x)$ Eq. (2):

$$L(x) = ln\frac{p(x = 0)}{p(x = 1)} \tag{2}$$

The transcription of the numerator and the denominator of the fraction in the argument of the natural logarithm function is, primarily the following – is expressed as a logarithm based on the Irrational Number e; which forms the basis of the following natural logarithms:

$$p(x = 1) = \frac{e^{-L(x)}}{1 + e^{-L(x)}} \qquad p(x = 0) = \frac{e^{L(x)}}{1 + e^{L(x)}} \tag{3}$$

In the next calculation, one can – with success, the Transcendent Function can be successfully used to convert these terms. The calculation is then simplified, when using one of the approximations of this function. In most cases, it is possible to use Linear Approximations in places. The Sum-Product Decoding algorithm is very similar to the bit-flipping algorithm, [6] - where Hard Decision Making is used for decoding individual symbols - except for the fact that, the calculation is iterative with Soft Decision Making and evaluates message symbols in a forward and backward manner; while the calculation is performed on the input symbol - as an a priori evaluation, and only after that, on the basis of further evaluations are decided upon ex-post. Which of the above-mentioned decoding methods will be prove suitable to choose from, is given by the application requirements, demands on speed, and decoding precision requirements, [11].

The dependence of the frequency of corrected errors on the energy efficiency E_b/N_0, in the figure above - taken from the literature, [11], shows how Error Correction Efficiency can be achieved by choosing Hard and Soft Decision-making measures and the use of regular and non-linear LDPC codes for BI-AWGN, (White Gaussian Noise Added to the Channel Input as a Binary Interference Signal).

The Dotted Line depicts the Dependence in the course of decoding the "Sum – Product" decoding algorithm. Graph Fig. 1, shows a dotted line depicting individual bits' error-rates, and a LDPC code Full-line Error Frequency Rate. The choice of Non-Regulatory codes can significantly improve LDPC code efficiency performance.

(The addition of ones in the Irregular Code Control Matrix are – then, used to decode columns with lower-weightings - with reference to a Non-linear LDPC Code Control Matrix.).

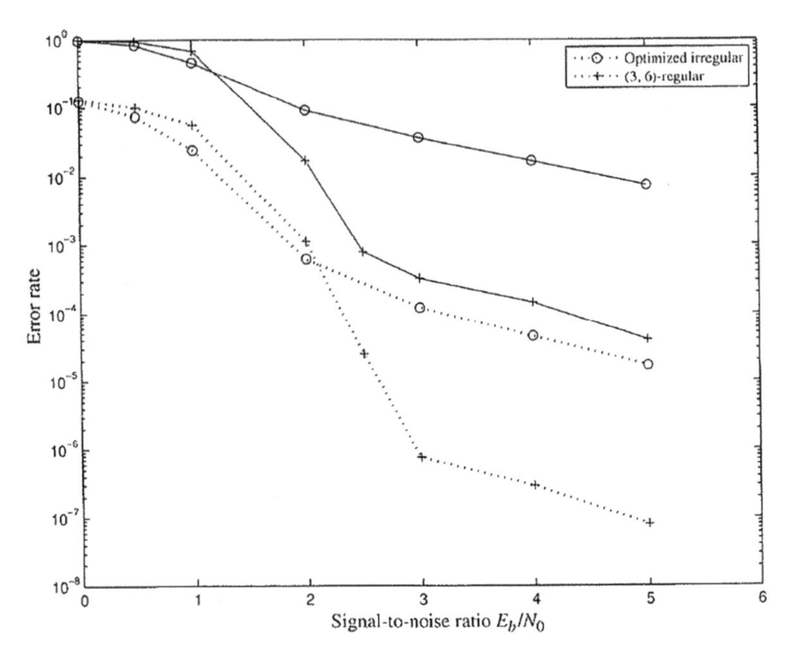

Fig. 1. The dependency of bit errors - bug error, (dotted line); and code word errors - (solid line); on energy effectivity in the course of white noise interference on the input of the transmission channel during decoding. [11]

5 Description of Transmission Channels' Empirical Properties

Wireless Message Transmission - recommended in 5G Mobile Access Networks to speed up Information Transfer uses higher frequencies; more susceptible - in particular, to higher signal attenuation in the course of transmitting information. The influence of changes to these Transmission Channel Parameters, [12], when assessing a transmitted message may impair the uniqueness of the individual message symbols. The message thereby, by changes to transmission environment parameters, becomes less readable.

In order to resolve this situation – it is, with reference to Shannon's Inequality, [6]; necessary to - either increase the transmitter power … or Code Security Measures which will ensure that the message will exhibit a low error rate – in the course of information transmission under existing current energy efficiency levels. For this purpose, standards based on LDPC Encoding are used. The applications field is very diverse now, not only: 10 Gb/s Ethernet (IEEE 803.3an), Wi-Fi (IEEE 802.11n), WiMAX (IEEE 802.16e), Wireless LAN Standards; but also their applicability to fields ranging from Optical Networks to Numerical Memory, they can also be used for repairing data flows burdened with information loss [13]. Empirical Model, [14] is used to transmit messages:

$$T_c \approx \sqrt{\frac{9}{16\pi f_m^2}} = \frac{0,423}{f_m} \tag{4}$$

Where, Time T_c is called Coherence Time, which is inversely proportional to Doppler Dispersal/Scattering. The maximum delay is related to high energy signals, [14, 15].

The investigation of empirical models describing wireless data transmission is a matter mainly addressed by telecommunication equipment manufacturers. The summarization of the most important properties of these models in a brief analysis is provide by the monograph [14]. The aim of this analysis is the modelling of transmission channels, and the consequent calculations to verify - to what extent, these models are close to the measurement results in the given environment. Of great significance for technical practice, is the most well-known Wireless Transmission Channel Model, in the ITU-R literature standards, under number M.1225 [14, 15].

6 New Observations Gained Using Differential Evolution for 5G Applications

Bit Interleaved Coded Modulation (BICM) was originally intended for transmission channels in third generation systems. It is clear that the demodulator calculates the vector that represents the output symbol according to the LLR for each bit b_k of the code, where $k \in \{1, 2, \ldots, m\}$ is 6:

$$L_e(b_k) = ln\frac{\Sigma_{x\in\chi_0^k}(y|x)}{\Sigma_{x\in\chi_1^k}(y|x)} \tag{5}$$

Where b_k denotes a subset of symbols $x \in \chi$ with a value of $b \in \{0, 1\}$ at the k-th position.

Using this approach, each bit of the received vector is decoded. After adjustments to the implementation, which brought higher data-rates and increased Spectral Efficiency, BICM also became applicable to fifth-generation mobile (5G) access networks, which require much greater data throughput. Therefore, coded modulations with BICM modulated LDPC codes are implemented, which provide flexible and simple circuit implementations for higher order modulation - (i.e. implementations with alphanumeric symbol applications [6]). By accepting this condition however, a new probability distribution of errors occurs due to the new probabilities of the occurrence of the new non-alphabet symbols.

By means of the use of Differential Evolution (DE), the authors of this study [17] have been able to find the modulation conditions, referred to in the acronym (BICM) - Bit Interleaved Coded Modulation [18]. These draw on the experience of circuit solutions that use LLRs when decoding - and proceed in an iterative manner [19]. Here, the use of the Gray Code in conjunction with BICM suggests itself, which is important for the practical application of Communication Systems Communications [17]. However, it is also necessary to respect the fact that the so-called Unequal Error Protection, (UEP) - which changes the position and meaning of the edges in the Tanner Graph - and thus also, the decryption algorithm of the respective LDPC code.

7 Conclusion

The lack of a clear algebraic structure of pseudo-random code words in constructing LDPC codes makes it difficult to determine the minimum code distance and this often requires the examination of all 2^K code-words to find the smallest Hamming Weights. While it is possible to arrive at a calculation for determining the smallest code distance for common codes; while for long code-words, such a calculation is very complicated. In principle, two methods based on qualified estimates are used to determine the minimum code distance, more in the monograph [11].

The Author wishes to express their thanks to Ing. Pavel Martinek, PhD. for his valuable advice, corrections and comments on the mathematical formulations used in the text. I especially appreciate his recommendation because he respects the practices used by the professional public from the mathematical profession.

References

1. Gallager, R.G.: Low density parity check codes. IRE Trans. Inf. Theory **IT-8**, 21–28 (1962)
2. Vlcek, K., Vorac, J., Mitrych, J.: Iterative decoder with very sparse matrices solution, In: Proceedings of the 7th IEEE Workshop on Design and Diagnostics of Electronic Circuits and Systems, Stará Lesná, Slovakia, pp. 203–206, 18–21 April 2004. ISBN 80-969117-9-1
3. Noor-A-Rahim, M., Nguyen, K.D., Lechner, G.: Finite length analysis of LDPC codes, arXiv:1309.7102v1, 27 September 2013
4. Noor-A-Rahim, M., Nguyen, K.D., Lechner, G.: Anytime characteristics of spatially coupled code. Accepted for Presentation at 51st Annual Allerton Conference on Communication, Control, and Computing, Allerton, Illinois, October 2013
5. Xilinx: Breakthrough UltraScale + Device Performance with SmartConnect Technology, WP478 (v1.0), pp. 1–11 (2016)
6. Knot, T., Vlček, K.: LDPC binary vectors coding enhances transmissions and memories reliability. In: Proceedings of the 6th Computer Science On-line Conference 2017 (CSOC 2017), vol. 2, pp. 434–443. https://doi.org/10.1007/978-3-319-57264-2, ISBN 978-3-319-57263-5
7. Olejník, R., Matyáš, J., Slobodian, P., Vlček, K.: Microwave antenna with integrated organic vapor sensor function. Czech Republic. Patent, 304850. Granted, 22 October 2014
8. Karami, A.R., Attari, M.A.: Novel LDPC Decoder via MLP Neural Networks. arXiv:1411.3425 [cs.IT]
9. European Cooperative in the Field of Science and Technical Research EURO-COST 231. Urban Transmission Loss Models for Mobile Radio in the 900 MHz and 1,800 MHz Bands. The Hague, the Netherlands, September 1991

10. ITU-R Recommendation M.1225: Guidelines for evaluation of radio transmission technologies for IMT 2000, February 1997. http://www.itu.int/rec/REC-R-M.1225/en. Accessed 20 April 2015
11. Johnson, S.J.: Iterative Error Correction. Cambridge University Press (2010). ISBN 978-0-521-87148-8
12. Grabner, M., Pechac, P., Valtr, P.: Analysis of propagation of electromagnetic waves in atmospheric hydrometeors on low-elevation paths. Radioengineering **27**(1), 29–33 (2018). https://doi.org/10.13164/re.2018.0029. ISSN 1210-2512
13. Shinkarenko, K.V., Vlček, K.: Design of erasure codes for digital multimedia transmitting. In: Proceedings of 2008 IEEE DDECS, pp. 30–33. IEEE, 16–18 April 2008 (2008). ISDN 978-1-4244-2276-0/08/
14. Kim, H.: Wireless Communications Systems Design, pp. 46–50. Wiley (2015). ISBN 9781118610152
15. Shokrollahi, A.: LDPC codes: an introduction. Digital Fountain, Inc., Technical report, p. 2 (2003)
16. Richardson, T.J., Shokrollahi, M.A., Urbanke, R.L.: Design of capacity-approaching irregular low-density parity-check codes. IEEE Trans. Inf. Theory **47**(2), 619–637 (2011)
17. Du, J., Zhou, L., Zhang, Z., Yang, L., Yuan, J.: Regular and irregular LDPC code design for bandwidth efficient BICM schemes. IEEE (2017). 978-1-5090-5019-2/17
18. Caire, G., Taricco, G., Biglieri, E.: Bit-interleaved coded modulation. IEEE Trans. Inf. Theory **44**(3), 927–946 (1998)
19. Zehavi, E.: 8-PSK trellis code for a Rayleigh cannel. IEEE Trans. Commun. **40**(5), 873–884 (1992)
20. ten Brink, S., Kramer, G., Ashikhmin, A.: Design of low-density parity-check codes for modulation and detection. IEEE Trans. Commun. **52**(4), 670–678 (2004)
21. Vlček, K.: SystemC – tools and framework for design System on Chip with mixed signals, 11 March 2014. (in Czech). http://www.utb.cz/file/44257_1_1/

Smart Thermostat as a Part of IoT Attack

Tibor Horák[1] and Ladislav Huraj[2(✉)]

[1] Institute of Applied Informatics, Automation and Mechatronics, Faculty of Materials Science and Technology in Trnava, Slovak University of Technology in Bratislava, Trnava, Slovakia
tibor.horak@stuba.sk

[2] Department of Applied Informatics, University of SS. Cyril and Methodius, Trnava, Slovakia
ladislav.huraj@ucm.sk

Abstract. The Internet of Things is a rapidly growing global network of devices. Due to a rapid growth, security issues are also associated with Distributed Denial of Service attacks on these devices. Nowadays, cyber-attacks are a chronic Internet affair. There are two views of attacks. In the first view, the goal is to disrupt the flow of service in any way. This type of attack is called Denial of Service. In the second view, the goal is to abuse Internet of Things devices for attacking so that the device sends a packet to a non-existent port, and a response with an error automatically sends the device to an unsuspecting victim. The Internet of Things thermostat, which is popular in both, home and industrial areas, has been selected to investigate these types of attacks. The article illustrates the possibility of attacking the Internet of Things device as well as integrating the Distributed Reflection Denial of Service targeted attack device into specific victims.

Keywords: Internet of Things devices · Distributed Denial of Service attack · Reflection attack · Thermostat

1 Introduction

Internet of Things (IoT) is introducing a rapidly growing global network of devices such as: manufacturing machines, network printers, intelligent lighting, waste bins, and others. These devices are equipped with a diverse variety or with simple electronic equipment, software, sensors, and network connectivity. Due to the rapid growth, there are also associated security issues that are linked with DDoS (Distributed Denial of Service) attacks on these devices. In the article, the focus is laid on a device which is oriented for industrial field and that is an intelligent heating control system with remote control of someone's building. Nowadays, the cyber-attacks are a chronic issue of Internet of Things. Any attack which intentions are to interfere, in any way, the smooth running of the service is considered as a Denial of Service attack. If there are multiple engines involved in this attack which is simultaneously attacking, usually even thousands of them, the attack is called distributed or, shortly, DDoS attack.

There are two views of the attacks. The first one is to disrupt a normal operation of the device or its networks in order to cause damage and create discomfort to the user.

R. Silhavy (Ed.): CSOC 2019, AISC 986, pp. 156–163, 2019.
https://doi.org/10.1007/978-3-030-19813-8_17

The second one is to misuse the device for attack the way that the attacker sends a packet to a non-existent port, and the answer with an error will be automatically forwarded to an unsuspecting victim.

Smart thermostats are IoT devices usable for intelligent households or industry and are responsible for the management of heating or air conditioning. They allow the user to plan day-to-day room temperature control, and additionally, they include other features such as sensors or Wi-Fi connection that allow remote communication via the mobile application as well. Nowadays, smart thermostats are becoming more and more a part of home automation [1].

For example, Shaikh et al. in [2] verified the infection status of 275,478 IoT devices worldwide where 4.5% of investigated devices were IoT thermostats. Moreover, they investigated global distribution of exploited IoT devices and smart thermostats also belong to this case. The role of this article is to take a closer look at the specific, intelligent thermostat Honeywell EvohomeTouch ATP921R3052 WiFi from the view of DDoS attacks conducted on this device as well as DDoS attacks using the intelligent thermostat.

The article is structured as follows: The Sect. 2 describes the importance and efficient use of home and industrial IoT devices as well as the background of IoT and DDoS attacks. The Sect. 3 explains design and topology of proposed experiments. The Sect. 4 describes the results of experimental attacks. The last Sect. 5 contains reports on the conclusions and possible goals and visions for the future.

2 Background

The article does not claim that the Internet of Things will cause another industrial revolution, even though it refers to the term Industry 4.0 or the fourth stage of the Industrial Revolution. The fact is that connecting objects to the Internet and interconnecting them with each is able to optimize existing processes. This also includes a regulator from Honeywell, which is part of the wireless control system. It is a control unit for heaters, floor heating, zone and mixing valves, and other different types of heating in domestic, industrial, or in office environment.

Thankfully to the wireless unit (BDR91), it is also possible to control the boiler, heat pump, or DHW (domestic hot water) cylinder. The control unit has a color display that can be controlled up to twelve independent zones. Each zone has its own weekly time schedule with the possibility of simple rewriting. This convenience allows people to set the desired thermal comfort remotely, even before entering the factory or office. The system is equipped with an optimization function that calculates the warm-up and cooling time, and so the system is offering the possibility to program the real heating needs. In addition, an adaptive regulation continually adjusts a current climate regulation and a heat load throughout the year. Such effective regulation can save up households and companies lots of money [3].

2.1 Internet of Things

The Internet of Things is the name for technologies that enable low-cost wireless connectivity and communication of various sensors and devices in order to automate, accelerate and streamline processes, distance measurement, remote control, comfort enhancement, to enable better quality of life, and many other uses such as agriculture, medical care, recycling of waste, producing an manpowered virtual reality, or motion in game industry [4–7]. The most common are sensors and small devices with low demands on the amount of data and on low data consumption. The sensors and devices can communicate with each other or with central systems via common or special types of wireless networks. There are being new ideas and projects of this kind created in the world, also new sensors and devices are being developed, specialized networks are being built, and a new sector is emerging. By typical present or future usage are monitoring and measurement sensors (for example: industry, farming, environment, and households) and a tracker movement, location of people or animals, transportation or goods. These areas are gradually replacing many of the devices and systems based on mobile networks that have been used so far [8, 9].

The specialized networks and communication protocols allow the operation and communication of ultra-low-power devices (the life of a battery for many months or years), thereby contributing to saving the costs and the growth of new possibilities for meaningful use. For the Internet devices, it is important to take care of security because they have the potential to run DDoS attacks of varying range, including application layers or encrypted attacks. Taking into an account the efficiency of IoT botnets, and, on the other hand, an increasing number of poorly protected IoT devices, it is quite clear that the number of such attacks will increase.

2.2 Attacks in IoT Environment

DDoS means denial of service. In fact, this is a certain form of denial of service. The networked device, on which the attack is executed, will, after all, refuse to provide its services. From the outside view, it simply look like the device stops communicating, and it becomes inaccessible. The DDoS attack does not serve to make anyone breaking into somewhere, but it serves to copy sensitive data from there. The DDoS attack has to overload the victim server only and thus prevent access to the requested service.

One of the most common and also the simplest attacks is the so-called SYN Flood attack. The TCP SYN Flood abuses the property of the TCP protocol, where the server after receiving TCP SYN has to send TCP response with features of SYN + ACK and has to keep the connection in memory until the timer expires.

There is an extremely simple and widespread attack when an attacker sends a huge amount of TCP SYN requests from a fake IP address source. The attacked server has to open a connection to each request and store that session. The attack is successfully terminated after system resources start to create new sessions, and the whole system crashes, making it inaccessible for a long period of time.

Other type of attack can be used as Reflected DoS (DRDoS). The attack consists in sending a TCP SYN packet to a victim with the wrong address of another machine on the network (ideally in the victim's network). Receiving a TCP SYN packet will cause the TCP SYN + ACK response to the source IP address. Since this machine did not start the connection, it cancels the connection by sending TCP RST. As a result, the server of the victim runs out of a source for generating a TCP SYN + ACK packet, while TCP RST packet will be overloaded again. TCP RST attack is an attack for legitimate connections of server users, where the attacker has to guess the sequential numbers of the TCP connection, and in case that the user guesses the sequential numbers right, the connection is terminated. Assuming that the attacker has a botnet, it is more than likely that such an attack may occur. For this type of attack, the client's IP address is needed; however, this makes it impossible to use the attack in many cases. However, when communicating while end nodes such as an electronic auction are agreed before hand, then the attack can be used.

The attack uses the PUSH sign in TCP. After receiving this message, the server immediately sends the contents of the buffer (cache) back to the client and confirms this action by sending an ACK sign. If the attacker can generate a large number of packets (for example, by using a botnet), it is possible to overload the server with many requests until the server is unable to process additional requests (which will deny the service for all users).

UDP flood is an invasive flood attack focused on the fourth (transport) layer of the OSI model. For this attack, the UDP protocol is used. An attacker sends a large amount of datagrams to the selected port of a victim, what can lead to overloading of the victim's line, which can be further increased by the victim's attempt to send back the ICMP Destination unreachable packet. This type of attacks was previously used in connection with the echo service (data that arrives at its port, are sent back) or Chargen (the service receives the data but returns the data randomly). If the source address is spoiled instead of the server's address of the provider on the specified Echo service port or the Chargen service and was routed to the Echo or Chargen address and port, provided with Echo or Chargen service, then it lead to an overflow of data between these servers.

The ICMP flood type of attack belongs among the invasive attacks oriented to the OSI's network layer model. The attacker overloads the targeted system ping with packets (ICMP Echo Requests,) which is further enhanced by the efforts of the victim to respond to them. The goal is to overload the victim's line, but at the same time, it also overloads the line of the attacker. For a successful attack, it is necessary that an attacker's line is faster than a victim's line [10–13].

3 Experiment Design

To evaluate the experiment the Honeywell Regulator was chosen as a representative of current IoT devices used in smart households and also used in industries. It is a control unit for heaters. A computer running Kali Linux was also used where the operating system is equipped with an hping3 toolkit for demonstrating DDoS and DRDoS attacks. In the article, an experimental network is also made up of a high speed router, ASUS-RT-AC66U, that can handle up to 1.75 Gbps of data rate and a four channel DVR recorder. The DVR recorder is also very popular in both, home and industry areas, which allow viewing of security cameras and remote, connect via browser or mobile app. There is no other communication between the devices during the experiment.

As shown in Fig. 1, a simple star topology with one ASUS 4-port router was used with a wirelessly connected IoT thermostat and a cable-connected DVR recorder, and a computer that served as an attack generator. The connection of IoT devices to router was secured at 54 Mbps. The computer was connected via Gigabit Ethernet interface, and the DVR recorder was connected via the 100 Mbps Ethernet interface.

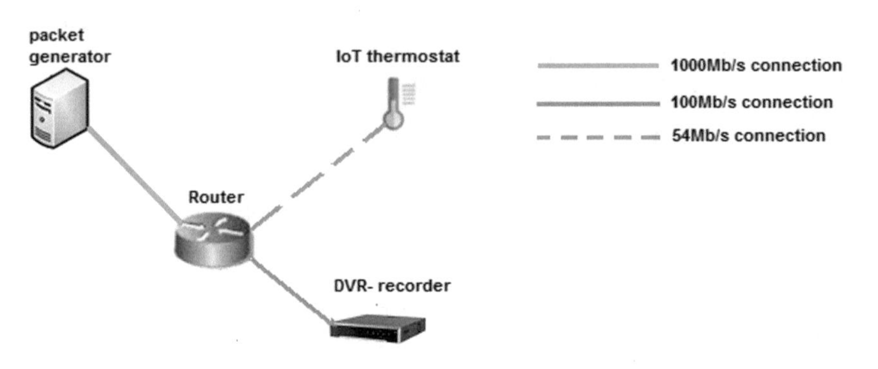

Fig. 1. The topology of the testing network

4 Performance Tests

The DDoS and DRDoS attacks were performed on a specific IoT device, in our case it is a thermostat for intelligent heat control. The TCP, UDP, and ICMP infiltrations were performed by Kali Linux through hping3 tool (http://hping.org). The Hping3 tool supports TCP, UDP, ICMP, and RAW-IP protocols. The control and generation of packets are performed through the command line. These tools are also used by security analysts in order to verify the safety of managed infrastructures.

The Fig. 2 illustrates outcomes of the first experiment where TCP flood and UDP flood packets were generated directly to the thermostat on a non-existent port 135 for 2 min. The sending of the packets has been set to the highest possible speed, regardless of whether the thermostat responds to the incoming packets or not. In the article, the smaller packet shifts in the graph can be seen in TCP SYN Flood and TCP ACK Flood attacks. This is because the attack is executed directly, so the device detects the attacker's IP address to which the packets are reflecting back. Therefore, it is better to use these attacks as reflected, where the IP address will be either listed randomly or in different station. During these direct attacks, the thermostat was unavailable only during generation of the packets.

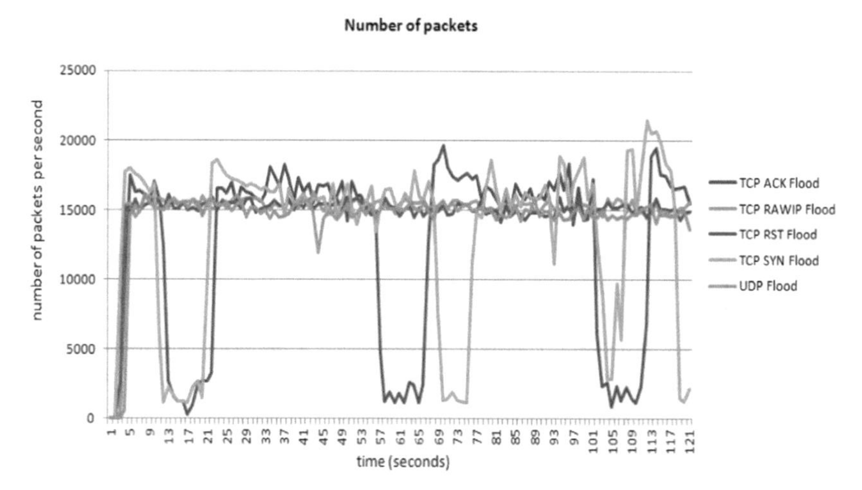

Fig. 2. The number of packets transmitted per second by various attacks to the network on the IoT thermostat.

The Fig. 3 illustrates the second test where three reflected attacks were performed and the packets were generated for 2 min. The sending of the packets has been set to the highest possible speed. The first ICMP echo flood attack is where the packet striker sends the packets to the router on a non-existent port 135. The router reflects these packets, containing an error message, to the false address that the thermostat had, as seen on the graph, after that the attack managed to get the thermostat out of the network permanently. The second attack was also the ICMP flood in which the attacker generated the packets to the thermostat on the non-existent port 135, and those responses with an error message the thermostat reflected on the router. The third attack was a TCP SYN flood, in which the attacker sends the packets to the thermostat and also to the nonexistent port 135. Then, the thermostat reflected the wrong answers to the IP address of the DVR recorder.

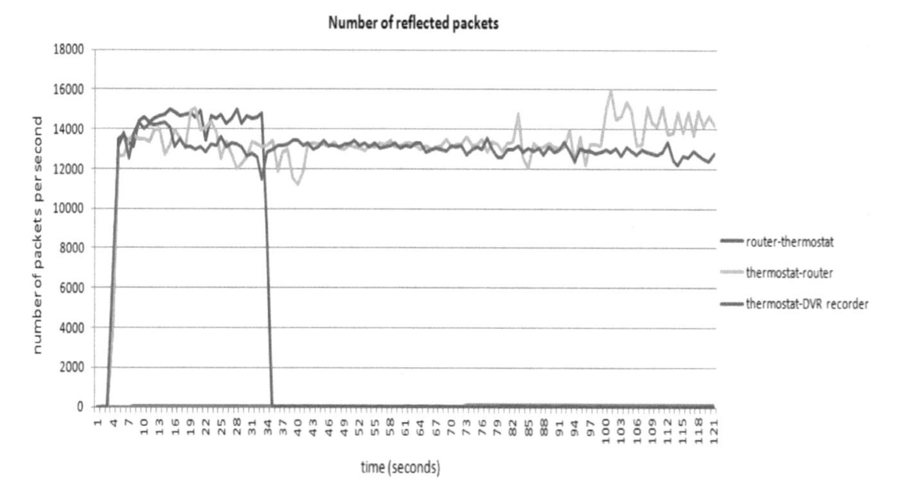

Fig. 3. Number of reflected packets per second transmitted to the network by the IoT devices

The Figs. 2 and 3 show the DDoS and DRDoS attacks on the IoT device, in this case the thermostat. As seen on Fig. 2, a direct DDoS attack on the thermostat was demonstrated for two minutes on the non-existent port 135. During this time, the thermostat was flooded by packets with various attack methods: TCP SYN flood, ACK flood, RST flood, RAWIP flood, and UDP flood at the highest possible speed rate. During this time, the thermostat became unavailable for another usage. After the attacks were completed, the thermostat became available again. For TCP SYN flood and TCP ACK flood attacks, slight fluctuations were noted in packets dispatching during the test, due to the fact that the thermostat recognized the attacker's IP address, so it started reflecting the packets back and thus slowed the flood caused by the attacker.

The Fig. 3 shows the DRDoS attack by the ICMP echo method where the attacker sent the packets to the non-existent port 135 on the router for two minutes by giving the spoiled IP address of the thermostat, on which the router reflects packets with an error message about the non-existent thermostat port. This attack on the thermostat was extremely effective because, after about 40 s, the thermostat stopped communicating on the network permanently even after the attack was completed, as can be seen on Fig. 3. So far, the DTC attacks have been demonstrated on the thermostat, either by DDoS attack or DRDoS attack. The Fig. 3 also shows how the thermostat can be misused for an attack, in a way, that the attacker sent packets to a thermostat that reflected packets to the router by the ICMP echo method and the TCP SYN flood on the DVR recorder.

The experimental results clearly proved that DDoS and DRDoS attacks are very dangerous for IoT devices as smart thermostats, and they can cause a great damage to network communication. Therefore, it is important to take care of their safety.

5 Conclusions and Future Work

The Internet of Things (IoT) is supposed to be another great opportunity for telecom operators, including electronics manufacturers, and it most likely will be. However, even this coin has two sides. The IoT is also a welcome opportunity even for nowadays cybercriminals. Although the machine communication of various sensors, intelligent devices, and other devices creates relatively small data streams, it can, with their huge numbers, generate extensive aggregate traffic. And since the network connectivity of IoT devices is often unsecured, they become an easy target for hackers. The hackers can steal not only data, but mainly use infected devices for DDoS attacks.

This case study demonstrates the type of attack on various heterogeneous categories of devices: a computer running Kali Linux, a router, a thermostat, and a DVR Recorder, as well as the first view of communication traffic during IoT device to attack. The experimental results show not only that the devices are easily flooded with DDoS attacks, but they can also be abused in DRDoS attacks and have the ability to cause severe problems in networks. Although, the phenomena discussed in this document are well known, its main contribution is to understand their importance as well as the actual implementation of the DRDoS attack on the IoT devices. The future extension of the case study requires further scrutiny of the DRDoS flood attacks on a larger number of other devices in the form of a large-scale simulation in the simulation environment.

Acknowledgements. The work was supported by the grant VEGA1/0272/18 *Holistic approach of knowledge discovery from production data in compliance with Industry 4.0 concept.*

References

1. Leporini, B., Buzzi, M.: Home automation for an independent living: investigating the needs of visually impaired people. In: ACM Proceedings of the Internet of Accessible Things. ACM, USA (2018)
2. Shaikh, F., et al.: Internet of malicious things: correlating active and passive measurements for inferring and characterizing internet-scale unsolicited IoT devices. IEEE Commun. Mag. **56**(9), 170–177 (2018)
3. Desmedt, Y.: Need to know to defend one's home, or should one buy a WiFi enabled thermostat? In: Proceedings of the Second International Conference on Internet of things and Cloud Computing. ACM (2017)
4. Horváthová, D., Siládi, V., Lacková, E.: Phobia treatment with the help of virtual reality. In: 13th International Scientific Conference on Informatics, pp. 114–119. IEEE (2015)
5. Ölvecký, M., Gabriška, D.: Motion capture as an extension of web-based simulation. Appl. Mech. Mater. **513**, 827–833 (2014)
6. Dirgová Luptáková, I., Šimon, M., Pospíchal, J.: Worst-case test network optimization for community detection method. In: Proceedings of the 8th Balkan Conference in Informatics. ACM (2017)
7. Hosťovecký, M., Novák, M., Horváthová, Z.: Problem-based learning: serious game in science education. In: ICEL 2017-Proceedings of the 12th International Conference on e-Learning, ACPI 2017, pp. 303–310 (2017)
8. Pervez, F., et al.: Wireless technologies for emergency response: a comprehensive review and some guidelines. IEEE Access **6**, 71814–71838 (2018)
9. Amadeo, M., et al.: Information-centric networking for the Internet of Things: challenges and opportunities. IEEE Netw. **30**(2), 92–100 (2016)
10. Šimon, M., Huraj, L., Čerňanský, M.: Performance evaluations of IPTables firewall solutions under DDoS attacks. J. Appl. Math. Stat. Inform. **11**(2), 35–45 (2015)
11. Šimon, M., Huraj, L., Hosťovecký, M.: IPv6 network DDoS attack with P2P grid. In: Creativity in Intelligent, Technologies and Data Science. Springer International Publishing, pp. 407–415 (2015)
12. Hofstede, R., Pras, A., Sperotto, A., Rodosek, G.D.: Flow-based compromise detection: lessons learned. IEEE Secur. Priv. **16**(1), 82–89 (2018)
13. Kolias, C., et al.: DDoS in the IoT: Mirai and other botnets. Computer **50**(7), 80–84 (2017)

Software Implementation of Spectral Correlation Density Analyzer with RTL2832U SDR and Qt Framework

Timofey Shevgunov[1,2](✉) and Evgeniy Efimov[1]

[1] Moscow Aviation Institute (National Research University), Moscow, Russia
shevgunov@gmail.com, omegatype@gmail.com

[2] National Research University Higher School of Economics, Moscow, Russia
tshevgunov@hse.ru

Abstract. This paper introduces the software implementation of spectral correlation density (SCD) estimator run on a personal computer equipped with low-cost quadrature demodulator RTL2832U. The proposed solution is based on software-defined radio concept and implements all processing in the software written in high-level programming language. The module architecture of the application is described with the scheme of the main subroutine conducting an iterative SCD estimator. The given outline of the main window reveals a graphical user interface designed with Qt framework which allows changing the estimation parameters during the run processing due to the multithreading; the scheme of interacting threads is also given. It is shown that the performance of a modern central processor unit is ample to implement near real-time processing of cyclostationary signals using the averaged cyclic periodogram method. The developed application offer to carry out a deeper cyclostationary-based analysis of electromagnetic signal in the desired frequency band at the rate of measured time samples being acquired.

Keywords: Cyclostationarity · Spectral correlation density · Software-defined radio · RTL2832U · Application design · Multithreading

1 Introduction

Cyclostationarity is an intrinsic property held by the majority of physical level signals used as media for radio communication [1] as well as by many other processes of natural or manmade origin [2]. The research being conducted over the last six decades has created the well-established theory and the bunch of original signal processing techniques which are still unknown to the majority of researches in various fields. The increasing interest to the cyclostationary methods over the last years [3] can be easily explained by the fact that analog systems have finally given way to digital ones and the latter are operating with digital signals exhibiting strong cyclostationary properties.

Basically, the cyclostationary models take into account the hidden correlated periodicities existing in random signals in the cases when the model of a deterministic periodic signal superimposed by an additive noise fails due to the randomness of the messages carried by the signal. On the contrary, the choice of the cyclostationary model

R. Silhavy (Ed.): CSOC 2019, AISC 986, pp. 164–173, 2019.
https://doi.org/10.1007/978-3-030-19813-8_18

rather than stationary can reveal more information about the signal although the model of the stationary process, which is primarily the filtered white noise, is widely spread among all branches of science and looks much simpler [4].

The cyclostationary signals can be described by their second-order characteristics which are basically functions depending on the pair containing time or frequency variables. There are four such characteristics [5], namely the two-dimensional auto-correlation function, the current (Wigner-Ville) spectrum, the cyclic auto-correlation function (CACF), and the spectral correlation function (SCF). For a given complex-valued signal $x(t)$ CACF is defined according to:

$$\mathcal{R}_x^{\alpha}(\tau) = \mathbb{E}\left\{ \lim_{B\to\infty} \frac{1}{B} \int_{-B/2}^{B/2} x(t+\tau/2)x^*(t-\tau/2)\exp(-j2\pi\alpha t)dt \right\}, \tag{1}$$

where τ is the time shift, $\mathbb{E}$ stands for the probabilistic expectation, superscript (*) denotes complex conjugation, and α is the cyclic frequency, which is the parameter of the Fourier transformation over the current time t; α takes values from a countable set.

The spectral correlation function as a function of the conventional frequency f is obtained via Fourier transform of CACF over the time shift variable:

$$\mathcal{S}_x^{\alpha}(f) = \int_{-\infty}^{+\infty} \mathcal{R}_x^{\alpha}(\tau)\exp(-j2\pi f\tau)d\tau. \tag{2}$$

A researcher interested in a deep analysis of measured signals should not avoid the estimation of their cyclostationary characteristics. Among four above-mentioned characteristics, the SCF usually provides the information in the most compact form. The first approach to build a SCF estimator was made in [6] where some important problems was highlighted standing it out on the background of the conventional one-dimensional spectral estimation [7]. The Fast Fourier Transform Accumulation Method (FAM) introduced in [8] in the early 1990s has become a de facto main technique for SCF estimation despite its drawbacks.

The novel approach based on cyclic periodograms [1] was shown in [9] and was successfully applied in radar [10] and electric fields estimation for electromagnetic compatibility measurements [11]. However, when one deals with a time-limited realization of a signal $x(t)$, the estimated characteristic is the density or distribution where the cyclic frequency α is assumed to be continuous. Basically, the averaged cyclic periodogram estimator could be written as

$$\hat{\mathcal{S}}_x(\alpha,f) = \frac{1}{K}\sum_{k=1}^{K} X_k(f+\alpha/2)X_k^*(f-\alpha/2), \tag{3}$$

where K is the number of averaged signal segments (generally overlapped) and

$$X_k(f) = \int_{t_k}^{t_k+T} w(t)x(t)\exp(-j2\pi ft)dt \tag{4}$$

is a short-time Fourier transform, or a current spectrum, T is the length of each segment, t_k is the starting point of the k-th segment, and $w(t)$ is a window or weighting function.

This paper deals with the near real-time software implementation of the SCD estimator based on the accumulation of cyclic periodograms. A promising approach involving the concept of software-defined radio consists in turning a versatile personal computer into a powerful and flexible signal processor. It acquires the processed data flows by means of an analog-to-digital converter (ADC) expanded by a quadrature demodulator receiving electromagnetic signals from an arbitrary frequency spectrum range. The range can be fully defined by its central frequency and bandwidth and the sampling rate of the ADC should be equal or greater than the bandwidth as far as complex envelopes are involved rather than real-valued modulated signals.

The primary goal of the paper is presenting the software prototype that successfully implements the algorithm of the spectral correlation density (SCD) estimator on the fly. The rest of the paper is organized as follows. The overview of the developed solution is presented in Sect. 2. The details of graphical user interface implemented by the main window of the Qt-based application are described in Sect. 3. The results of carried out experiments revealing the screenshots of the developed software are shown in Sect. 4. The paper ends with the Conclusion where a brief summary is given.

2 System Overview

The solution developed in the current work based on a very popular low-cost USB receiver based on Realtek RTL2832U [12]. The user-ready device depicted in Fig. 1 is a small printed circuit board that can be plugged to a PC by the standard USB 2.0 (or higher) interface. An antenna must be attached to the device to obtain the desired gain in the chosen frequency band. The vendor provides consumers with the software package including drivers for most popular operating systems. The PC equipped with this receiver is a versatile hardware platform to start further prototyping.

Fig. 1. SDR USB-receiver based on RTL2832U

The interactive graphical user interface (GUI) implements the following functionality of the presented system: setting up the parameters of signal acquisition and further signal processing; displaying the evaluated estimations and its integral characteristic; controlling the signal acquisition unit. The development process benefits significantly from usage of free open-source technologies and libraries. The software prototype has been written in Python 3 and targets a GNU/Linux operating system. In order to account for a possible cross-platform application, GUI has been implemented with Qt graphical framework [13] which also provides a rather simple possibility to implement the multithreading approach for the software design.

The necessary parameters of the signal acquisition unit must be defined during the configuration stage prior to the acquisition. They are the central frequency F_c which the complex envelope is taken from, the sampling rate F_s that is equal to the bandwidth of the acquired signal located symmetrically around the central frequency, the buffer size N that is the number of samples defining the observation period. The latter is usually chosen to be a multiple of 1024.

The acquisition unit RTL2832U is equipped with the quadrature demodulator returning digital signals $i[n]$ and $q[n]$ with the sampling period T_s reciprocal to the sampling rate: $T_s = 1/F_s$:

$$i[n] = i(nT_s), \quad q[n] = q(nT_s). \tag{5}$$

These signals are merged to assemble the complex envelope as follows:

$$x[n] = i[n] + jq[n] = x(nT_s). \tag{6}$$

Thus, the actual real-valued signal which complex envelope is given can be restored:

$$s(t) = |x(t)| \cos(2\pi F_c t + \varphi_0 + \arg(x(t))), \tag{7}$$

where $|\bullet|$ and $\arg(\bullet)$ stand for the absolute value and argument of the complex number correspondingly, and φ_0 is the initial phase, either a negligible or nuisance parameter.

The following parameters are to be defined in order to perform the estimations of the spectral correlation density that is one of the cyclic characteristics of the signal. They are the base length of the FFT N_{FFT} that will point out the frequency resolution, the number of the intervals N_α to split cyclic bandwidth into, and the cyclic frequency range as the closed interval $[\alpha_{\min}, \alpha_{\max}]$ which boundaries ought to be chosen according to:

$$2F_c - F_s \leq \alpha_{\min} \leq \alpha_{\max} \leq 2F_c + F_s. \tag{8}$$

The procedure of the spectral correlation density estimation is roughly described in [9] and consists in splitting the acquired signal into a series of segments and using then the window function in order to weight each signal segment. Therefore, the parameters of the segmentation and weighting have to be defined; they are the length of each segment equal to the window length T_W, the overlapping coefficient κ, and the type of the window.

Once the signal is acquired, the developed software prototype will calculate the spectral density estimation matrix and visualize it in a 2D plot alongside with the integral characteristic function $P(\alpha)$ [14].

The main modules of the application presented in Fig. 2 are following: `rtlssdr.py` is a library for the interaction with RTL2832U unit that acquires the signals according to the defined parameters and returns it to Python as a `numpy` array; `qscdanalysis.py` is a GUI library that holds all user interface components implemented within Qt; `scdestimation.py` returns the SCD estimation as a complex-valued matrix. The latter implements the spectral correlation density (SCD) estimator described in [9] as Python class `ScdEstimator`. The sequence of actions required to obtain the estimation is shown in Fig. 3.

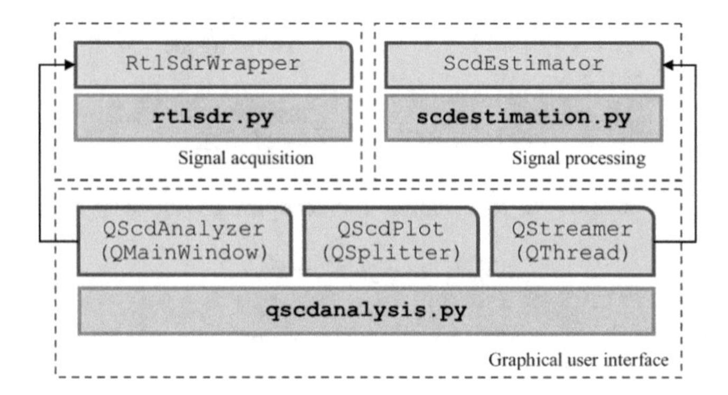

Fig. 2. Design scheme of the main software components and libraries

As far as the SCD estimator is concerned, the implementation of the estimation procedure in Python can be optimized by using vectorized `numpy` operations. The sampled signal of the length N is split into R segments of the length T_w/T_s each:

$$R = \left\lfloor \frac{(NT_s - T_w\kappa)}{(T_w - T_w\kappa)} \right\rfloor, \tag{9}$$

where the half-brackets stands for rounding down to the nearest integer number. The small portion of the signal less than the length of a segment may therefore be neglected during the analysis of the long time series.

The evaluation of the SCD estimation within the target cyclic frequency band requires the tensor of the shape $N_\alpha \times (T_w/T_s) \times R$ to be processed. For instance, the typical parameters $N_\alpha = 1024$, $T_w/T_s = 1024$, $R = 128$ imply that approximately 2048 MB is required to store it in the RAM. In practice, these parameters tend to increase so the required amount of the memory may rocket.

In order to lift the memory constraint, the tensor must be split into a series of smaller tensors which will be processed sequentially. That separation may be carried out alongside one of its axes. If the splitting occurs alongside the cyclic frequency axis then the tensor of shape $T_w/T_s \times R$ must be processed for each N_α-th cyclic frequency α

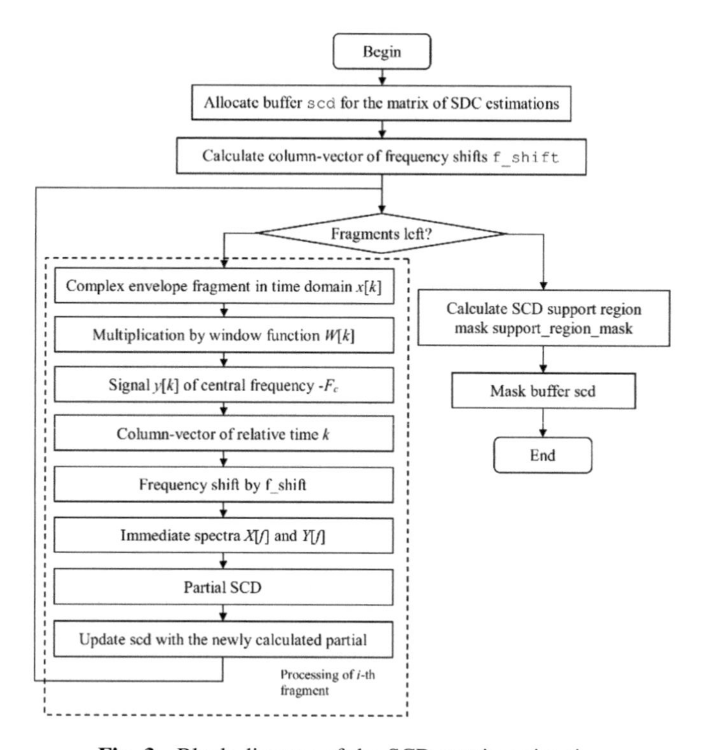

Fig. 3. Block diagram of the SCD matrix estimation

fragment. In that case, the resulting estimation matrix is calculated row-by-row. An alternative approach involves the processing of each signal segment T_w sequentially and yields a partial SCD estimation matrix each time a new segment of the signal is acquired. In the whole, R operations with the tensor of the shape $N_\alpha \times T_w/T_s$ is required.

The authors have chosen the latter strategy to design the algorithm shown in Fig. 3. The advantage of the approach is that there is no need to know in advance the length of the full signal. Therefore, one may begin to visualize the SCD estimation matrix once the first segment has been processed and then go on updating the plot as the matrix accumulates more and more processed segments.

3 Graphical User Interface

GUI library `qscdanalysis.py` is developed as the essential part of the prototype software and consists of three main classes that are `QScdAnalyzer`, `QScdPlot` and `QStreamer`. The library aims to perform the following tasks. The first one is a control panel that allows a user to set up the signal acquisition and processing parameters. The second task is illustrating the current SCD estimation matrix in form of 2D plot as it was introduced earlier in [14]. Finally, the third task is the isolation of the estimation subroutine in a separate thread that releases the thread of the main program to interact with the user and do not let the main window freeze.

The design of the main application window is shown in Fig. 4 where one can see three main sections. The parameter control tree is on the left-hand side of the window, the area of the SCD matrix visualization is in the middle of the window while the integral characteristic $P(\alpha)$ visualization block is on the right-hand side of the window. It is important to notice, that the vertical axis limits of the plot depicting the integral characteristic have to be linked to the corresponding limits of the vertical axis of SCD matrix estimation plot; this makes their further rescaling matched.

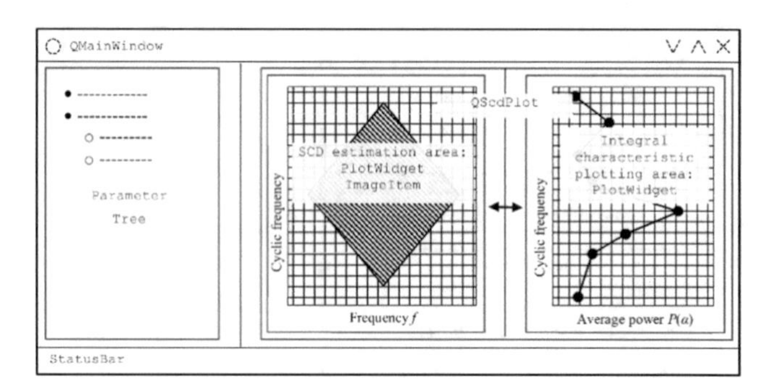

Fig. 4. Design of the main window of the program

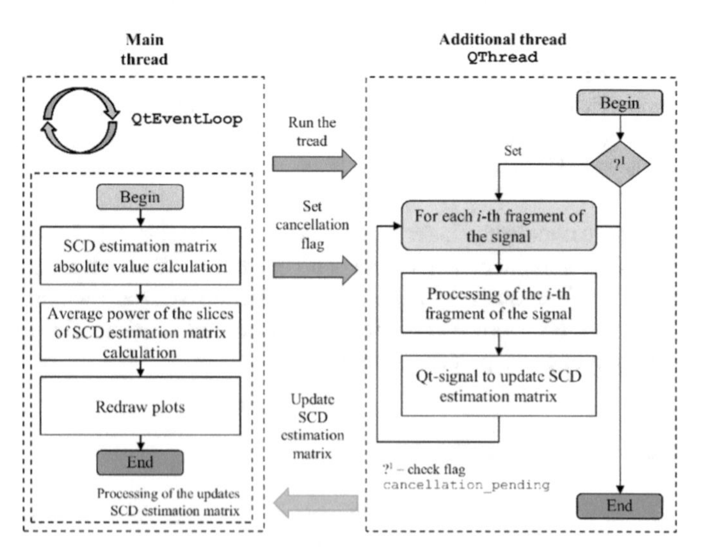

Fig. 5. Multithreading model of the program: the main thread with the Qt event loop, an additional thread and the communication signals between them

The key feature of the developed software prototype is its ability to asynchronously calculate and improve on the original SCD estimations matrix as newer signal segments obtained from the acquisition unit. The structural scheme of the intercommunicating threads is shown in Fig. 5 which is facilitated by 'signals and slots' mechanism provided by Qt. Thus, the SCD estimation matrix is visualized and updated in parallel with the data acquisition and processing. The main thread holds an event loop, processes the user input and draws plots representing the SCD estimation matrix and the integral characteristic $P(\alpha)$. As soon as a newer signal segment yields an update to the resulting SCD estimation matrix, the matrix is first being updated and only then the flag to redraw it is set high. The additional thread `QThread` in Fig. 4 acts in accordance with the SCD estimation procedure shown in Fig. 3. The iterative nature of the algorithm allows users to assess the result of the SCD estimation up to the moment and make it possible for them to adjust the processing parameters or conclude the acquisition when segments have been accumulated enough to the accurate estimation.

4 Experimental Results

In order to access the proposed approach, the set of experiments was conducted. The following signal acquisition and processing parameters are set as follows: the central frequency F_c = 102 MHz, the sampling frequency Fs = 312.5 kHz, the size of signal buffer N = 16384. The obtained complex envelope of the signal was then being processed by the developed application with the parameters: the FFT base N_{FFT} = 1024; the cyclic frequency band limits: $\alpha_{\min} = 2F_c - 2$ MHz, $\alpha_{\max} = 2F_c + 600$ kHz; the number of fragments alongside cyclic frequency axis N_α = 2048. The Hanning window [15] of 256-sample length was used to weight each segment. The overlapping coefficient was set to $\kappa = 0.8$. The results of the processing are shown in Fig. 6.

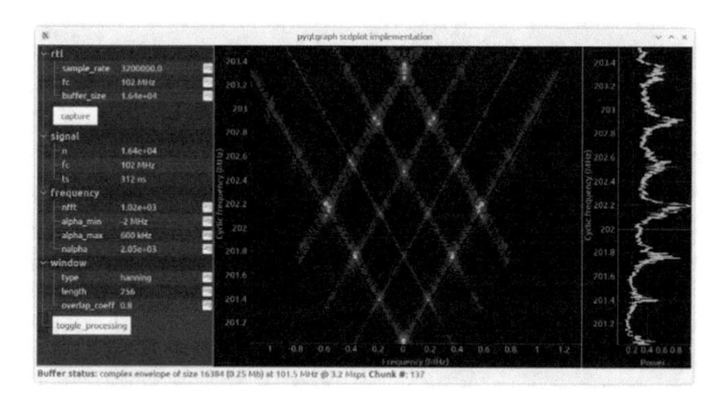

Fig. 6. The main window with the plots of the SCD estimated matrix and the integral characteristic function

Figure 7 demonstrates the area of the SCD matrix illustrating the cyclostationary properties of the signal vividly. For the detailed analysis, FFT base was doubled: N_{FFT} = 2048, the cyclic frequency bandwidth limits was narrowed: $\alpha_{\min} = 2F_c$, $\alpha_{\max} = 2F_c + 600$ kHz, the number of fragments alongside cyclic frequency axis was halved: $N_{\alpha} = 1024$, the size of signal buffer N remained unchanged. One could note that the obtained characteristic inherits the visual form of a rhombus with two generating stripes. This is a typical behavior of the SCD estimation for signals exhibiting some degree of the second-order cyclostationarity. The deeper theoretical reasoning for the estimated SCD to exhibit such a picture in the bispectral plane can be found in [1].

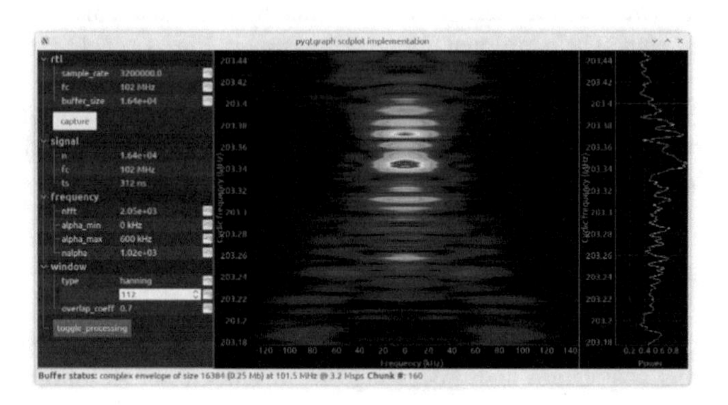

Fig. 7. The area of interest of the SCD estimation matrix and its integral characteristic function

5 Conclusion

The developed solution utilizing a relatively cheap software-defined radio USB receiver, which is cost about $20 per genuine unit, has clearly demonstrated that signal processing performed in accordance with the SCD estimation technique designed within the cyclostationary model could be carried out in near-real time using a modern PC workstation equipped with Core i5 CPU and 8 GB RAM. This result has significantly refined the statement made in [16] about 30 years ago when it was thought that the SCD estimation in the full bispectral band involving both frequency and cyclic frequency axis would inevitably employ a computer with a highly-parallel hardware organization rather than a versatile PC. As one can see, the rapid progress in semiconductor chips and the CPU architecture, especially in multithreading, has overcome the limit so that large arrays containing 2048×2048 and grater floating point complex numbers could be processed in the time enough to support up to 1 M Sample per second rate of the ADC data acquisition. This solution generally paves the way for further development of more sophisticated cyclostationary measurements of electromagnetic signals although the higher speed ADC are bound to be tested in the further development.

Acknowledgement. The work was supported by state assignment of the Ministry of Education and Science of the Russian Federation (project 8.8502.2017/BP).

References

1. Gardner, W.A.: Cyclostationarity in Communications and Signal Processing. IEEE Press, New York (1994)
2. Gardner, W.A., Napolitano, A., Paura, L.: Cyclostationarity: half a century of research. Signal Process. **86**(4), 639–697 (2006). https://doi.org/10.1016/j.sigpro.2005.06.016
3. Napolitano, A.: Cyclostationarity: new trends and applications. Signal Process. **120**(3), 385–408 (2016). https://doi.org/10.1016/j.sigpro.2015.09.011
4. Napolitano, A.: Generalizations of Cyclostationary Signal Processing: Spectral Analysis and Applications. Wiley/IEEE Press (2012)
5. Gardner, W.A.: The spectral correlation theory of cyclostationary time-series. Signal Process. **11**(1), 13–36 (1986). https://doi.org/10.1016/0165-1684(86)90092-7
6. Gardner, W.A.: Measurement of spectral correlation. IEEE Trans. Signal Process. **34**(5), 1111–1123 (1986). https://doi.org/10.1109/TASSP.1986.1164951
7. Kay, S.M.: Modern Spectral Estimation: Theory and Application. Prentice Hall, Upper Saddle River (1988)
8. Brown, W.A., Loomis, H.H.: Digital implementations of spectral correlation analyzers. IEEE Trans. Signal Process. **41**(2), 703–720 (1993). https://doi.org/10.1109/78.193211
9. Shevgunov, T., Efimov, E., Zhukov, D.: Averaged absolute spectral correlation density estimator. In: Proceedings of Moscow Workshop on Electronic and Networking Technologies (MWENT), pp. 1–4 (2018). https://doi.org/10.1109/MWENT.2018.8337271
10. Shevgunov, T., Efimov, E.: Artificial neural networks implementing maximum likelihood estimator for passive radars. In: Advances in Intelligent Systems and Computing, vol. 764. Springer, Cham (2019). https://doi.org/10.1007/978-3-319-91189-2_15
11. Efimov, E., Shevgunov, T., Kuznetsov, Y.: Time delay estimation of cyclostationary signals on PCB using spectral correlation function. In: Proceedings of Baltic URSI Symposium, Poznan, pp. 184–187 (2018). https://doi.org/10.23919/URSI.2018.8406726
12. RTL-SDR (RTL2832U) and SDR news and projects. http://www.rtl-sdr.com/
13. Qt Documentation. Signals & Slots. http://doc.qt.io/qt-5/signalsandslots.html/
14. Efimov, E., Shevgunov, T., Kuznetsov, Y.: Cyclic spectrum power density estimation of info-communication signals. Trudy MAI **97**, 14 (2017)
15. Harris, F.J.: On the use of windows for harmonic analysis with the discrete Fourier transform. IEEE Proc. **66**(1), 51–83 (1978). https://doi.org/10.1109/PROC.1978.10837
16. Roberts, R.S., Brown, W.A., Loomis, H.H.: Computationally efficient algorithms for cyclic spectral analysis. IEEE Signal Process. Mag. **8**(2), 38–49 (1991). https://doi.org/10.1109/79.81008

Historical 3D Visualisations of Brumov Castle in Different Time Periods

Pavel Pokorný[✉] and Monika Vatalová

Department of Computer and Communication Systems,
Faculty of Applied Informatics, Tomas Bata University in Zlín,
Nad Stráněmi 4511, 760 05 Zlín, Czech Republic
pokorny@utb.cz, monika.vatalova@seznam.cz

Abstract. The main task of this paper is to describe a visualisation method for Brumov Castle. This castle was probably founded in the first half of the 13th Century and grew rapidly until the end of 16th Century. Brumov Castle was an important residence of the local aristocracy in this period. At the beginning of the 18th Century, the castle lost its importance and was gradually abandoned and the construction was dismantled for building material. Research by historians has made it possible to create sketches and paintings of the castle in different centuries. Five 3D complex models of the castle were created - in the 13th, 14th, 15th and 16th Centuries, based on the appearance of the present ruins and its vicinity. The last model corresponds to its current appearance. All buildings and accessories were separately modelled and textured using UV mapping techniques. The visualisation outputs are performed by rendered images and animations. The Blender software creation suite was used for visualisation purposes.

Keywords: 3D visualisation · Historical visualization · Modelling

1 Introduction

Visualisation is the process of representing information, data, or knowledge in a visual form in order to support their presentation, exploration, confirmation, and understanding. It is used as a communication mechanism. A single picture can contain a wealth of information, and can be then processed much more quickly than a comparable page full of words. Pictures can also be independent of local language - just as a graph, chart or map may be understood by a group of people with no common tongue [1].

Visualisation research at SCI [2] has focused on applications spanning computational fluid dynamics, medical imaging and analysis, biomedical data analysis, healthcare data analysis, weather data analysis, poetry, network and graph analysis, financial data analysis, etc. Research involves novel algorithm and technique development to build tools and systems that assist in the comprehension of massive amounts of (scientific) data.

A typical example for creating successful visualisations is the Computer Graphics field. Along with the development and improved performance of computer technologies, the limits and possibilities of computer graphics continue to increase. The

R. Silhavy (Ed.): CSOC 2019, AISC 986, pp. 174–184, 2019.
https://doi.org/10.1007/978-3-030-19813-8_19

consequences of this trend are 3D visualisations - based on creating 3D models, the rendered image outputs can offer better quality [3], and the area of the visualised environment that can be show at the same moment is increasing [4].

One of these fields is historical visualisations. Based on historical documents, drawings, maps, photographs or plans, one can create 3D models of objects that no longer exist. Chronological sequences of computer-implemented visualisation methods of historic events are described in [5]. Modern approaches to 3D reconstruction solutions and visualisations of historic buildings can be found in [6].

This article describes a 3D visualisation method used for Brumov Castle. Using modern 3D software modelling tools, we tried to achieve a very credible appearance of the castle buildings and its vicinity [7].

Fig. 1. Brumov Castle artwork from the mid-13th Century; Author: Radim Vrla [8]

2 A Short History of Brumov Castle

The history of surveys of Brumov Castle are difficult to determine since these archives were lost in the 18th or 19th centuries. Because of this, all information is based on knowledge and other archaeological research archives [9].

The actual foundation of the castle is dated - from the available evidence, from somewhere in the years 1210–1220. It was founded by Margrave Vladislav Jindřich during the reign of Přemysl Otakar I. in the late Romanesque style. The building was probably dominated by a massive square tower and a part thereof was probably the opulent castle chapel - as evidenced by findings of rich architectural decorations, like tympanum heads, the portal, and other details [10].

From its very inception, Brumov Castle fulfilled important guard, defensive and residential functions when entering the Vlar Pass, which has always served as a natural connection between the Váh and Morava Rivers.

Fig. 2. Brumov Castle artwork from the mid-16th Century; Author: Radim Vrla [8]

The first written record comes from Opava in a document dating back to 1224, where Svěslav - a bourgeois in Brumov, is mentioned. At the beginning of the 14th Century, the castle was attacked by Matůš Čák and his Trenčanský army. The castle was besieged twice and captured by King Zikmund's army, in the course of years. In 1342, Brumov Castle changed its status to a royal castle.

Bernard from Cimburk dominated Brumov in the years 1445–1461. He used this well-fortified and armed castle for military assaults on the surrounding estates and on Hungary. On December 7, 1503, the castle - roughly after 300 years, passed from royal to noble hands, when it was sold to Štěpán Podmanický and Jiřík Tarczy [9].

East Moravia was exposed to many wars, conflicts, and plundering by rebels and the invasions of Asian invaders in the 16th Century and the beginning of the 17th Century. Thus, possession of the Brumov Manor was therefore, a particularly difficult and dangerous task. During the devastating Thirty Years War, Brumov Castle was occupied by Wallachians, who improved its defences with stone bastions and further fortifications.

When the Tatar and Turks invaded the Brumov Estate in 1663, it was devastated and burned; only the castle was preserved. After a fire in 1760, the castle was temporarily repaired; but, after another fire in 1820, it was finally abandoned and became a local source of cheap building materials.

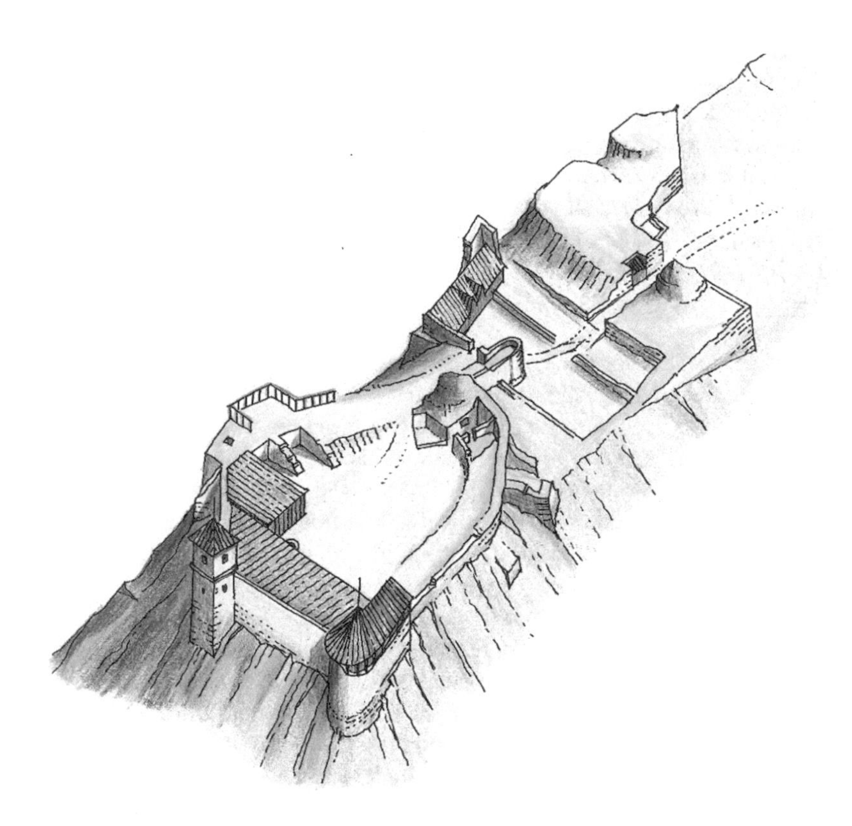

Fig. 3. Brumov Castle artwork from the present-day; Author: Radim Vrla [8]

3 Used Resources and Software

The first phases that needed to be performed were to collect all suitable and available historical materials of Brumov Castle and to select adequate software for visualisation creation.

3.1 Acquiring Resources

The overall progress of this work was initiated by the collation of available historic materials and information about Brumov Castle down through the centuries. The main resources were found in the State District Archive in Zlín [11], the National Heritage Institute in Kroměříž [8], the castle´s webpage [10], as well as a book on this subject [9]. Attention was mainly focused on drawings and sketches.

The most useful drawings obtained were sourced from Radim Vrla, an employee of the National Heritage Institute in Kroměříž. Based on archaeological excavations, he created nice drawings of Brumov Castle´s appearance in different periods – in the 13th, 14th, 15th and 16th Centuries. The last drawing corresponds to its current appearance. Some of these drawings are shown in Figs. 1, 2 and 3.

The created 3D models of the castle were mainly based on these drawings. Since there is no exact form of the castle from the past and these drawings capture only the main shapes, the details of these shapes were a matter of improvisation. However, to avoid historical inaccuracies, all works on these models were consulted with the above-mentioned historian - Radim Vrla.

3.2 Software Used

In order to select applications that we wanted to use, we compiled the requisite works that are required for creating high-quality visual outputs. The whole process includes 3d model creation - (wide spectrum of modelling tools are suitable for this), correct texture mapping, and then to create animations. The final phase comprises the rendering process, which generates rendered images and animations based on the set rendering parameters. In addition, we wanted to use free-to-use software.

The GIMP software was used for preparing, (drawing), and textures [12]. All other partial processes were made in the Blender software suite [13].

GIMP stands for GNU Image Manipulation Program - one of the most successful applications in the free software world. It is freely distributed to anybody, and anybody can look at its contents and its source-code and can add features or fix problems. The GIMP suite features also include many selection, drawing, transformation, layer and masking tools, as well as simple and complex filters and brushes [14].

Blender is a fully integrated creation suite under the GNU GPL license. It is based on cross-platform OpenGL technology and offers essential tools for modelling, texturing, UV mapping, lighting, skinning, scripting, camera work, rendering, animation, rigging, particles and other simulations. Blender also allows one to perform compositing, post-production, and game creation [15].

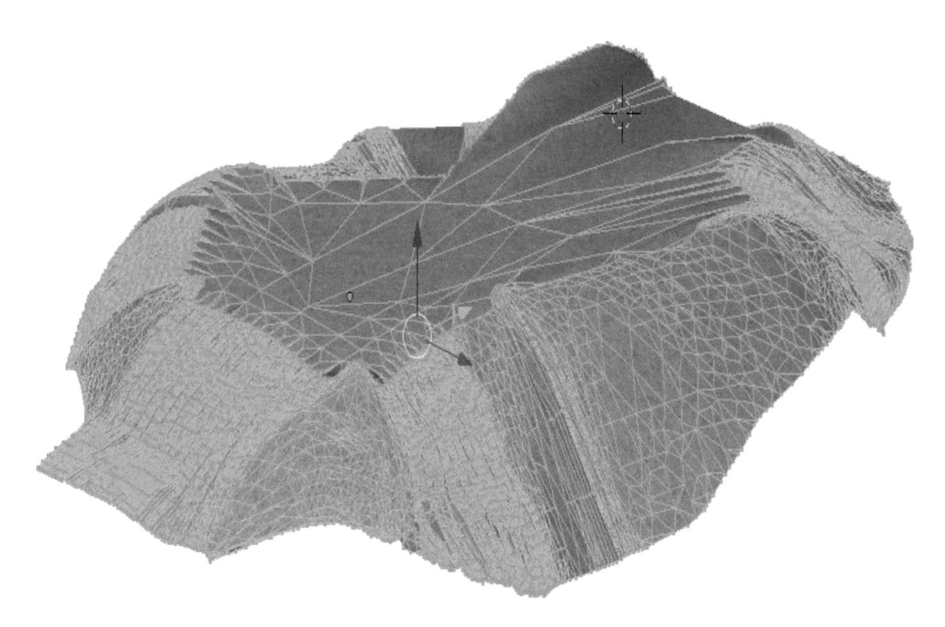

Fig. 4. A landscape mesh model (the ground under Brumov Castle)

4 Modelling

When the collation of all of suitable resources was completed, the modelling phase was started in Blender. To begin with, a landscape model was created; and used together with all five castle models. Each castle model was created in 3D scenes separately.

Only polygonal representation was used for all the models. It is performed with mesh objects in Blender.

4.1 A Landscape Model

In Blender, a plane object was added and shaped into a rectangle corresponding to the ground plan of the castle and its vicinity. After that, the Subdivide toll was used several times, which split the rectangle into smaller parts and added new vertices. In this way, we obtained a simple model with about 5,000 vertices.

The next step was to modify the (spatial), z-coordinates of all of the vertices based on altitude on a scale corresponding to the model size and reality ratio. To achieve this, we used web tools found on the DaftLogic homepage [16], which offer the possibility of determining the altitude - (elevation) for a specific latitude and longitude for most places in the world. On the basis of the obtained altitude values, we performed the modification of the z-coordinates for each vertex.

The last part was to reduce, (optimise), the number of vertices in the landscape model in order to speed up the rendering process. This operation is performed by a Decimate modifier in the Blender environment. With this tool, the number of vertices was reduced by about half. With this, we obtained the final landscape model (Fig. 4).

4.2 Modelling the Castle

As mentioned above, five historical sketches of Brumov Castle were obtained. It was decided that five different 3D complex castle models based on these sketches be made, because these sketches are the most important resources. Each model was made in a separate scene, but used the same techniques.

The same landscape model was used for each castle model. It was only modified a little - where necessary, in order to correspond to the actual castle model.

The first important resource of Brumov Castle is the top-down view of the present-day castle. This was acquired from the DaftLogic web page [16]. This image was loaded into the Blender environment and placed on the background screen. This image also contained the border of the current main castle shape. We also added a plane object in the Blender 3D scene and, using simple editing tools, modified it to this castle´s shape. For the castle models from the 13th, 14th, 15th and 16th Centuries, we compared this shape with sketches, and where necessary, modified this shape. After that, the shapes were extruded into the third dimension.

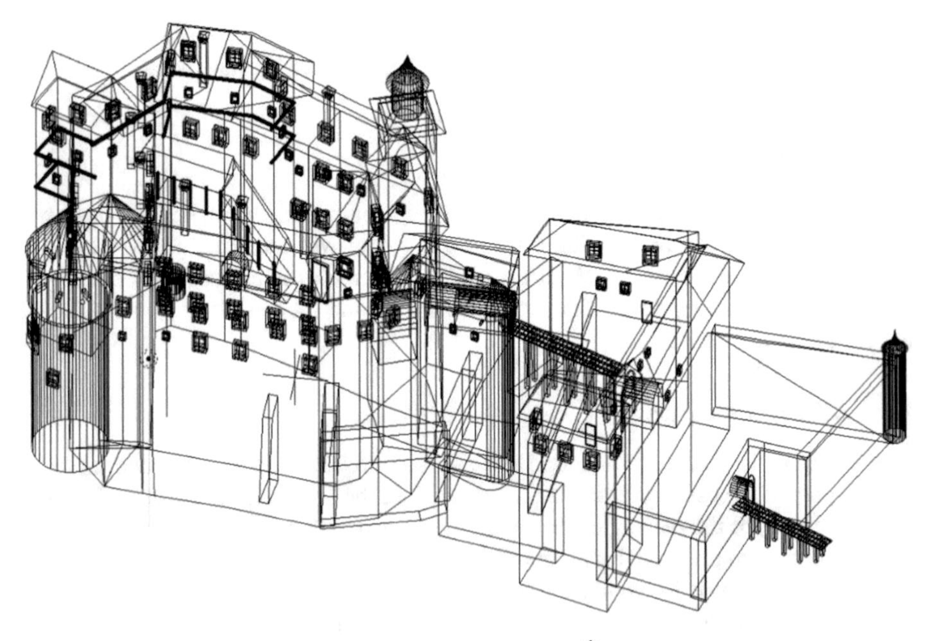

Fig. 5. The mesh model of Brumov Castle from the 16th Century, (wireframe shading)

The advantage was that the castle buildings usually have a box shape - so modelling was not difficult for this reason. The extruded 3D model was then modified by standard modelling tools (basic editing commands, transformations, modifiers, Extrude, Subdivide, Knife, Loop Cut and Slide, etc. [15]) in order to obtain the main appearance.

In the following steps, we modelled details – doors, windows, passages, gateways, stairs, roofs, chimneys and bridges. We also used typical mesh modelling tools. While modelling, we paid attention to always have "pure models", i.e. that the models do not have multiple or unnecessary vertices, edges and shapes. This eliminated potential problems for the next modelling phases - especially texturing and rendering.

The example of one completed model - (displayed in wireframe shading), is shown in Fig. 5. The whole model, including landscape, contains approx. 32,000 vertices.

Fig. 6. The model of Brumov Castle with landscape from the 16th Century, with textures in Blender

5 Texturing

All textures used had two sources. The first was our own. We visited Brumov Castle and took many photographs of its current appearance. Some of them could later be used as textures. The second resource was the webpage [17]. It is possible to find thousands of different textures here. After registration, it is possible to download and use them under the given conditions (free for use with the 3D models).

In order use them as textures, they had to be prepared and modified. All operations were made in GIMP software. The first step was to cut off unnecessary image parts. Then we removed undesirable objects - (like grass in front of walls) with the Clone tool, and unified brightness and contrast in order that the images have corresponding colours.

We used UV mapping techniques for texture mapping in Blender. This process starts with the decomposition of each object into 2D sub-surfaces - (e.g. a UV map). Blender supports the Unwrap tool with several settings for these purposes. The UV map created in this way was saved into the .png raster graphic format, (in order to achieve precision mapping, this requires the use of a lossless compression algorithm). The resolutions: 512×512 or 1024×1024 pixels, were used for the UV maps, (to increase the rendering process speed).

All UV maps were opened in GIMP and - using its common tools, textures were prepared and transformed into the correct parts of these maps. Once this process was completed, all of the created textures were saved back into the same files and opened in Blender. This way, correct texture mapping was performed on the appropriate 3D models.

The mesh model of Brumov Castle, dating from the 16th Century - including mapped textures, is shown in Fig. 6.

6 Rendering and Animations

After finishing the modelling and texture phases, the whole scene was completed. In order to attain high-quality render outputs - (raster images), it is also necessary to set other suitable parameters like the surroundings, lighting and camera. The surrounding area setting parameters were performed by using the Blender World window. There, is possible to set simple colours for zenith and horizon, and make a gradient with them, or to use the internal - (procedural), or external texture, (any bitmap file). An external sky texture was used for the background in all our 3D scenes.

The scene lighting can be realised in several ways in the Blender environment. It is possible to use global influences - (i.e. Environment Light, Ambient Light, Ambient Occlusion and Indirect Light); or light objects - (i.e. Point, Spot, Area, Sun and Hemi) for local lighting. We only used light objects in these projects – one global Sun object with high energy followed by some Point objects with low energy - used for lightning parts that are not lightened enough by the global light.

The last step before rendering the images was to select a suitable position for the camera. Because it is appropriate to have pictures from different positions, we used more cameras, correctly oriented in order to capture the best graphic images of the whole scene. It is possible to make simple switches among them and to make different image outputs. The image resolution was set to 1920×1080 pixels, (Full HD).

Other render outputs can be realised with the help of animation. In order to attain the animations that we required, this could be performed with the help of animation curves that can be freely shaped and transformed, in Blender. The camera can then follow this in a defined time, and - with this, it is possible to set the animation speed. The animation curve was transformed into a spiral shape with 2 turns, in order to capture the castle models in this way from all directions and different zooms.

The Render command is used to perform the rendering calculation process in the Blender environment. Additionally, users can set many other parameters for images or animations. The decision was made to use Blender's internal renderer with allowed ray-tracing; and, as mentioned above, the resolution was set to 1920×1080 pixels.

Animation parameters were set to 25 frames per second, and to the MPEG-2 output. Figure 7 shows a sample of a rendered image of Brumov Castle from the 16th Century.

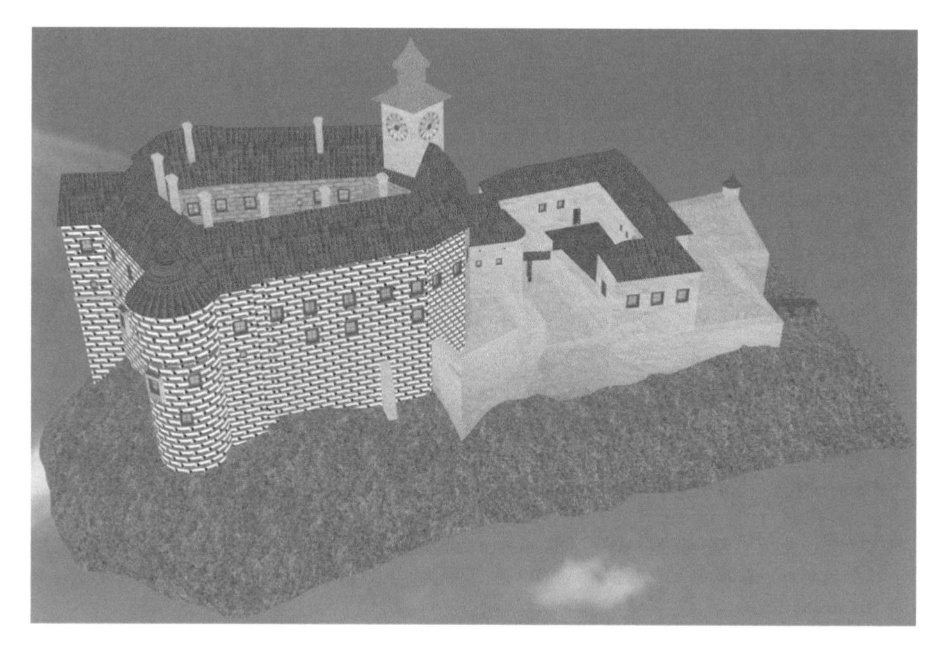

Fig. 7. A rendered image of Brumov Castle from the 16th Century

7 Conclusion

This article briefly describes the visualisation method used for depicting Brumov castle in different time-periods. Five more complex 3D models were completed - based on historical materials - mainly drawings and sketches. These models are from 13th, 14th, 15th and 16th Centuries; the last model corresponds to its current appearance.

All models were modelled, textured and animated in the Blender software suite and the final complex scenes were performed with rendered images and animations. These outputs present the probable appearances of the castle in the centuries mentioned above.

The future goal is to improve and expand these models. This process will include modelling more detailed parts of the castle - like doors, windows and courtyards. The higher quality of some models can also be improved using high-quality or more-layer textures. Further improvements can be performed using interactive web application outputs. Blender can be extended with a Blend4Web plug-in, which allows users to export the whole 3D scene into a single .html file that can be opened and controlled in any modern web browser.

Acknowledgments. This work arose from collaboration with the Czech National Heritage Institute and the Kroměříž Regional Council, which provides both methodological and professional assistance to heritage-site owners and managers in the Zlín Region.

References

1. Ward, M.: Interactive Data Visualization: Foundations, Techniques, and Applications, 1st edn. A K Peters Ltd., India (2010). ISBN 978-1568814735
2. Visualization. http://www.sci.utah.edu/research/visualization.html. Accessed 12 Jan 2019
3. Qiu, H., Chen, L., Qiu, G., Yang, H.: Realistic simulation of 3D cloud. WSEAS Trans. Comput. **12**(8), 331–340 (2013)
4. Wettel, R., Lanza, M.: Codecity: 3D visualization of large-scale software. In: Companion of the 30th International Conference on Software Engineering, Leipzig, Germany, pp. 921–922. ACM (2008)
5. Boice, G.L., Black, J.L.: System and method for animated computer visualization of historic events. U.S. Patent No. 7,999,810, 16 August 2011
6. Ragia, L., Sarri, F., Mania, K.: 3D reconstruction and visualization of alternatives for restoration of historic buildings: a new approach. In: 2015 1st International Conference on Geographical Information Systems Theory, Applications and Management (GISTAM). IEEE (2015)
7. Vatalová, M.: A 3D visualization of Brumov Castle in different time periods. http://digilib.k.utb.cz/bitstream/handle/10563/43430/vatalov%C3%A1_2018_dp.pdf?sequence=1&isAllowed=y. Accessed 12 Jan 2019
8. National Heritage Institute in Kroměříž. https://www.npu.cz/cs/uop-kromeriz. Accessed 12 Jan 2019
9. Obadal, M., Vrla R.: Hrad Brumov: historie a stavební vývoj. Brumov-Bylnice, Město Brumov-Bylnice (2013)
10. The history of the castle – Official website of Castle Brumov. http://www.hradbrumov.cz/historie-hradu/. Accessed 12 Jan 2019
11. State District Archive in Zlín, Moravian Provincial Archives in Brno. http://www.mza.cz/zlin/. Accessed 12 Jan 2019
12. GIMP - The GNU image manipulation program. http://www.gimp.org. Accessed 12 Jan 2019
13. Blender.org. http://www.blender.org. Accessed 12 Jan 2019
14. Lecarme, O., Delvare, K.: The Book of GIMP: A Complete Guide to Nearly Everything, 1st edn. No Starch Press (2013). ISBN 978-1593273835
15. Fisher, G.: Blender 3D Basics, 2nd edn. Packt Publishing (2014). ISBN 978-1783984909
16. Daftlogic – Google maps find altitude. https://www.daftlogic.com/sandbox-google-maps-find-altitude.htm. Accessed 12 Jan 2019
17. Textures – Textures for 3D, graphic design and Photoshop!. https://www.textures.com/. Accessed 12 Jan 2019

LCBC-XTEA: High Throughput Lightweight Cryptographic Block Cipher Model for Low-Cost RFID Systems

R. Anusha(✉) and V. Veena Devi Shastrimath

Department of Electronics and Communication Engineering,
NMAM Institute of Technology, NITTE, Udupi, India
anu4research@gmail.com

Abstract. The RFID Technology is widely used in many authentications and a sophisticated application, still having many prospective issues includes privacy and security issues. To resolve these issues, this paper presents an efficient Lightweight cryptographic Block cipher algorithm. The hardware architecture of Tiny Encryption Algorithm (TEA) is designed and which is simple, flexible, fewer computations required and simple key scheduling. To overcome the security attacks in key scheduling on TEA, an Extended TEA (XTEA) is designed, which is having pipelined architecture with parallel computation to improve the throughput and provide better security. The proposed XTEA is in reconfigurable nature, by changing the mode to process encryption or decryption. The TEA and XTEA simulation results are obtained from Xilinx ISE tool on ModelSim 6.5f simulator and implemented on FPGA Platform-Artix-7 with resource constraints like Area, time and power are tabulated. The proposed XTEA is compared with similar existing research approaches like AES-8bit, TinyXTEA1, and Tiny XTEA-3 with an improvement of Area, throughput and Efficiency on same FPGA platform. The proposed XTEA works at the high throughput of 81 Mbps and Efficiency of 0.34 Mbps/Slice.

Keywords: RFID systems · TEA · XTEA · Lightweight · Block cipher · Security · Key scheduling · Encryption · Decryption · FPGA

1 Introduction

Security is an important issue to authenticate the information in communication systems and other applications because of its ubiquitous nature. To prevent user's privacy, trust, and future problems, the systems needs provides an appropriate security engine depends on applications viewpoint [1]. For high-speed applications and low power applications like RFID tag-reader and wireless sensor nodes, an effective less area and energy consumed Block ciphers are needed. The TEA (Tiny Encryption Algorithm) is a block cipher algorithm, and its goal is to produce a simple, flexible and does not depend on larger pre-computations or lager Tables. The TEA uses simple algebraic functions includes addition, shifting and XOR with small operation code [2–4,10]. The TEA block cipher with low power applications is designed with three approaches

R. Silhavy (Ed.): CSOC 2019, AISC 986, pp. 185–196, 2019.
https://doi.org/10.1007/978-3-030-19813-8_20

which include sequential, structural and subprogram methods. The sequential method provides the best resource utilization than other two ways [9].

The extended TEA (XTEA) is having simple Feistel structure with fastest and most efficient light-weight Cryptographic Block Cipher (LCBC) which is having 64-bit block data and 128-bit key size and has 64- rounds of operations and XTEA is used in most of the real-time cryptographic applications [5]. The XTEA is extended to provide security implementation with GEN2 protocol, which includes the tag identification layer with command detection module (CDM) response module (RM) and control unit with the state machine. In between the control unit and the response module, the XTEA is introduced [6]. The solve privacy and security issues in RFID systems, a robust security protocol is designed, which includes back-end server model with reader and tag model. In a tag model, XTEA key is set which will automatically be updated when communicating with a reader. What the system can apply with different attacks includes replay attack, cloning attack, the man in the middle attack, Denial-of-service [DoS] attack, and Active scanning attack [ASA] [7]. The MMA (meet-in-the-middle attack) on XTEA with reduced round is introduced in [8,11] with 15, and 23 rounds attacks are described in detail using XTEA. The Secured RFID authentication protocol with key updating method is explained [12] which includes symmetric key technique, with server-based tag-reader authentication. The security level analysis with different attacks is analyzed with computational cost, computation load for the tag and storage requirement with improvements.

The previous work towards TEA and XTEA and its approach towards Security applications are explained in Sect. 2. The Problem statement of the block cipher is highlighted in Sect. 3. Section 4 described the Hardware architecture of TEA and proposed XTEA with description. Section 5 analyzes the simulation results of TEA and XTEA of encryption and decryption and tabulate the hardware constraints and comparison of XTEA with existing Block ciphers with improvements. Whereas the overall work is highlighted in conclusion Sect. 7.

2 Related Work

Venugopal et al. [13] present high throughput models of cryptographic algorithms include TEA and XTEA on GPU (Graphical processing units) and FPGA (Field programmable gate array). The designs work at a high level of parallelism with coarse-grained architecture with optimization. Hardware acceleration tool (CHAAT) for benchmarking with automated fashion to select the cryptographic algorithms and using these tool Performance, cost, and security issues are resolved inefficient manner. FPGA design results give better for smaller data and GPU results for more substantial data sizes. Huang et al. [14] describe the key scheduling of Lightweight Block Ciphers (LBC). Overcome the weakness of many attacks in a key schedule by using a suitable tool to find the computation chains which covers the actual key information (AKI) for key scheduling iteration in LBC's. The novel-ness is shown in many applications includes TWINE-80 as a case study. Mozaffari-Kermani et al. [15] have presented a fault recovery light-weight Cryptographic Block Ciphers (LCBC) for securing the embedded devices. The LCBC includes XTEA, SIMON and PRESENT algorithms with error detection schemes to achieve the reliability, better accuracy and overhead than AES (Advanced Encryption Systems). The error detection schemes are incorporate with LCBC with RERO (Recomputing with Rotated Operands (RERO) manner to detect the permanent faults easily. The differential cryptanalysis of Reduced round LBC's like TEA and XTEA are described in Hajari et al. [16]. The TEA and XTEA are done by bit-wise operations and by using early halt technique, removing the not suitable pairs in early stages for impossible differential attacks, which results from 19 rounds for TEA and 25 rounds for XTEA with analysis of time, memory and data complexities.

The Hussain et al. [17] presents the TEA Block cipher on FPGA based implementation with different design approaches. The various design approaches include combinational, sequential and hybrid (combination of combinational and sequential) approaches for TEA block cipher. The hybrid approach used only four working clock cycles to complete the encryption and decryption process and gives better area slices, power and throughput than the other two methods. The Tian et al. [18] describes the Research study of Bitsum attack on TEA block cipher. The Bitsum attack is the same and similar to the dictionary attack, which includes the messages with a quality of '1'. The Bitsum is generated from the input messages, and message information cannot generate via Bitsum. The data is always not secure under Bitsum attack by using large key sizes. The Almeer et al. [19] has the XTEA lightweight-BC engine in hardware–software co-design on FPGA platform. The software approach gives average throughput using XTEA hardware accelerator on NIOS II soft-core processor and in Hardware approach, even higher throughput obtained for XTEA on FPGA. The review of lightweight block ciphers (LBC) is explained in detail by Hatzivasilis et al. [20] and cryptographic algorithms for IOT by Surendran et al. [22] and the comparison of LBC's with hardware constraints like throughput, power, efficiency, energy consumption, and technology based-CMOS and also software approaches to LBC's with constraints and pitfalls are listed. The modified version of TEA to improve the security is presented by Rajak et al. [21], the new version of key scheduling with S-BOX to allocate the round keys to TEA is approached with time improvement. Anusha and Devi Shastrimath [23] have demonstrated qualitative calculation on efficiency of Security Approaches towards preservation NFC Devices & Services. Anusha and Devi Shastrimath [24] have presented Tag-Reader Mutual Authentication (TRMA) - technique of RIFD tag-reader with the modified form of PadGen. The modified Pad-Gen design employs XOR operation for the RIFD-TRMA protocol.

3 Problem Statement

The wireless devices can exchange information via internet or satellite links through gateways to the external world. These wireless devices have limited resource constraints like area (space in a device), and power consumption. But gateways have no memory and power constraints in assumption. So, the hacker can easily access all the information from these wireless devices. Secure the gateway with encryption and decryption is needed. But the response time is enormous to operate in the gateway with a secured manner. Motivated by this issue, FPGA based lightweight cryptographic block cipher (LCBC) with XTEA is used in gateways, to speed up the encryption and decryption and also most of the existing researches on LCBC are software-based approaches and very less work on Hardware-based approaches which are not flexible and reconfigurable nature. The obtained resource constraints from the previous works are not standardized yet to process in Low-cost RFID wireless systems with better security.

4 Lightweight Block Cipher (LBC) Algorithms

To enhance the security on low cost embedded devices, the lightweight Block ciphers are introduced which includes Tiny Encryption Algorithm (TEA) and Extended Tiny Encryption Algorithm (XTEA). The hardware architecture of these two algorithms is elaborated in the below section.

4.1 Tiny Encryption Algorithm (TEA)

David Wheeler developed the TEA in 1994, which is the fastest encryption algorithm with minimal operation code. The TEA is a Lightweight Block Cipher (LBC) which includes 64-bit block size (Plain text) and 128-bit key size. The key size is enough to adapt to the modern security requirements. The 128-bit Key is divided into four parts each is 32-bit wide; i.e., K0, K1, K2, and K3 and perform both encryption and decryption Process. When TEA process starts, for encryption key scheduling, Counter initiates to reset with zero (count [0] = 0), if Yes, K0/K1 and if not, K2/K3 for the functional key scheduling process Fk0/Fk1 generation (as sub key) for round function operation is represented in Fig. 1(a). For Decryption key scheduling, Counter initiates to zero (count [0] = 0), if Yes, K2/K3 and if not, K0/K1 for functional key process Fk0/Fk1 generation for round function operation is represented in Fig. 1(b). The round function operation is in the form of a Feistel structure with 64 rounds for encryption and decryption Process. Two Feistel rounds perform 32- iterations of operation parallelly. The round operation is represented in Fig. 1(c). The 32-bit input ‘x’ parallelly shifts left to 4 times then added by sub key ‘Fk0’ and shift right to 5times then added by ‘Fk1’ and Xor’ed the added results to form the Round function output ‘F(x)’.

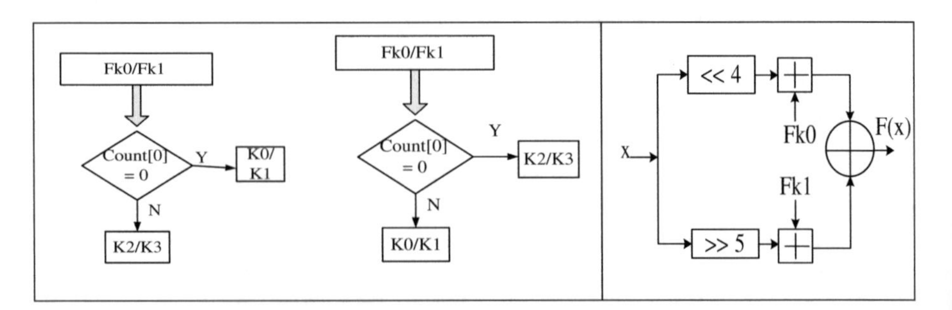

Fig. 1. (a) Key scheduling-encryption, (b) Key scheduling-decryption (c) Round function operation for TEA.

The Plain Text (PT) has 64-bit and is separated by two 32-bits wide, i.e. V0 and V1. For encryption process, the plaintext is divided to V0 [31:0] as first 32-bits, and V1 [63:31] is next 32-bits. Similarly, for decryption process, the plaintext is divided into V1 [31:0] as first 32-bits, and V0 [63:31] is next 32-bits. The 7-bit counter is used to count till 64 then reset to zero which is controlled by the clock.

The TEA uses Constants in 1st round: for Encryption- Delta ($\Delta1 = 9E3779B9_H$) and alpha ($\alpha1 = 9e3779b9_H$) are added ($\alpha1 + \Delta1$), when counter (count [0]) is one. Similarly, for Decryption-Delta ($\Delta2 = 9E3779B9_H$) and alpha ($\alpha2 = C6EF3720_H$) are subtracted ($\alpha2 - \Delta2$), when counter (count [0]) is one.

The TEA encryption and decryption process are represented in Fig. 2(a) and (b) respectively. When Counter (count [6] = 1) reaches 63, The O_V1 [63:32] and next 64trh count, O_V0 [31:0] is generated to form the 64-bit cipher text for encryption and plain recovered text for decryption. In TEA, the key scheduling is simple and weak, so it is suffered from third-party attack in most of the cases. So, these drawbacks are overcome by XTEA algorithm.

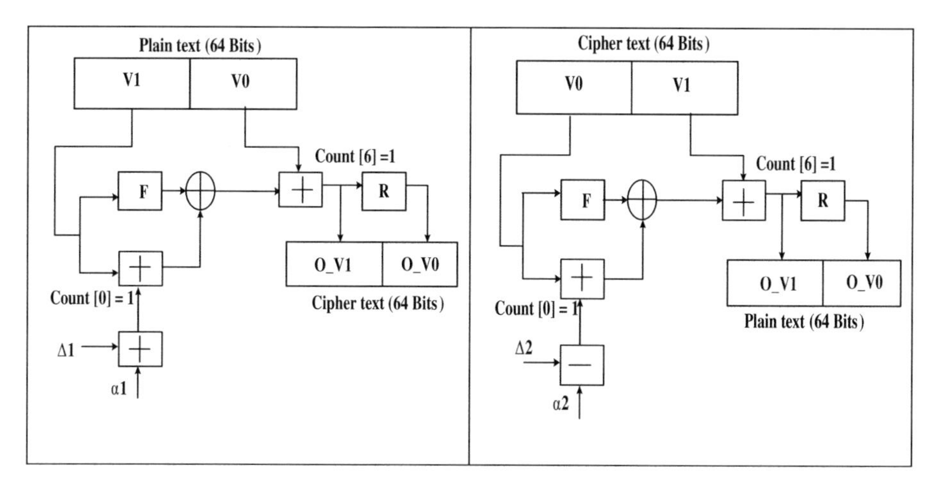

Fig. 2. (a) Encryption operation, (b) Decryption operation of TEA

5 Extended TEA (XTEA)

The TEA is extended to correct the Key scheduling weakness by many attacks and to enhance the security efficiently, The Extended TEA (XTEA) is a Lightweight cryptographic block cipher (LCBC), and it has the 128-bit security key to perform encryption and decryption for 64-bit block data (plain text).

The key scheduling of encryption and decryption for XTEA is represented in Fig. 3 (a) and (b) respectively and it is working based on the counter method. The key scheduling is operated, only when count = 00 or 10, for both Encryption and decryption operation because of Feistel structure. The 128-bit key is divided to four 32-bit sub-keys which are store in 4 memory locations i.e. K [0], K [1], K [2] and K [3]. The XTEA uses Constants in initial round: for Encryption- Delta ($\Delta1 = 9E3779B9_H$)

and alpha ($\alpha 1 = 0$) are added ($\alpha 1 + \Delta 1$), when counter (count = 00) is zero. Similarly, for Decryption-Delta ($\Delta 2 = 9E3779B9_H$) and alpha ($\alpha 2 = 8DDE6E40_H$) are subtracted ($\alpha 2 - \Delta 2$), when counter (count = 0) is zero. For encryption, the constants are added and then perform the (S + K [S & 3]) and (S + K [S $\gg$ 11 & 3]) operations for count = 00 and count = 10 respectively to generate the Key output (KO). For Decryption, the constant values are subtracted and then perform the (S + K [S $\gg$ 11 & 3]) and (S + K [S & 3]) operations for count = 00 and count = 10 respectively to generate the Key output (KO) which is input to Round function of XTEA.

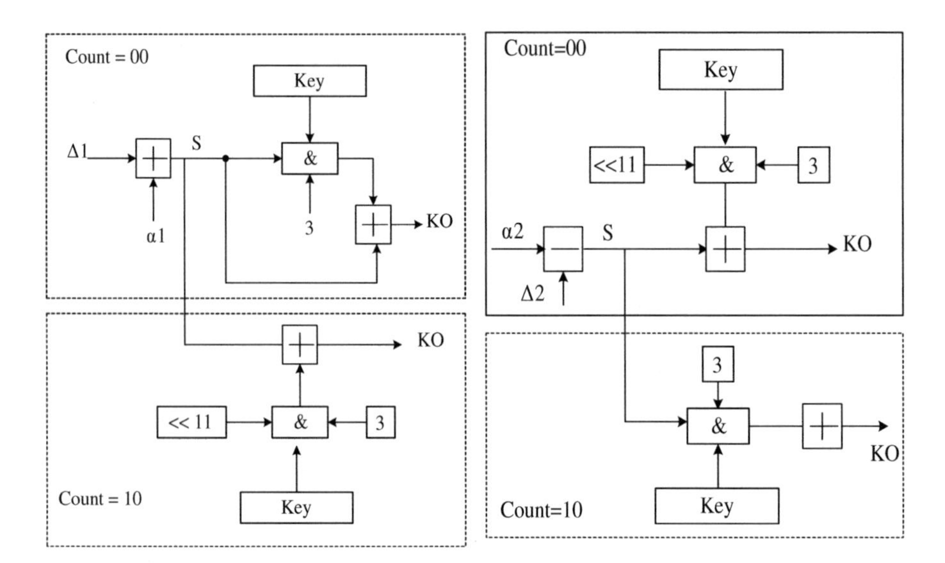

Fig. 3. (a) Key scheduling-encryption, (b) Key scheduling-decryption for XTEA.

The round function operation of XTEA is in the form of a Feistel structure with 64 rounds for encryption and decryption. Two Feistel rounds perform 32- iterations of operation parallelly. Each iteration uses four counts. The round operation of XTEA is represented in Fig. 4. The 32-bit input 'Ri' perform Arithmetic operations to generate the Rounded function Output as Ro = KO ^ ((Ri $\ll$ 4 ^ Ri $\gg$ 5) + Ri).

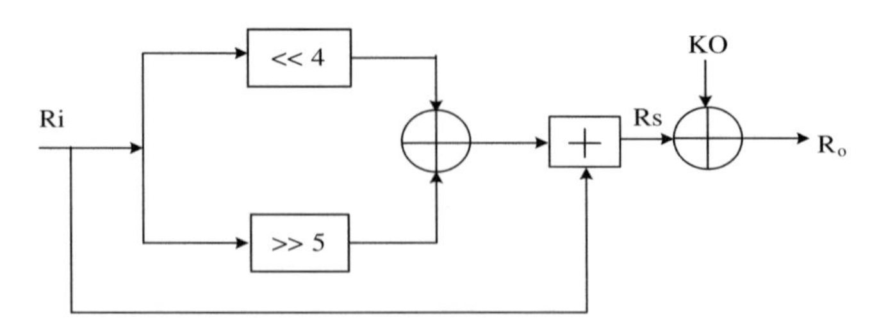

Fig. 4. Round function of XTEA

The Complete design flow of XTEA operation is represented in Fig. 5. The control signal 'enc' used to perform when '0' for encryption and '1' for decryption. The Plain Text (PT) has 64-bit and is separated by two 32-bits parts, i.e. V0 and V1. For enc = 0, the plain text (PT) is divided to V0 [31:0] as first 32-bits and V1 [63:31] is next 32-bits. Similarly, for enc = 1, the plaintext is divided into V1 [31:0] as first 32-bits, and V0 [63:31] is next 32-bits.

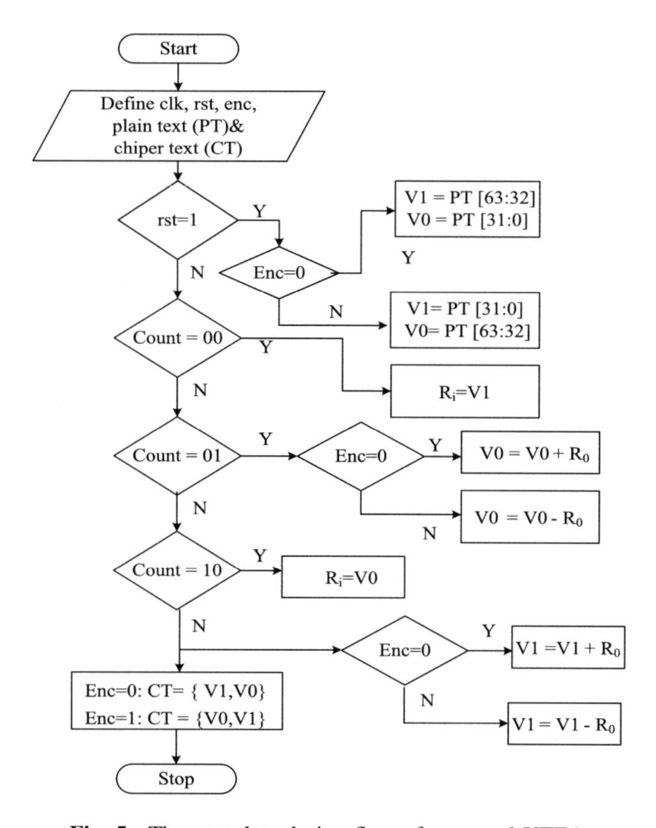

Fig. 5. The complete design flow of proposed XTEA

When Count = 00 or 10, The Round input is selected to V1 or V0 respectively. When count = 01 and for enc = 0, updated V0 = V0 + Ro and enc = 1, updated V0 = V0 − Ro. When count = 11 and for enc = 0, updated V1 = V1 + Ro and enc = 1, updated V1 = V1 − Ro. Finally, concatenate V0 and V1 according to the enc = 0, The 64-bit Cipher text is {V1, V0} and for enc = 1, The Cipher text is {V0, V1}.

The Proposed XTEA used the pipelined architecture with parallel processing using the counter to improve the throughput. The proposed XTEA is reconfigurable architecture, by changing the Control signal 'enc' to 0, the full architecture work as XTEA encryption process and enc to 1, for the XTEA Decryption process. The TEA and XTEA results are analyzed with simulation and performance metrics in the next section.

6 Results and Analysis

The Tiny Encryption Algorithm (TEA) and Extended TEA (XTEA) Design results are analyzed in the below section. The TEA and XTEA are modeled over Xilinx ISE platform using Verilog-HDL and simulation is performed using Modelsim 6.5f environment and implemented on FPGA- Artix-7.

The Simulation results of TEA Encryption and Decryption (TEA-ED) is represented in Fig. 6. For Encryption operation, Once Global clock signal is activated with low asynchronous reset (rst), the control signal 'enables' set high to reset the 'sum' data and make it low. The 128-bit key (key_in) is set and 64-bit input plain text is set to 64'haaaabbbb12345678. Perform the TEA-Encryption operation, after 64 clock cycles, the done signal becomes high to generate the output 64-bit ciphertext, 64'h024e85614871a30a. Similarly, for Decryption Process, the key will be same and encrypted output as an input, Perform the TEA-Decryption operation, after 64 clock cycles, done signal set high and generates the same plaintext as output 64'haaaabbbb12345678.

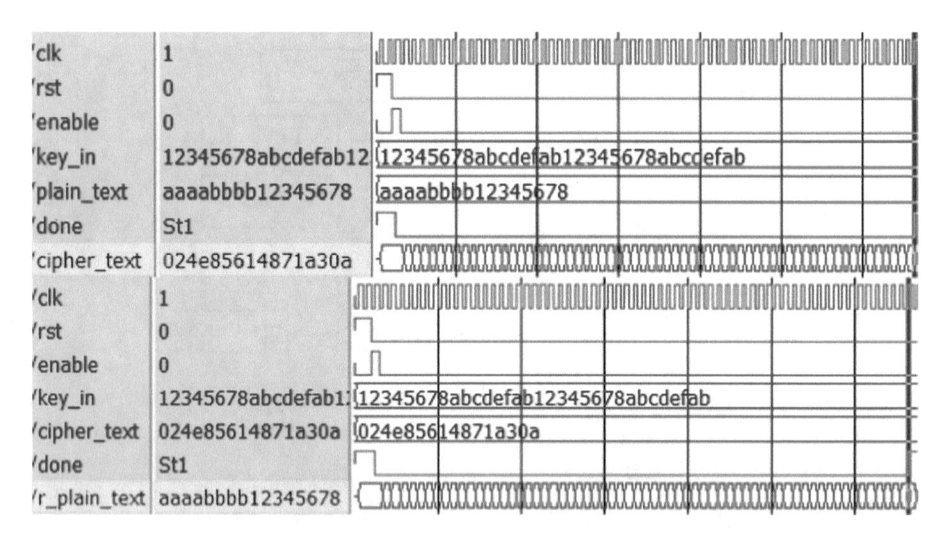

Fig. 6. Simulation results of TEA encryption and decryption

The Simulation results of XTEA Encryption and Decryption (XTEA-ED) is represented in Fig. 7. For Encryption operation, The Global clock signal is activated with low asynchronous reset (rst), the control signal 'enc' set to zero for the encryption process. The 128-bit key (key_in) is set and 64-bit input plain text is set to 64'haaaabbbb12345678. Perform the XTEA-Encryption operation, after 64 clock cycles, 'done' signal becomes high and the output 64-bit cipher-text, 64'h500171e538e098f2 is obtained. Similarly, for Decryption Process, the control signal 'enc' set to High, the key remains same and XTEA encrypted output as an input to Perform the XTEA-Decryption operation, after 64 clock cycles, done signal set high and generates the same plaintext as output 64'haaaabbbb12345678. The XTEA Architecture is flexible and works for encryption or decryption operation in design by changing the 'enc' signal to '0' or '1' respectively.

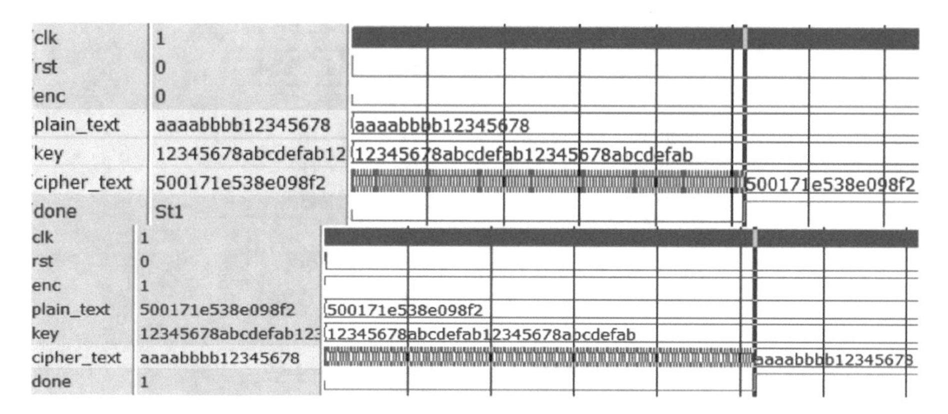

Fig. 7. Simulation results of XTEA encryption and decryption

The TEA-ED and XTEA-ED designs are simulated through test bench using ModelSim simulator. The Three test scenarios include for plain text and key for TEA and XTEA designs with 64-bit cipher texts are tabulated in Table 1. The TEA and XTEA designs are verified for both encryption and decryption and also working correctly for different test cases.

Table 1. Different test cases with TEA and XTEA ciphers

Test cases	TEA-64-bit Cipher text	XTEA-64-bit Cipher text
Plain text: 00000000_00000000	94baa94041ea3a0a	4ad0ed50fc924d12
Key: 00000000000000000000000000000000		
Plain text: aaaabbbbccccdddd	0f2d9bc17accd022	a6547a18fb5dcae2
Key: abcdefabcdefabcdefabcdefabcdefab		
Plain text: aaaabbbb12345678	024e85614871a30a	500171e538e098f2
Key: 12345678abcdefab12345678abcdefab		

The TEA-ED and XTEA-ED designs are synthesized, and resource constraints like area, timing, and power are obtained after place and routing operation and are tabulated with numerical values in Table 2. The area utilization concerning slice registers, slice LUT's and LUT-FF Pairs for TEA-ED is 212, 446 and 200 and for XTEA-ED utilized 316, 641 and 243 respectively. The Timing utilization of TEA and XTEA operates with the maximum frequency of 258.88 MHz and 263.762 MHz respectively on Artix-7 FPGA. The power consumption of TEA-ED and XTEA-ED using X-Power-analyzer after place and route is noted with the clock frequency of 100 MHz. The total power consumption of 0.194 W for TEA-ED and 0.222 W for XTEA-ED obtained and also a dynamic power of 0.111 W and 0.139 W for TEA and XTEA is utilized respectively.

Table 2. Resource constraints of TEA and XTEA designs

Resources	TEA-ED	XTEA-ED
Area		
Number of slice registers	212	316
Number of slice LUTs	446	641
LUT-FF pairs	200	243
Time		
Minimum period (ns)	3.863	3.791
Maximum frequency (MHz)	258.88	263.762
Power		
Dynamic power (W)	0.111	0.139
Total power (W)	0.194	0.222

The TEA-ED utilizes fewer resources includes Area and power, because of simple architecture and possible implementation. But For XTEA-ED is an extended version of TEA and architecture is complicated than TEA in Key scheduling and internal modules. So XTEA consumes more Resources than TEA on a hardware level.

The comparison of XTEA cipher with similar existing work [4] with numerical values regarding hardware constraints is tabulated in Table 3.

Table 3. Comparison of XTEA cipher with existing work [1]

Resources	AES-8bit [4]	TinyXTEA 1 [4]	TinyXTEA 3 [4]	Proposed XTEA
Block size (Bits)	128	64	64	64
Key size (Bits)	128	128	128	128
Clock cycles	3900	240	112	128
Maximum delay (ns)	14.93	13.87	15.97	6.216
Area (Slices)	522	266	254	238
Throughput (Mbps)	2	19	36	80.43
Efficiency (Mbps/Slice)	0.01	0.07	0.14	0.34
FPGA device	xc2s15-6	xc3s50-5	xc3s50-5	xc3s50-5

The Plaintext (PT) of 64-bits and the key size of 128-bits are used for the below block ciphers. The existing block ciphers include AES-8bit, TinyXTEA-1, and TinyXTEA-3 over proposed XTEA are compared on same FPGA Platform. The AES-8-bit block cipher resource utilization is huge than other block ciphers. The XTEA cipher uses 128 clock cycles over TinyXTEA-1 uses 240 clock cycles with an improvement of 46.66%. The Maximum path delay of XTEA is 6.216 ns over TinyXTEA-3 is 15.97 ns with an improvement of 61.07%. The Area (Slices) of XTEA is 238 over TinyXTEA-3 is 254 with an improvement of 6.2%. The throughput of XTEA is 80.43 Mbps over TinyXTEA-3 is 36 Mbps with an improvement of 55%. The efficiency (Throughput/Slice) of XTEA is 0.34 Mbps/slice over TinyXTEA-3 is 0.14 Mbps/slice with an improvement of 58%. Similar to that, XTEA has been improved over TinyXTEA-1 and AES-8bit block ciphers regarding all the hardware resource constraints.

7 Conclusion

The hardware architecture of lightweight cryptographic Block cipher (LCBC) algorithms includes TEA and XTEA with encryption and decryption are designed for low-cost RFID applications. The TEA block cipher is a simple and easy key scheduling process with less coding to secure information, which may attack by the hacker because of simple key function. To resolve the issues, proposed XTEA is introduced which includes Pipelined hardware architecture with parallel computation for key scheduling to improve the speed of the process with enhanced security. The TEA and XTEA design simulation results are presented with waveforms for different test cases. The Hardware constraints include area, time and power for TEA and XTEA designs are tabulated. TEA gives better utilization than XTEA because of simple architecture. But lags with speed and security. The proposed XTEA is compared with existing similar architecture TinyXTEA-3 with area improvement of 6.2%, throughput improvement of 55% and efficiency improvement of 58% on the same FPGA platform. In the future, by using the proposed XTEA design with RFID systems, designing an efficient security protocol will be implemented on a low-cost FPGA Platform.

References

1. Israsena, P.: On XTEA-based encryption/authentication core for wireless pervasive communication. In: 2006 International Symposium on Communications and Information Technologies, ISCIT 2006. IEEE (2006)
2. Castro, C.H., Vinuela, P.: New results on the genetic cryptanalysis of TEA and reduced-round versions of XTEA. New Gener. Comput. **23**(3), 233–243 (2005)
3. Israsena, P., Wongnamkum, S.: Hardware implementation of TEA-based lightweight encryption for RFID security. RFID Security, pp. 417–433. Springer, Boston (2008)
4. Kaps, J.-P.: Chai-Tea, cryptographic hardware implementations of XTEA. In: International Conference on Cryptology in India. Springer, Heidelberg (2008)
5. Lu, J.: Related-key rectangle attack on 36 rounds of the XTEA block cipher. Int. J. Inf. Secur. **8**(1), 1–11 (2009)
6. Seshabhattar, S., Jagannatha, S.K., Engels, D.W.: Security implementation within GEN2 protocol. In: 2011 IEEE International Conference on RFID-Technologies and Applications (RFID-TA). IEEE (2011)
7. Yu, J., Khan, G., Yuan, F.: XTEA encryption based novel RFID security protocol. In: 2011 24th Canadian Conference on Electrical and Computer Engineering (CCECE), pp. 000058–000062. IEEE (2011)

8. Sekar, G., Mouha, N., Velichkov, V., Preneel, B.: Meet-in-the-middle attacks on reduced-round XTEA. In: Cryptographers' Track at the RSA Conference, pp. 250–267. Springer, Heidelberg (2011)
9. Ruhan Bevi, A., Malarvizhi, S.: Performance analysis of TEA block cipher for low power applications. Commun. Comput. Inf. Sci. **1**(292), 605–610 (2012)
10. Geetha, G.: On the security of reduced key tiny encryption algorithm. In: 2012 International Conference on Computing Sciences (ICCS), pp. 322–325. IEEE (2012)
11. Isobe, T., Shibutani, K.: Security analysis of the lightweight block ciphers XTEA, LED and Piccolo. In: Australasian Conference on Information Security and Privacy, pp. 71–86. Springer, Heidelberg (2012)
12. Khan, G.N., Zhu, G.: Secure RFID authentication protocol with key updating technique. In: 2013 22nd International Conference on Computer Communications and Networks (ICCCN), pp. 1–5. IEEE (2013)
13. Venugopal, V., Shila, D.M.: High throughput implementations of cryptography algorithms on GPU and FPGA. In: 2013 IEEE International Instrumentation and Measurement Technology Conference (I2MTC). IEEE (2013)
14. Huang, J., Vaudenay, S., Lai, X.: On the key schedule of lightweight block ciphers. In: International Conference in Cryptology in India, pp. 124–142. Springer, Cham (2014)
15. Mozaffari-Kermani, M., et al.: Fault-resilient lightweight cryptographic block ciphers for secure embedded systems. IEEE Embed. Syst. Lett. **6**(4), 89–92 (2014)
16. Hajari, M., Azimi, S.A., Aref, M.R.: Impossible differential cryptanalysis of reduced-round TEA and XTEA. In: 2015 12th International Iranian Society of Cryptology Conference on Information Security and Cryptology (ISCISC), pp. 58–63. IEEE (2015)
17. Hussain, M.A., Badar, R.: FPGA based implementation scenarios of TEA Block Cipher. In: 2015 13th International Conference on Frontiers of Information Technology (FIT), pp. 283–286. IEEE (2015)
18. Tian, J., Gao, X.: A Bitsum attack research based on TEA. In: 2016 IEEE Advanced Information Management, Communicates, Electronic and Automation Control Conference (IMCEC), pp. 1721–1724. IEEE (2016)
19. AlMeer, M.H.: FPGA implementation of a hardware XTEA light encryption engine in co-design computing systems. In: 2017 Seventh International Conference on Innovative Computing Technology (INTECH), pp. 26–30. IEEE (2017)
20. Hatzivasilis, G., Fysarakis, K., Papaefstathiou, I., Manifavas, C.: A review of lightweight block ciphers. J. Cryptogr. Eng. **8**, 1–44 (2017)
21. Rajak, C.K., Mishra, A.: Implementation of modified TEA to enhance security. In: International Conference on Information and Communication Technology for Intelligent Systems, pp. 373–383. Springer, Cham (2017)
22. Surendran, S., Nassef, A., Beheshti, B.D.: A survey of cryptographic algorithms for IoT devices. In: 2018 IEEE Long Island on Systems, Applications and Technology Conference (LISAT), pp. 1–8. IEEE (2018)
23. Anusha, R.: Qualitative Assessment on Effectiveness of Security Approaches towards safeguarding NFC Devices & Services. Int. J. Electr. Comput. Eng. (IJECE) **8**(2), 1214–1221 (2018)
24. Anusha, R., Veena Devi Shastrimath, V.: TRMA: an efficient approach for mutual authentication of RFID wireless systems. In: Computer Science On-line Conference, pp. 290–299. Springer, Cham (2018)

Type-2 Fuzzy Controller for Stability of a System

D. Nagarajan[1(✉)], M. Lathamaheswari[1], and J. Kavikumar[2]

[1] Department of Mathematics, Hindustan Institute of Technology and Science, Chennai 603 103, India
dnrmsu2002@yahoo.com, lathamax@gmail.com
[2] Department of Mathematics and Statistics, Faculty of Applied Science and Technology, Universiti Tun Hussein Onn, Malaysia, Parit Raja, Malaysia
kavi@uthm.edu.my

Abstract. Type-2 Fuzzy sets (T2FSs) handle a greater modeling and uncertainties which exist in the real world applications especially control systems. To avoid mathematical complexity, interval T2FSs (IT2FSs) have been pertained in majority of the fields. One of the important components which influence the fuzzy controller is the t-norm. For obtaining the stability of a control system, t-norm operator can be preferred for better results and in this paper the minimum and maximum operations have been used to simplify the work of the system with Gaussian interval type-2 membership function (GIT2MF). Also proposed Gaussian interval type-2 weighted geometric (GIT2WG) operator and mathematical properties of aggregation operator have been proved using the proposed operator. The goal of this work is to analyze the stability of an inverted pendulum using interval type-2 fuzzy logic controller (IT2FLC) and the results are compared with Proportional Integrated Derivative (PID) Controller. It is observed that IT2FL controller gives the better stability.

Keywords: Aggregation properties · Control systems · Gaussian membership function · T Norm · Type-2 fuzzy sets · PID controller · Interval type-2 fuzzy logic controller · Inverted pendulum

1 Introduction

Fuzzy Control System (FCS) is reflects the human maturity for using linguistic rules with vague implication in order to develop control behavior. Triangular norms play an important role in control systems. Fuzzy logic controller (FLC) consists of linguistic IF-THEN rules. In the comparison of type-1 (T1) and type-2 (T2) categories of fuzzy logic, T1FL system has the difficulties in imitate and decrease the effect of uncertainties. Whereas in the case of T2 Fuzzy logic system, at least one T2FS must be taken and should be characterized by fuzzy membership grades.

An important part of T2FS is that Foot Print of Uncertainty (FOU), which constitutes the uncertainties in nature and posture of T1FS. Also it is an extra mathematical aspect provided by the focusable area and this area contains T1 membership function

R. Silhavy (Ed.): CSOC 2019, AISC 986, pp. 197–213, 2019.
https://doi.org/10.1007/978-3-030-19813-8_21

(T1MF) and has crisp membership grades. Simply FOU is the area between upper and lower MFs and which tells about the level of uncertainty of the information. Also it has the possibility of outperforming their T1 counterparts. Type reducer will convert the type-2 fuzzy outputs into crisp outputs. Moreover in IT2FSs, every element of FOU has a secondary membership grade as unity.

Under fuzzy based Control design, membership functions and rule base are the essential things and usually it is difficult to determine. In this work, the antecedent part of the rule base has been designed using IT2FS, whereas for consequent part T1FS has been applied. Type reduction process is differentiate T2 from T1, since for each fired rules the outputs are T2FS and this should be done prior to the defuzzifier is manipulate to provoke a crisp output. One of the type reducer is center of sets. This will incorporate all the type-2 fuzzy outputs and produce type-1 fuzzy set, which is the type reduced set.

FL controller is usually designed by T1FS which is known as T1FL controller and it has been applied in many of the fields, specifically in controlling complex non-linear systems where the researchers faced difficulties in designing and handling uncertainties. The disadvantage is failing to catch all the feature of a certain plant. Generally, controller which handles more uncertainties is preferable. It has been noted that interval type-2 fuzzy logic controllers have been applied for controlling the stability of mobile robot quality, sound speakers and admission in ATM networks [13]. The most applicable membership functions in control system are triangular and trapezoidal are the most applied membership function in control system. Due to their poor approximation, Gaussian membership function (GMF) is chosen as it gives actual representation at each point. Since MF virtually expresses the fuzziness, its characterization is the main aspect of the fuzzy operation [8].

In fuzzy inference theory, MFs, triangular norm operators, defuzzification methods and types of input to the controller are the main components. Though there are many MFs to represent a linguistic value, researchers found that GMF is the ideal one. There is a possibility of getting negative effects while getting the information from sensors corrupted by noise and this is the reason why fuzzy inputs are introduced with the computational complexity. For an efficient system, an influence of the operators must be taken care before implementation to get an enforced achievement level. Therefore, the selection of T Norm, defuzzifier and GMF has the greatest influence of the fuzzy controller [10].

Classical control designs are based on point to point whereas FL controller is either range to point or range to range i.e. FL controller is a function from an input data vector to a scalar output [1, 6]. Using MFs, outputs of the FL controller by fuzzifying both input and outputs are obtained. T2FLSs distinguished by fuzzy IF-THEN rules and T2 fuzzy values are used for antecedent and consequent parts as the parameters. In GIT2FSs, uncertainties may be incorporated with the mean or standard deviation and therefore consider either GIT2FS with uncertain mean or uncertain standard deviation. In this work, GIT2MF with uncertain mean and standard deviation is considered. In the case of T2FS, the antecedent and consequent parts are T2 or any one of the two. Usually consequent part is taken as T1 [2, 3, 7, 11, 12].

In this work the performance of PID controller and IT2FL controller is compared. PID controller has been used in many applications but it gives poor performance as it has poor knowledge of input and output parameters and therefore tuning is very difficult in this case. Neuro fuzzy controller works better than PID controllers as it performs better with reduced fluctuations and settling time faster. The performance of the controllers can be improved by increasing the number of combinations of input and output. From these observations it is observed that FL controller, which handles imprecision and uncertainty, is a good replacement to PID controller.

The lower and upper MFs can be captured effectively in IT2FL which is the collection of T1 fuzzy models. The FOU processes the stability analysis of the system [4, 5] and [9]. The operators may be chosen according to the characteristic properties and then the operations for minimum and maximum can be applied [13, 14]. Norms play as the synthesize operators for which these maximum and minimum operators are just an exclusive choice [15–17]. In many of the real world problems it is necessary to have a MV itself fuzzy instead of crisp value which is called T2FS [18, 19]. The parameter η in Yager triangular norms, accepts for tuning the norm between the other norms [20–22]. Image processing based on T2 Fuzzy system has been considered [23, 24]. Every pixel has some number of bits that determines available number of various gray levels [25–30]. Interval type-2 fuzzy set has been applied in image processing (edge detection and feature extraction), traffic control management to handle more uncertainties exists in the image and in the data successfully [31–36].

In this paper, the rest of the part is organized as follows. In Sect. 2, basic concepts have been given for better understanding of the paper. In Sect. 3, operational laws have been derived using Gaussian interval type-2 fuzzy numbers. In Sect. 4, the mathematical properties have been proved using proposed operator namely Gaussian interval type-2 weighted geometric operator. In Sect. 5, basic concepts of control system is given. In Sect. 6, stability analysis has been done for inverted pendulum using interval type-2 fuzzy logic controller and the result is compared with PID controller. In Sect. 7, conclusion of the work is given with future direction.

2 Basic Concepts

2.1 Gaussian Membership Function

Gaussian membership function for a Fuzzy set is defined by

$\psi_{\overline{D}}(x) = \exp\left[-\frac{1}{2}\left(\frac{x-m}{\sigma}\right)\right]$, $-\infty < x < \infty$, where m is the mean and σ is the standard deviation.

2.2 GMF with Type-2 Fuzzy Set

Here two different cases are considered for GITMF, according to the nature of the parameters namely mean (m) and standard deviation (σ) namely GIT2MF with fixed mean and uncertain standard deviation (FM&USD) and fixed standard deviation and uncertain mean (FSD&UM) as follows (Figs. 1 and 2).

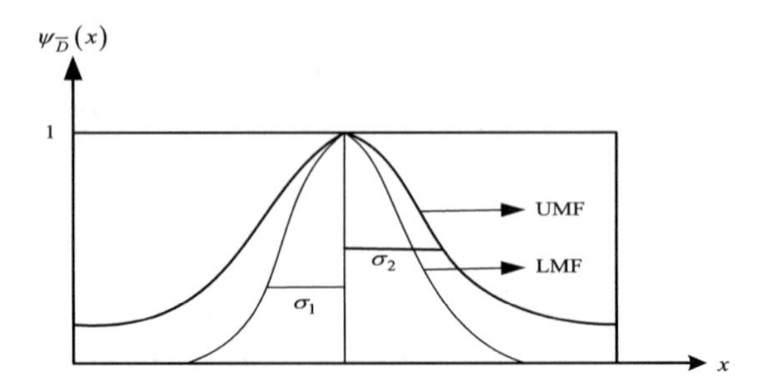

Fig. 1. GIT2MF with FM&USD

and is defined by, $$\psi_{\overline{D}}(x) = \exp\left[-\frac{1}{2}\left(\frac{x-m}{\sigma}\right)^2\right],\ \sigma \in [\sigma_1, \sigma_2] \tag{1}$$

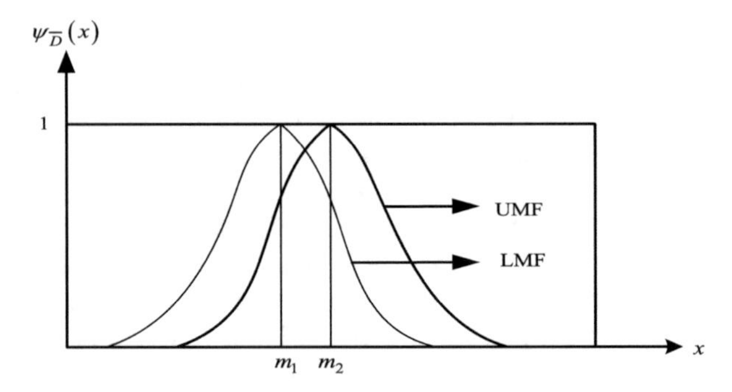

Fig. 2. GIT2MF with FSD&UM

and is defined by $$\psi_{\overline{D}}(x) = \exp\left[-\frac{1}{2}\left(\frac{x-m}{\sigma}\right)^2\right],\ m \in [m_1, m_2] \tag{2}$$

Where LMF is the lower membership function and UMF is the upper membership function in the pictorial representations.

2.3 Triangular Norms Used

Consider Dubois Prade (DP) triangular norms as defined below.

$$\text{DP T Norm: } T(x, y) = \frac{xy}{\max(x, y, \upsilon)} \tag{3}$$

$$\text{DP T Conorm: } TC(x, y) = 1 - \frac{(1-x)(1-y)}{\max[(1-x), (1-y), (1-\upsilon)]} \tag{4}$$

In this paper T Norm is used as it is preferable for control systems with min and max operations and T conorm will be used in the stage of defuzzification in the control system with uncertain parameters.

3 Operational Laws

Let $\overline{D}_1, \overline{D}_2, \ldots, \overline{D}_n,\ n = 1, 2, 3, \ldots, n$ be three Gaussian interval type-2 fuzzy numbers and the parameter $\vartheta \in [0, 1]$, then the following operations are hold.

3.1 Addition Operation

$$\overline{D_1} \oplus \overline{D_2} = 1 - \left[\frac{\left(1 - \exp\left[-\frac{1}{2}\left(\frac{x_1 - m_1}{\sigma_1}\right)^2\right]\right)\left(1 - \exp\left[-\frac{1}{2}\left(\frac{x_2 - m_2}{\sigma_2}\right)^2\right]\right)}{\max\left(\left(1 - \exp\left[-\frac{1}{2}\left(\frac{x_1 - m_1}{\sigma_1}\right)^2\right]\right), \left(1 - \exp\left[-\frac{1}{2}\left(\frac{x_1 - m_1}{\sigma_1}\right)^2\right]\right), (1 - \vartheta)\right)} \right] \tag{5}$$

3.2 Multiplication Operation

$$\overline{D_1} \otimes \overline{D_2} = \frac{\left(\exp\left[-\frac{1}{2}\left(\frac{x_1 - m_1}{\sigma_1}\right)^2\right]\right)\left(\exp\left[-\frac{1}{2}\left(\frac{x_2 - m_2}{\sigma_2}\right)^2\right]\right)}{\max\left(\left(\exp\left[-\frac{1}{2}\left(\frac{x_1 - m_1}{\sigma_1}\right)^2\right]\right), \left(\exp\left[-\frac{1}{2}\left(\frac{x_1 - m_1}{\sigma_1}\right)^2\right]\right), \vartheta\right)} \tag{6}$$

3.3 Multiplication by an Ordinary Number and Power

$$p.\overline{D} = 1 - \left[\frac{\left(1 - \exp\left[-\frac{1}{2}\left(\frac{x - m}{\sigma}\right)^2\right]^p\right)}{\max\left(\left(1 - \exp\left[-\frac{1}{2}\left(\frac{x - m}{\sigma}\right)^2\right]^p\right), (1 - \vartheta)\right)} \right] \tag{7}$$

$$\text{and} \quad \overline{D}^p = \frac{\left(\exp\left[-\frac{1}{2}\left(\frac{x - m}{\sigma}\right)^2\right]^p\right)}{\max\left(\left(\exp\left[-\frac{1}{2}\left(\frac{x - m}{\sigma}\right)^2\right]^p\right), \vartheta\right)} \tag{8}$$

4 Proposed Theorems

The below theorems are constituting the mathematical properties of aggregation operator (AO) namely triangular norms and shows the role of their properties in control system. Here the theorems of first, Idempotency, associativity and stability represents the facts that a control system can have any number of inputs (finite), unanimity of the system, the system can extend the process without ambiguity and the strength of the system respectively.

4.1 Theorem

Let $\overline{D}_i = \left(\exp\left[-\frac{1}{2}\left(\frac{x_i - m_i}{\sigma_i}\right)^2\right]\right)$, $i = 1, 2, 3, \ldots,$n be a collection of GIT2FNs then their aggregated value by

GIT2WG operator is still a GIT2FN and

$$GIT2WG_{\varpi}(\overline{D}_1, \overline{D}_2, \ldots, \overline{D}_n) = \frac{MOT\left(\exp\left[-\frac{1}{2}\left(\frac{x_i - m_i}{\sigma_i}\right)^2\right]\right)^{\varsigma_i}}{\max\left(\left(\exp\left[-\frac{1}{2}\left(\frac{x_1 - m_1}{\sigma_1}\right)^2\right]\right)^{\varsigma_1}, \left(\exp\left[-\frac{1}{2}\left(\frac{x_2 - m_2}{\sigma_2}\right)^2\right]\right)^{\varsigma_2}, \ldots, \left(\exp\left[-\frac{1}{2}\left(\frac{x_n - m_n}{\sigma_n}\right)^2\right]\right)^{\varsigma_n}, \vartheta\right)} \tag{9}$$

Proof:
By the method of mathematical induction. For $n = 2$, using Power operation

$$\overline{D}^{\varsigma_1} = \frac{\left(\exp\left[-\frac{1}{2}\left(\frac{x - m}{\sigma}\right)^2\right]\right)^{\varsigma_1}}{\max\left(\left(\exp\left[-\frac{1}{2}\left(\frac{x - m}{\sigma}\right)^2\right]\right)^{\varsigma_1}, \vartheta\right)} \quad \text{and} \quad \overline{D}^{\varsigma_2} = \frac{\left(\exp\left[-\frac{1}{2}\left(\frac{x - m}{\sigma}\right)^2\right]\right)^{\varsigma_2}}{\max\left(\left(\exp\left[-\frac{1}{2}\left(\frac{x - m}{\sigma}\right)^2\right]\right)^{\varsigma_2}, \vartheta\right)}$$

Here $i = 1, 2$, MOT = Multiplication Of Terms

$$GIT2WG_{\varpi}(\overline{D}_1, \overline{D}_2) = \frac{MOT\left(\exp\left[-\frac{1}{2}\left(\frac{x_i - m_i}{\sigma_i}\right)^2\right]\right)^{\varsigma_i}}{\max\left(\left(\exp\left[-\frac{1}{2}\left(\frac{x_1 - m_1}{\sigma_1}\right)^2\right]\right)^{\varsigma_1}, \left(\exp\left[-\frac{1}{2}\left(\frac{x_2 - m_2}{\sigma_2}\right)^2\right]\right)^{\varsigma_2}, \vartheta\right)}$$

$$= \frac{MOT\left(\exp\left[-\frac{1}{2}\left(\frac{x_i - m_i}{\sigma_i}\right)^{2\varsigma_i}\right]\right)}{\max\left(\left(\exp\left[-\frac{1}{2}\left(\frac{x_1 - m_1}{\sigma_1}\right)^{2\varsigma_1}\right]\right), \left(\exp\left[-\frac{1}{2}\left(\frac{x_2 - m_2}{\sigma_2}\right)^{2\varsigma_2}\right]\right), \vartheta\right)}$$

For $n = k$,

$$GIT2WG_{\varpi}(\overline{D}_1, \overline{D}_2, \ldots, \overline{D}_k) = \frac{MOT\left(\exp\left[-\frac{1}{2}\left(\frac{x_i - m_i}{\sigma_i}\right)^2\right]\right)^{\varsigma_i}}{\max\left(\left(\exp\left[-\frac{1}{2}\left(\frac{x_1 - m_1}{\sigma_1}\right)^2\right]\right)^{\varsigma_1}, \left(\exp\left[-\frac{1}{2}\left(\frac{x_2 - m_2}{\sigma_2}\right)^2\right]\right)^{\varsigma_2}, \ldots, \left(\exp\left[-\frac{1}{2}\left(\frac{x_k - m_k}{\sigma_k}\right)^2\right]\right)^{\varsigma_k}, \vartheta\right)}$$

For $n = k + 1$, $GIT2WG_{\varpi}(\overline{D}_1, \overline{D}_2, \ldots, \overline{D}_k, \overline{D}_{k+1})$

$$= \frac{MOT\left(\exp\left[-\frac{1}{2}\left(\frac{x_i-m_i}{\sigma_i}\right)^2\right]\right)^{\varsigma_i}}{\max\left(\left(\exp\left[-\frac{1}{2}\left(\frac{x_1-m_1}{\sigma_1}\right)^2\right]\right)^{\varsigma_1}, \left(\exp\left[-\frac{1}{2}\left(\frac{x_2-m_2}{\sigma_2}\right)^2\right]\right)^{\varsigma_2}, \ldots, \left(\exp\left[-\frac{1}{2}\left(\frac{x_k-m_k}{\sigma_k}\right)^2\right]\right)^{\varsigma_k}, \vartheta\right)}$$

$$\otimes \frac{\left(\exp\left[-\frac{1}{2}\left(\frac{x_{k+1}-m_{k+1}}{\sigma_{k+1}}\right)^2\right]\right)^{\varsigma_{k+1}}}{\max\left(\left(\exp\left[-\frac{1}{2}\left(\frac{x_{k+1}-m_{k+1}}{\sigma_{k+1}}\right)^2\right]\right)^{\varsigma_{k+1}}, \vartheta\right)}$$

$$= \frac{MOT\left(\exp\left[-\frac{1}{2}\left(\frac{x_i-m_i}{\sigma_i}\right)^2\right]\right)^{\varsigma_i}}{\max\left(\left(\exp\left[-\frac{1}{2}\left(\frac{x_1-m_1}{\sigma_1}\right)^2\right]\right)^{\varsigma_1}, \left(\exp\left[-\frac{1}{2}\left(\frac{x_2-m_2}{\sigma_2}\right)^2\right]\right)^{\varsigma_2}, \ldots, \left(\exp\left[-\frac{1}{2}\left(\frac{x_{k+1}-m_{k+1}}{\sigma_{k+1}}\right)^2\right]\right)^{\varsigma_{k+1}}, \vartheta\right)}$$

Hence, the result is true for all the values of n.

4.2 Theorem (Idempotency)

Let $\overline{D}_i = \left(\exp\left[-\frac{1}{2}\left(\frac{x_i-m_i}{\sigma_i}\right)^2\right]\right)$, $i = 1, 2, 3, \ldots,$ n be a collection of GIT2FNs. If be a collection of GIT2FNs. If $\forall \overline{D}_i,\ i = 1, 2, 3, \ldots, n$ are equal i.e., $\overline{D}_i = \overline{D}$ then

$$GIT2WG_{\varpi}\left(\overline{D}_1, \overline{D}_2, \ldots, \overline{D}_n\right) = \overline{D} \tag{10}$$

Proof:
Using Theorem 4.1, $GIT2WG_{\varsigma}\left(\overline{D}_1, \overline{D}_2, \ldots, \overline{D}_n\right)$

$$= \frac{MOT\left(\exp\left[-\frac{1}{2}\left(\frac{x_i-m_i}{\sigma_i}\right)^2\right]\right)^{\varsigma_i}}{\max\left(\left(\exp\left[-\frac{1}{2}\left(\frac{x_1-m_1}{\sigma_1}\right)^2\right]\right)^{\varsigma_1}, \left(\exp\left[-\frac{1}{2}\left(\frac{x_2-m_2}{\sigma_2}\right)^2\right]\right)^{\varsigma_2}, \ldots, \left(\exp\left[-\frac{1}{2}\left(\frac{x_n-m_n}{\sigma_n}\right)^2\right]\right)^{\varsigma_n}, \vartheta\right)}$$

$$= \frac{\left(\exp\left[-\frac{1}{2}\left(\frac{x_i-m_i}{\sigma_i}\right)^2\right]\right)^{\sum_{i=1}^{n}\varsigma_i}}{\max\left(\left(\exp\left[-\frac{1}{2}\left(\frac{x_1-m_1}{\sigma_1}\right)^2\right]\right)^{\varsigma_1}, \left(\exp\left[-\frac{1}{2}\left(\frac{x_2-m_2}{\sigma_2}\right)^2\right]\right)^{\varsigma_2}, \ldots, \left(\exp\left[-\frac{1}{2}\left(\frac{x_n-m_n}{\sigma_n}\right)^2\right]\right)^{\varsigma_n}, \vartheta\right)}$$

$$= \frac{\left(\exp\left[-\frac{1}{2}\left(\frac{x_i-m_i}{\sigma_i}\right)^2\right]\right)}{\max\left(\left(\exp\left[-\frac{1}{2}\left(\frac{x_1-m_1}{\sigma_1}\right)^2\right]\right), \left(\exp\left[-\frac{1}{2}\left(\frac{x_2-m_2}{\sigma_2}\right)^2\right]\right), \ldots, \left(\exp\left[-\frac{1}{2}\left(\frac{x_n-m_n}{\sigma_n}\right)^2\right]\right), \vartheta\right)}$$

$$= \overline{D}$$

Hence, the theorem holds.

4.3 Theorem (Associativity)

If $\overline{D}_1, \overline{D}_2$ and $\overline{D}_3$ are the three GIT2FNs then the following result

$$\left(\overline{D}_1 \otimes \overline{D}_2 \otimes \overline{D}_3\right) = \left(\overline{D}_1 \otimes \overline{D}_2\right) \otimes \overline{D}_3 \text{ is hold.} \tag{11}$$

Proof:
Using associativity property we have,

$$\left(\overline{D}_1 \otimes \overline{D}_2 \otimes \overline{D}_3\right) = \left(\overline{D}_1 \otimes \overline{D}_2\right) \otimes \overline{D}_3.$$

Consider, $\left(\overline{D}_1 \otimes \overline{D}_2\right) \otimes \overline{D}_3$

$$= \left[\left(\exp\left[-\frac{1}{2}\left(\frac{x_1 - m_1}{\sigma_1}\right)^2\right]\right) \otimes \left(\exp\left[-\frac{1}{2}\left(\frac{x_2 - m_2}{\sigma_2}\right)^2\right]\right)\right] \otimes \left(\exp\left[-\frac{1}{2}\left(\frac{x_3 - m_3}{\sigma_3}\right)^2\right]\right)$$

$$= \frac{\left[\left(\exp\left[-\frac{1}{2}\left(\frac{x_1-m_1}{\sigma_1}\right)^2\right]\right) \otimes \left(\exp\left[-\frac{1}{2}\left(\frac{x_2-m_2}{\sigma_2}\right)^2\right]\right)\right] . \left(\exp\left[-\frac{1}{2}\left(\frac{x_3-m_3}{\sigma_3}\right)^2\right]\right)}{\max\left[\left(\exp\left[-\frac{1}{2}\left(\frac{x_1-m_1}{\sigma_1}\right)^2\right]\right) \otimes \left(\exp\left[-\frac{1}{2}\left(\frac{x_2-m_2}{\sigma_2}\right)^2\right]\right), \left(\exp\left[-\frac{1}{2}\left(\frac{x_3-m_3}{\sigma_3}\right)^2\right]\right), \upsilon\right]}$$

$$= \frac{\dfrac{\left(\exp\left[-\frac{1}{2}\left(\frac{x_1-m_1}{\sigma_1}\right)^2\right]\right) . \left(\exp\left[-\frac{1}{2}\left(\frac{x_2-m_2}{\sigma_2}\right)^2\right]\right)}{\max\left[\left(\exp\left[-\frac{1}{2}\left(\frac{x_1-m_1}{\sigma_1}\right)^2\right]\right), \left(\exp\left[-\frac{1}{2}\left(\frac{x_2-m_2}{\sigma_2}\right)^2\right]\right), \upsilon\right]} . \left(\exp\left[-\frac{1}{2}\left(\frac{x_3-m_3}{\sigma_3}\right)^2\right]\right)}{\max\left[\dfrac{\left(\exp\left[-\frac{1}{2}\left(\frac{x_1-m_1}{\sigma_1}\right)^2\right]\right) . \left(\exp\left[-\frac{1}{2}\left(\frac{x_2-m_2}{\sigma_2}\right)^2\right]\right)}{\max\left[\left(\exp\left[-\frac{1}{2}\left(\frac{x_1-m_1}{\sigma_1}\right)^2\right]\right), \left(\exp\left[-\frac{1}{2}\left(\frac{x_2-m_2}{\sigma_2}\right)^2\right]\right), \upsilon\right]}, \left(\exp\left[-\frac{1}{2}\left(\frac{x_3-m_3}{\sigma_3}\right)^2\right]\right), \upsilon\right]}$$

$$= \frac{\overline{D}_1.\overline{D}_2.\overline{D}_3}{\max\left[\overline{D}_1, \overline{D}_2, \upsilon\right] \max\left[\frac{\overline{D}_1.\overline{D}_2}{\max\left[\overline{D}_1, \overline{D}_2, \upsilon\right]}, \overline{D}_3, \upsilon\right]}$$

$$= \frac{\left(\exp\left[-\frac{1}{2}\left(\frac{x_1-m_1}{\sigma_1}\right)^2\right]\right) . \left(\exp\left[-\frac{1}{2}\left(\frac{x_2-m_2}{\sigma_2}\right)^2\right]\right) . \left(\exp\left[-\frac{1}{2}\left(\frac{x_3-m_3}{\sigma_3}\right)^2\right]\right)}{\max\left[\left(\exp\left[-\frac{1}{2}\left(\frac{x_1-m_1}{\sigma_1}\right)^2\right]\right), \left(\exp\left[-\frac{1}{2}\left(\frac{x_2-m_2}{\sigma_2}\right)^2\right]\right), \left(\exp\left[-\frac{1}{2}\left(\frac{x_3-m_3}{\sigma_3}\right)^2\right]\right), \upsilon\right]}$$

$$= \frac{\overline{D}_1.\overline{D}_2.\overline{D}_3}{\max\left[\overline{D}_1, \overline{D}_2, \overline{D}_3, \upsilon\right]} = \left(\overline{D}_1 \otimes \overline{D}_2 \otimes \overline{D}_3\right)$$

Hence, this result also holds for all the values of n.

4.4 Theorem (Stability)

Let $\overline{D}_i = \left(\exp\left[-\frac{1}{2}\left(\frac{x_i - m_i}{\sigma_i}\right)^2\right]\right)$, $i = 1, 2, 3, \ldots,$ n be a collection of GIT2FNs. If $p > 0$ and $\overline{D}_{n+1} = \left(\exp\left[-\frac{1}{2}\left(\frac{x_{n+1} - m_{n+1}}{\sigma_{n+1}}\right)^2\right]\right)$ is a GIT2FN on the set X then,

$$\begin{aligned} & GIT2WG_{\varpi}\left(\overline{D}_1^p \otimes \overline{D}_{n+1}, \overline{D}_2^p \otimes \overline{D}_{n+1}, \ldots, \overline{D}_n^p \otimes \overline{D}_{n+1}\right) \\ & = \left[GIT2WG_{\varpi}\left(\overline{D}_1, \overline{D}_2, \ldots, \overline{D}_n\right)\right]^p \otimes \overline{D}_{n+1} \end{aligned} \tag{12}$$

Proof:
Using Power operation of GIT2FN,
We know that,

$$\begin{aligned} & GIT2WG_{\varpi}\left(\overline{D}_1^p \otimes \overline{D}_{n+1}, \overline{D}_2^p \otimes \overline{D}_{n+1}, \ldots, \overline{D}_n^p \otimes \overline{D}_{n+1}\right) \\ & = GIT2WG_{\varpi}\left(\overline{D}_1, \overline{D}_2, \ldots, \overline{D}_n\right) \otimes \overline{D}_{n+1} \end{aligned} \tag{13}$$

$$\overline{D}_i \otimes \overline{D}_{n+1} = \frac{\underset{j=\{i,n+1\}}{MOT}\left[\left(\exp\left[-\frac{1}{2}\left(\frac{x_j - m_j}{\sigma_j}\right)^2\right]\right)\right]}{\max\left[\left(\exp\left[-\frac{1}{2}\left(\frac{x_j - m_j}{\sigma_j}\right)^2\right]\right), \upsilon\right]}$$

$$\begin{aligned} & GIT2WG_{\varpi}\left(\overline{D}_1^p \otimes \overline{D}_{n+1}, \overline{D}_2^p \otimes \overline{D}_{n+1}, \ldots, \overline{D}_n^p \otimes \overline{D}_{n+1}\right) \\ & = \frac{\underset{j=1}{\overset{n}{MOT}}\left[\exp\left[-\frac{1}{2}\left(\frac{x_j - m_j}{\sigma_j}\right)^2\right]\right]\left(\exp\left[-\frac{1}{2}\left(\frac{x_{n+1} - m_{n+1}}{\sigma_{n+1}}\right)^2\right]\right)}{\max\left[\exp\left[-\frac{1}{2}\left(\frac{x_j - m_j}{\sigma_j}\right)^2\right], \exp\left[-\frac{1}{2}\left(\frac{x_{n+1} - m_{n+1}}{\sigma_{n+1}}\right)^2\right], \upsilon\right]} \end{aligned} \tag{14}$$

$$\begin{aligned} & GIT2WG_{\varpi}\left(\overline{D}_1, \overline{D}_2, \ldots, \overline{D}_n\right) \otimes \overline{D}_{n+1} \\ & = \frac{\underset{j=1}{\overset{n}{MOT}}\left[\exp\left[-\frac{1}{2}\left(\frac{x_j - m_j}{\sigma_j}\right)^2\right]^{\varpi_j}\right]}{\max\left[\exp\left[-\frac{1}{2}\left(\frac{x_1 - m_1}{\sigma_1}\right)^2\right], \exp\left[-\frac{1}{2}\left(\frac{x_2 - m_2}{\sigma_2}\right)^2\right], \ldots, \exp\left[-\frac{1}{2}\left(\frac{x_n - m_n}{\sigma_n}\right)^2\right], \upsilon\right]} \otimes \exp\left[-\frac{1}{2}\left(\frac{x_{n+1} - m_{n+1}}{\sigma_{n+1}}\right)^2\right] \\ & = \frac{\underset{j=1}{\overset{n}{MOT}}\left[\exp\left[-\frac{1}{2}\left(\frac{x_j - m_j}{\sigma_j}\right)^2\right]^{\varpi_j}\right] . \exp\left[-\frac{1}{2}\left(\frac{x_{n+1} - m_{n+1}}{\sigma_{n+1}}\right)^2\right]}{\max\left[\underset{j=1}{\overset{n}{MOT}}\left[\exp\left[-\frac{1}{2}\left(\frac{x_j - m_j}{\sigma_j}\right)^2\right]^{\varpi_j}\right], \exp\left[-\frac{1}{2}\left(\frac{x_{n+1} - m_{n+1}}{\sigma_{n+1}}\right)^2\right], \upsilon\right]} \end{aligned} \tag{15}$$

From (14) and (15), $GIT2WG_{\varpi}\left(\overline{D}_1^p \otimes \overline{D}_{n+1}, \overline{D}_2^p \otimes \overline{D}_{n+1}, \ldots, \overline{D}_n^p \otimes \overline{D}_{n+1}\right) = GIT2WG_{\varpi}\left(\overline{D}_1, \overline{D}_2, \ldots, \overline{D}_n\right) \otimes \overline{D}_{n+1}$.

Also we have, $GIT2WG_{\varpi}\left(\overline{D}_1^p, \overline{D}_2^p, \ldots, \overline{D}_n^p\right) = \left(GIT2WG_{\varpi}\left(\overline{D}_1, \overline{D}_2, \ldots, \overline{D}_n\right)\right)^p$ (16)

$$GIT2WG_{\varpi}\left(\overline{D}_1^p, \overline{D}_2^p, \ldots, \overline{D}_n^p\right)$$

$$= \frac{\underset{j=1}{\overset{n}{MOT}}\left[\exp\left(\left[-\frac{1}{2}\left(\frac{x_j-m_j}{\sigma_j}\right)^2\right]^p\right)^{\varpi_j}\right]}{\max\left[\exp\left(\left[-\frac{1}{2}\left(\frac{x_1-m_1}{\sigma_1}\right)^2\right]^p\right)^{\varpi_1}, \exp\left(\left[-\frac{1}{2}\left(\frac{x_2-m_2}{\sigma_2}\right)^2\right]^p\right)^{\varpi_2}, \ldots \exp\left(\left[-\frac{1}{2}\left(\frac{x_n-m_n}{\sigma_n}\right)^2\right]^p\right)^{\varpi_n}, \upsilon\right]}$$

$$= \frac{\underset{j=1}{\overset{n}{MOT}}\left[\exp\left(\left[-\frac{1}{2}\left(\frac{x_j-m_j}{\sigma_j}\right)^2\right]^{p\varpi_j}\right)\right]}{\max\left[\exp\left(\left[-\frac{1}{2}\left(\frac{x_1-m_1}{\sigma_1}\right)^2\right]^{p\varpi_1}\right), \exp\left(\left[-\frac{1}{2}\left(\frac{x_2-m_2}{\sigma_2}\right)^2\right]^{p\varpi_2}\right), \ldots \exp\left(\left[-\frac{1}{2}\left(\frac{x_n-m_n}{\sigma_n}\right)^2\right]^{p\varpi_n}\right), \upsilon\right]} \tag{17}$$

Also since, $\left(GIT2WG_{\varpi}\left(\overline{D}_1, \overline{D}_2, \ldots, \overline{D}_n\right)\right)^p$

$$= \frac{\underset{j=1}{\overset{n}{MOT}}\left[\exp\left(\left[-\frac{1}{2}\left(\frac{x_j-m_j}{\sigma_j}\right)^2\right]^{\varpi_j}\right)\right]^p}{\max\left[\exp\left(\left[-\frac{1}{2}\left(\frac{x_1-m_1}{\sigma_1}\right)^2\right]^{\varpi_1}\right)^p, \exp\left(\left[-\frac{1}{2}\left(\frac{x_2-m_2}{\sigma_2}\right)^2\right]^{\varpi_2}\right)^p, \ldots, \exp\left(\left[-\frac{1}{2}\left(\frac{x_n-m_n}{\sigma_n}\right)^2\right]^{\varpi_n}\right)^p, \upsilon\right]}$$

$$= \frac{\underset{j=1}{\overset{n}{MOT}}\left[\exp\left(\left[-\frac{1}{2}\left(\frac{x_j-m_j}{\sigma_j}\right)^2\right]^{p\varpi_j}\right)\right]}{\max\left[\exp\left(\left[-\frac{1}{2}\left(\frac{x_1-m_1}{\sigma_1}\right)^2\right]^{p\varpi_1}\right), \exp\left(\left[-\frac{1}{2}\left(\frac{x_2-m_2}{\sigma_2}\right)^2\right]^{p\varpi_2}\right), \ldots \exp\left(\left[-\frac{1}{2}\left(\frac{x_n-m_n}{\sigma_n}\right)^2\right]^{p\varpi_n}\right), \upsilon\right]} \tag{18}$$

From (17) and (18), we get $GIT2WG_{\varpi}\left(\overline{D}_1^p \otimes \overline{D}_{n+1}, \overline{D}_2^p \otimes \overline{D}_{n+1}, \ldots, \overline{D}_n^p \otimes \overline{D}_{n+1}\right) = \left[GIT2WG_{\varpi}\left(\overline{D}_1, \overline{D}_2, \ldots, \overline{D}_n\right)\right]^p \otimes \overline{D}_{n+1}$.
Hence the theorem.

5 Basics of Control System

5.1 Components of Fuzzy Inference System (FIS)

Rule Base, database and reasoning Mechanism are the components of FIS used for selecting fuzzy rules, to define membership function and to derive sensible conclusion based on the rule of fuzzy reasoning respectively.

5.2 Components of Fuzzy Logic System

Rule base, fuzzy inference engine (FIE), fuzzifier and defuzzifier are the four components used to choose Fuzzy rule which shows the human thinking, judgment and perception, to combine rules for developing a scaling from crisp inputs to type-2 fuzzy outputs, Gaussian fuzzifier to simplify the computation in the FIE when the MFs in the IF-THEN rules are Gaussian and a mapping from fuzzy set to crisp point and calculates the crisp Output respectively.

5.3 Role of T-Norm in Control System

The role of triangular norms plays a key role in fuzzy control system, especially in getting an output. The T Norms are expresses differently and come out with different properties as proved by the theorems.

6 Application

The pendulum moves vertically, the force F is the control input of the cart which moves horizontally and the angular position of the pendulum θ and the horizontal position of the cart x are the outputs. Also N is the reaction force [5] (Fig. 3).

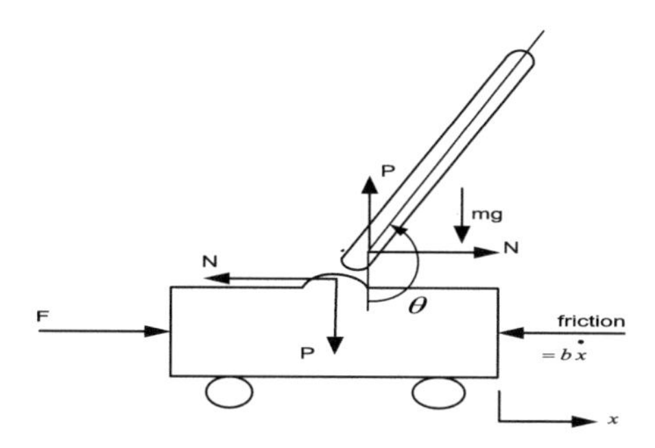

Fig. 3. Inverted Pendulum

The motion in the cart is defined by

$$M\ddot{x} + b\dot{x} + N = F \tag{19}$$

The motion in the pendulum is

$$(M + m)\ddot{x} + b\dot{x} + ml\ddot{\theta}\cos\theta - ml\dot{\theta}^2\sin\theta = F \tag{20}$$

$$\left(I + ml^2\right)\ddot{\theta} + mgl\sin\theta = -ml\ddot{x}\cos\theta \tag{21}$$

The system is to be linearized.
The two linearized motion of the equations are

$$\left(I + ml^2\right)\ddot{\phi} - mgl\,\phi = ml\ddot{x} \tag{22}$$

$$(M + m)\ddot{x} + b\dot{x} - ml\ddot{\phi} = u \tag{23}$$

The transfer function of the linearized system is

$$\frac{\Phi(s)}{U(s)} = \frac{\frac{ml}{q} s^2}{s^4 + \frac{b(I+ml^2)}{q} s^3 - \frac{(M+m)mgl}{q} s^2 - \frac{bmgl}{q} s} \tag{24}$$

$$\text{where, } q = \left[(M+m)(I+ml^2) - (ml)^2\right] \tag{24.1}$$

And the state space equation of the system is

$$\begin{bmatrix} \dot{x} \\ \ddot{x} \\ \dot{\phi} \\ \ddot{\phi} \end{bmatrix} = \begin{bmatrix} 0 & 1 & 0 & 0 \\ 0 & \frac{-(I+ml^2)}{I(M+m)+Mml^2} & \frac{m^2gl^2}{I(M+m)+Mml^2} & 0 \\ 0 & 0 & 0 & 1 \\ 0 & \frac{-mlb}{I(M+m)+Mml^2} & \frac{mgl(M+m)}{I(M+m)+Mml^2} & 0 \end{bmatrix} \begin{bmatrix} x \\ \dot{x} \\ \phi \\ \dot{\phi} \end{bmatrix} + \begin{bmatrix} 0 \\ \frac{(I+ml^2)}{I(M+m)+Mml^2} \\ 0 \\ \frac{ml}{I(M+m)+Mml^2} \end{bmatrix} u \tag{25}$$

$$y = \begin{bmatrix} 1 & 0 & 0 & 0 \\ 0 & 0 & 1 & 0 \end{bmatrix} \begin{bmatrix} x \\ \dot{x} \\ \phi \\ \dot{\phi} \end{bmatrix} + \begin{bmatrix} 0 \\ 0 \end{bmatrix} u \tag{26}$$

Here the nonlinear plant is an Inverted Pendulum (IP) subject to parameter uncertainty without considering the cart movement for demonstration process. The proposed fuzzy controller is engaged for stabilizing the IP with IT2FLC. The dynamical equation of an IP is defined as follows.

$$\ddot{\theta}(t) = \frac{g\ \sin(\theta(t)) - a\, m_p L \dot{\theta}(t)^2 \sin(2\theta(t))/2 - a \cos(\theta(t)) u(t)}{4L/3 - a\, m_p L \cos^2(\theta(t))} \tag{27}$$

where $\theta(t)$ is the angular displacement of the pendulum, g is the acceleration due to gravity, m_p is the mass of the pendulum, $m_p \in \left[m_{p_{\min}}\ m_{p_{\max}}\right]$, $a = \frac{1}{(m_p + M_c)}$, $M_p \in \left[M_{c_{\min}}\ M_{c_{\max}}\right]$, M_c is the mass of the cart, $2L = 1m$ is the length of the pendulum, $u(t)$ is the force applied to the cart and m_p, M_c are regarded as the parameter uncertainties.

6.1 Interval Type-2 Fuzzy Logic Controller (IT2FLC)

To fix the position of the input MFs and uniformly distributed between −1 and +1. Limit these inputs to a minimum and maximum values using two saturation blocks Saturation 1 and saturation consecutively. The fuzzy controller is tuned by scaling gains. Control the spread of the input MFs by the input gains ‘Gain 1’ and ‘Gain 2’. To rescale the axes we can change gains. The MFs are uniformly spread out and contracted for the gains, which is less than 1 and greater than 1 respectively. The spread of the output MFs controlled by the output gain 'Gain' and the changes in it will lead to scale the vertical axis of the controller surface. If we increase the gains 'Gain 1' and 'Gain 2' then the proportional gain and the derivative gain in a PD controller will be increased respectively. If the proportional gain is increased then the system respond will be faster (Fig. 4).

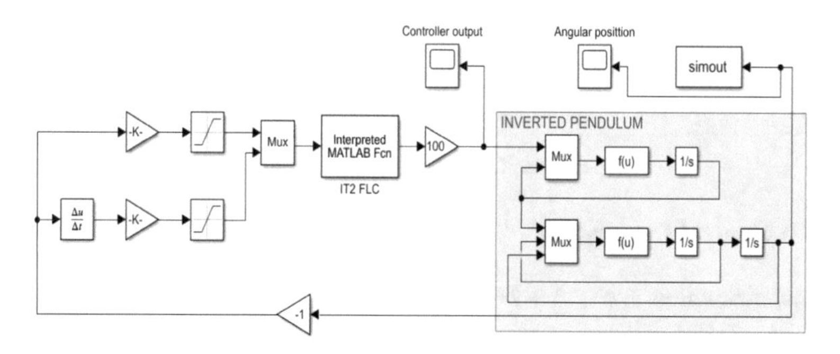

Fig. 4. IT2FLC

6.2 Controller Output

It shows the optimized control output of IT2FLC system (Fig. 5).

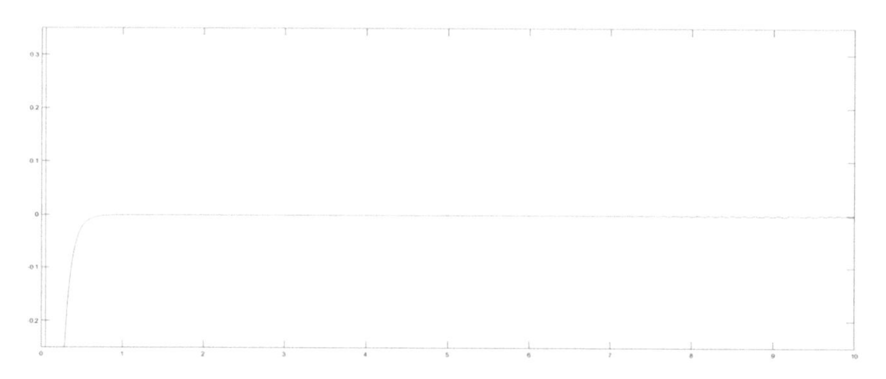

Fig. 5. Chart for controller output

6.3 Angular Position

It shows its varying between the angular positions from 0 to 1 (Fig. 6).

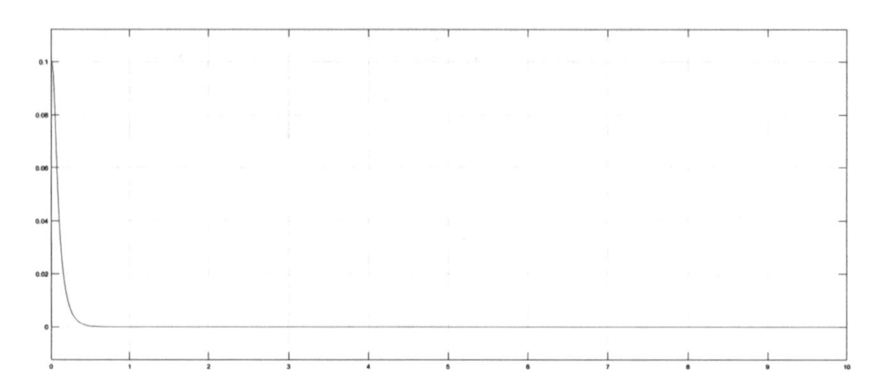

Fig. 6. Chart for angular position

6.4 PID Control System

The Transfer function is $P + \frac{I}{S} + D\left(\frac{N}{1+\frac{N}{S}}\right)$ (Fig. 7).

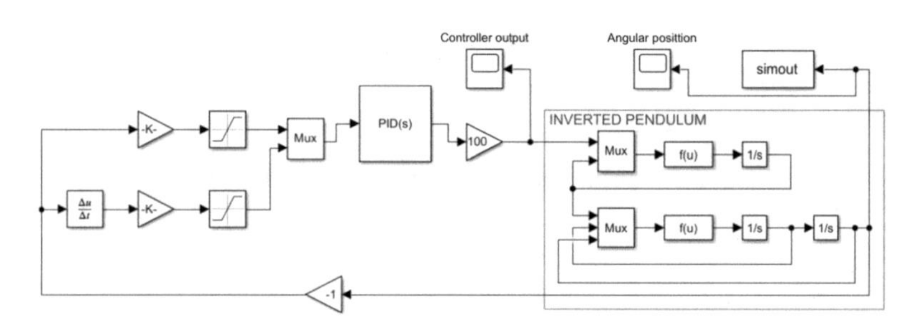

Fig. 7. PID control system

6.5 Controller Output

It shows the poor response (Fig. 8).

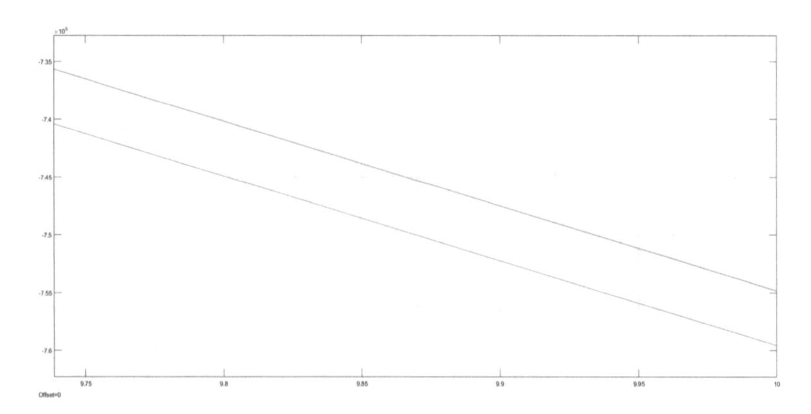

Fig. 8. Chart for controller output

P = 286.7, I = 733.234, D = 10081, Filter N = 269.93.
The control output is not stationary and it's getting decaying.

6.6 Angular Position

See Fig. 9.

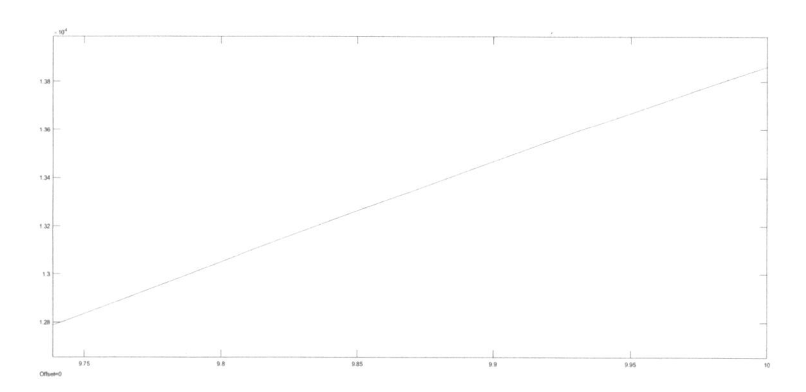

Fig. 9. Chart for angular position

6.7 Comparison of Type-2 Fuzzy Controller and PID Controller

In the field of control system, PID controller has been widely used but it has poor capability of handling the relation between input and output parameters, whereas Type-2 fuzzy logic controller has a very good capability of analyzing stability of the system with uncertain parameters. From this present work, it is proved that IT2FLC gives better stability than PID controller.

7 Conclusion

The results reveal that Gaussian membership function gives the exact result with smoothness and T2FSs with interval MF, handles more uncertainties and less mathematical complexity than Type-1. Therefore GIT2MF is used and the mathematical properties of aggregation operator using IT2GWG operator are proved. These properties play an important role in control system for the characteristics like continuity, robustness and stability. In this research, IT2FLC is used to check the stability for an inverted pendulum and compared the result with PID controller which proved IT2FLC is better than PID controller for this system. In future work, stability analysis can be done using neutrosophic controller and the comparative study can be made with the PID controller and fuzzy controller.

References

1. Bai, Y., Wnag, D.: Fundamentals of fuzzy logic control-fuzzy sets fuzzy rules and defuzzification. In: Advances Fuzzy Logic Technologies in Industrial Applications, pp. 17–36 (2006)
2. Dubois, D., Prade, H.: On the use of aggregation operations in information fusion processes. Fuzzy Sets Syst. **142**, 143–161 (2004)
3. Franzoi, L., Sgarro, A.: Linguistic classification: T-norms, fuzzy distances and fuzzy distinguishabilities. Procedia Comput. Sci. **112**, 1168–1177 (2017)
4. Gaurav, A.K., Kaur, A.: Comparison between conventional PID and fuzzy logic controller for liquid flow control: performance evaluation of fuzzy logic and PID controller by using MATLAB/Simulink. Int. J. Innov. Technol. Explor. Eng. (IJITEE) **1**(1), 84–88 (2012). ISSN 2278-30 75
5. Ghosh, A., Krishnan, R., Subudhi, B.: Robust PID compensation of an inverted cart pendulum system. An experimental study. IET Control Theory Appl. **6**(8), 1145–1152 (2012)
6. Kayacan, E., Kaynak, O.: Design of an adaptive interval type-2 fuzzy logic controller for the position control of a servo system with an intelligent sensor. In: IEEE World Congress on Computational Intelligence, pp. 18–23 (2010)
7. Klement, E.P., Mesiar, R., Pap, E.: Triangular norms. Position paper III: continuous t-norms. Fuzzy Sets Syst. **145**, 439–454 (2004)
8. Kobersi, I.S., Finaev, V.I.: Control of the heating system with fuzzy logic. World Appl. Sci. J. **23**(11), 1441–1447 (2013)
9. Lam, H.K., Seneviratne, L.D.: Stability analysis of interval type-2 fuzzy model-based control systems. IEEE Trans. Syst. Man Cybern.-Part-B: Cybern. **38**(3), 617–628 (2008)
10. Rojas, I., Ortega, J., Pelayo, F.J., Prieto, A.: Statistical analysis of the main parameters in the fuzzy inference process. Fuzzy Sets Syst. **102**, 157–173 (1999)
11. Tan, D.W.W.W.: A simplified type-2 fuzzy logic controller for real time control. ISA Trans. **45**(4), 503–516 (2006)
12. Volosencu, C.: Stabilization of fuzzy control systems. WSEAS Trans. Syst. Control **3**(10), 879–896 (2009)
13. Wu, D., Tan, W.W.: A type-2 fuzzy logic controller for the liquid-level process. In: Conference Proceedings, IEEE International Conference on Fuzzy Systems, vol. 2, no. 2, pp. 953–958 (2004)
14. Chiu, C.H., Wang, W.J.: A simple computation of MIN and MAX operations for fuzzy numbers. Fuzzy Sets Syst. **126**(2), 273–276 (2002)
15. Gera, Z., Dombi, J.: Exact calculations of extended logical operations on fuzzy truth values. Fuzzy Sets Syst. **159**(11), 1309–1326 (2008)
16. Zhou, S., Chiclana, F., John, R., Garibaldi, J.: Type-1 OWA operators for aggregating uncertain information with uncertain weights induced by type-2 linguistic quantities. Fuzzy Sets Syst. **159**(24), 3281–3296 (2008)
17. Franke, K., Koppen, M., Nickolay, B.: Fuzzy image processing by using Dubois and Prade fuzzy norms. In: Pattern Recognition, pp. 1–4 (2000)
18. Kiaei, A.A., Shouraki, S.B., Khasteh, S.H., Khademi, M., Samani, A.R.G.: New S-norm and T-norm operators for active learning method. In: Fuzzy Optimization and Decision Making, pp. 1–11 (2010)
19. Dombi, J.: A general class of fuzzy operators, the DeMorgan class of fuzzy operators and fuzziness measures induced by fuzzy operators. Fuzzy Sets Syst. **8**(2), 149–163 (1982)

20. Yao, Y., Wang, J.: Interval based uncertain reasoning using fuzzy and rough sets. In: Advances in Machine Intelligence and Soft-Computing, pp. 1–20 (1996)
21. Franke, K., Koppen, M.: A computer-based system to support forensic studies on handwritten documents. Int. J. Doc. Anal. Recogn. **3**(4), 218–231 (2001)
22. Hirota, K., Nobuhara, H., Kawamoto, K., Yoshida, S.I.: On a lossy image compression/reconstruction method based on fuzzy relational equations. Iran. J. Fuzzy Syst. **1**(1), 33–42 (2004)
23. Hidalgo, M.G., Torres, A.M., Aguilera, D.R., Sastre, J.T.: Image analysis applications of morphological operators on uninorms. Soft Comput. Tech. Uncertain. Manag. Image Process. 630–635 (2009)
24. Chaira, T.: Medical Image Processing: Advanced Fuzzy Set Theoretic Techniques. CRC Publications, Boca Raton (2015)
25. Castillo, O., Sanchez, M.A., Gonzalez, C.I., Martinez, G.E.: Review of recent type-2 fuzzy image processing applications. Information **8**(97), 1–18 (2017)
26. Sheikh, I., Khan, K.A.: Simulation of image edge detection using fuzzy logic in MATLAB. Int. J. Comput. Math. Sci. **6**(5), 19–22 (2017)
27. Kenjharayoobchandio, Y.: Fuzzy logic based digital image edge detection. Int. J. Electr. Electron. Data Commun. **6**(2), 18–22 (2018)
28. Dutta, A.K.: Intuitionistic fuzzy logic implementation in image fusion technique. Asian J. Res. Comput. Sci. **1**(1), 1–7 (2018)
29. Privezentsev, D.G., Zhiznyakov, A.L., Astafiev, A.V., Pugin, E.V.: Using fuzzy fractal features of digital images for the material surface analysis. J. Phys. **944**, 1–5 (2018)
30. Maini, R., Aggarwal, H.: Study and comparison of various image edge detection techniques. Int. J. Image Process. **13**(1), 1–12 (2009)
31. Lathamaheswari, M., Nagarajan, D., Udayakumar, A., Kavikumar, J.: Review on type-2 fuzzy in biomedicine. Indian J. Publ. Health Res. Dev. **9**(12), 322–326 (2018)
32. Lathamaheswari, M., Nagarajan, D., Kavikumar, J., Phang, C.: A review on type-2 fuzzy controller on control system. J. Adv. Res. Dyn. Control Syst. **10**(11), 430–435 (2018)
33. Nagarajan, D.: Three dimensional visualization of brain using machine learning. Int. J. Pure Appl. Math. **117**(7), 459–466 (2017)
34. Nagarajan, D., Lathamaheswari, M., Sujatha, R., Kavikumar, J.: Edge detection on DIOM image using triangular norms in type-2 fuzzy. Int. J. Adv. Comput. Sci. Appl. **9**(11), 462–475 (2018)
35. Nagarajan, D., Lathamaheswari, M., Kavikumar, J., Hamzha: A type-2 fuzzy in image extraction for DICOM image. Int. J. Adv. Comput. Sci. Appl. **9**(12), 352–362 (2018)
36. Nagarajan, D., Lathamaheswari, M., Broumi, S., Kavikumar, J.: A new perspective of traffic control management using type 2 fuzzy and interval neutrosophic sets. Oper. Res. Perspect. (in press)

The Importance of Safe Coding Practices and Possible Impacts on the Lack of Their Application

Gleidson Sobreira Leite(✉) and Adriano Bessa Albuquerque

Universidade de Fortaleza, Fortaleza, CE, Brazil
gleidson.sleite@gmail.com, adrianoba@unifor.br

Abstract. A significant change in the way organizations and companies use software has been brought by the evolution of the Internet over the time, resulting in an increasing presence of Information Systems in Web environments and, consequently, an increase in security vulnerabilities and threats. In this context, secure application development has become a crucial component for information systems in the market. This paper aims to contribute to a greater awareness of the importance of secure software coding by exposing possible impacts examples resulting from the lack of safe coding practices, as well as to present some existing recommendations from specialized agencies to avoid or treat occurrences of exploitation of vulnerabilities by malicious agents.

Keywords: Secure coding · Information security · Information systems · Web applications vulnerabilities

1 Introduction

Over time, technical, academic, business and also the society as a whole, has been impacted by the evolution and influence of the Internet as we increasingly use online services to access information or work, allowing the Web to consolidate itself as one of the main way of communication, relationships and business [1].

Due the ease and convenience provided, many internal business systems are migrating to Web, where, over time, they evolved rapidly from simple web sites, intended to enable online browsing of information, to complex information systems.

However, as noted in the Symantec Corporation report [2], in which 78% of web applications have at least one vulnerability, and 15% of them are critical, this universe of digital content is subject to various forms of threats that affect the security of people and their information by exploiting vulnerabilities.

Attacks on web applications are, according to Verizon [3], the leading cause of confirmed data breaches and are increasing in a rapid pace.

Although becoming increasingly essential to society, the Internet, according to OWASP [4], is an infrastructure inherently insecure and dependent on a large number of programs developed by people (software) that define the rules of functioning of computers. As well as all human activity, software development is subject to errors that

R. Silhavy (Ed.): CSOC 2019, AISC 986, pp. 214–224, 2019.
https://doi.org/10.1007/978-3-030-19813-8_22

can generate failures as security, leading to exponential impacts to companies or individuals.

With the increase of available and transacted information in the Web, a greater reflection on vulnerabilities is necessary in order to avoid fraud, invasion of intruders, propagation of viruses or means that circumvent the principles of information security (Confidentiality, Availability, Authenticity, Integrity and Non-repudiation) [5].

According to [6], sometimes the companies recognize the importance of guaranteeing the security of the information late and after effectiveness of attacks, losses or destruction of information.

That is, only after the application has been developed that secure modules are included, which increase the time and resources spent on the project as a whole, increasing the risk of threat materializations, as well as the possibility of not being able to address all identified vulnerabilities as it is very costly to undertake a number of changes, which can lead to high risk and cost increases for organizations and companies.

Motivated by this scenario, there is a growing need to increase the awareness of the importance of information security in Web applications and seek the minimization of vulnerabilities in software coding.

In this context, this paper aims to contribute to a greater awareness of the importance of secure software coding by exposing possible examples of impacts resulting from the lack of safe coding practices, as well as to present some existing recommendations (proactive controls) from specialized agencies to avoid or treat occurrences of exploitation of vulnerabilities by malicious agents.

This work also intends to be a contribution to the studies related to security in information systems, regarding the treatment of vulnerabilities with the intention to minimize the action of malicious agents.

This article is organized as follow: the methodology and procedures performed in the work (Sect. 2); the security breaches impacts examples (Sect. 3); the work related to the study about the recommendations (Sect. 4); and finally, the selection of proactive controls (Sect. 5), followed by the final considerations (Sect. 6).

2 Research Methodology

To accomplish the objective of this work, four specific actions were carried out. They were:

(1) Research and selection of data security breaches examples that occurred around the world with big companies or organizations, illustrating occurrences of cyber-attacks and their impacts, with the aim to contribute to a greater awareness of the importance of secure software coding.
(2) Bibliographic research and selection of the main international organizations specialized in conceptualization, treatment and exploitation of information security, followed by the selection of works focused on the treatment of vulnerabilities in Web information systems.

(3) From the organizations selected (OWASP, CWE and WASC), a comparative analysis, mapping and union of their work results were made. It was created a category list of main risks, programming errors and security vulnerabilities in Web applications.
(4) Finally, from the result of item 3, a selection of proactive treatment for vulnerabilities and risks by the developers of information systems in Web environments was made to support software development, based in proactive actions (recommendations) of previous works of the organizations selected.

3 Security Breaches Impacts

Just as the evolution of technology has made a major breakthrough in access to information, how vulnerabilities (failures or weaknesses) are exploited by malicious agents has also evolved, causing various cyber-attacks that have impacted several companies around the world.

The increase in new technologies has also increased risks and new vulnerabilities for companies, so they are subject to new types of attacks that exploit these new technologies.

Unfortunately, the number of data breach victims, which occurs when a cybercriminal successfully infiltrates a data source and extracts sensitive information, has increased, including the participation of large companies in this list of victims.

Table 1 list some examples of big breach incidents reported [7] that evolve millions of records stolen, order by the number of records stolen, illustrating that even big companies have been victims of cybercriminals.

Table 1. Big breach incidents examples

Company/Organization	Number of records stolen	Date of breach
Yahoo	3 billion	August, 2013
Equifaz	145.5 million	July, 2017
eBay	145 million	May, 2014
Heartland Payment Systems	134 million	March, 2008
Target	110 million	December, 2013
TJX Companies	94 million	December, 2006
JP Morgan & Chase	83 million	July, 2014
Uber	57 million	November, 2017
U.S. Office of Personnel Management (OPM)	22 million	Between 2013 and 2014
Timehop	21 million	July, 2018

According to Akamai research [8], attacks on Web applications increased by 69 percent comparing its state of the internet reports from 2016 to 2017, and the number one attack vector continues to be SQL Injection.

Data breaches can result in the loss of a large quantity of private records and sensitive data, affecting not just the breached organization, but also everyone whose personal information may have been stolen.

Thus, due to the exponential damage to people and companies caused by attacks by malicious agents or even several failures, the importance of information security in the corporate world is growing.

4 Related Works

According to [9], currently there are several methodologies and practices in information security because of the difficulties of technology companies in updating their models and control structures due to the constant change of technologies.

In order to assist organizations with these difficulties, there are worldwide-recognized standards such as NBR ISO/IEC 27001 and NBR ISO/IEC 27002, which together [10] are considered the most complete standard of information security management. They guide the creation of secure systems based on controls defined by international standards and good practices. Besides, there is the NBR ISO/IEC 15408 (Common Criteria), which provides a set of criteria to ensure the safety of the application through safety specifications based on the environment's characteristics.

Even after the definition and implementation of security policies in a company based on the adoption of standards and good practices of security management, software vulnerabilities can become a gateway to attacks that may lead, according to [11], to unauthorized access to privileged data, destruction of critical data, or even destruction of an entire system.

According to [12], the initial focus of software development is fundamentally to meet functional requirements that meet customer needs, where security requirements are often dealt with late, during testing and validation of software, because of the lack of preparation of customers and developers to address the issue in advance. This can produce a high negative impact on the project and even on companies.

Therefore, it is essential to adequately complement the concepts of information security in the specification and coding of information systems, especially in open environments such as Web, in order to minimize the possibility of greater occurrences of system failures or attacks.

For achieve this goal, some international agencies and institutions have carried out studies and published work results, defining ranks of the main vulnerabilities, weaknesses (insufficient controls) and risks (possibilities of threats) to support de software industry, as well as recommendations (proactive controls) to mitigate or avoid them.

4.1 OWASP Top 10 – Most Critical Web Application Security Risks

The Open Web Application Security Project group is a global, nonprofit organization created to enable the improvement of software development companies to develop trusted Web-centric applications, increasing the visibility of application security, helping individuals from whole world to make better decisions regarding software security risks [13].

The work products produced by OWASP are available to the entire international community and used as a reference by entities and organizations in the area of technology and security, such as the US Defense Information Systems Agency (DISA), the US Federal Trade Commission and PCI Council [14].

According to [15], OWASP Top 10 groups the highest risks of critical attacks exploited by vulnerabilities in Web applications. It has periodic updates based on research data and statistics on the major attacks around the world.

Table 2 lists the major security risks of applications made available in Web environments, according to OWASP study [16], ranked in order of increasing prevalence and criticality, in combination with consensus related to estimates of exploration, detection and impact.

Table 2. Owasp Top 10-2017

Rank	Name
A1	Injection
A2	Broken Authentication
A3	Sensitive Data Exposure
A4	XML External Entities (XXE)
A5	Broken Access Control
A6	Security Misconfiguration
A7	Cross-Site Scripting (XSS)
A8	Insecure Deserialization
A9	Using Components with Known Vulnerabilities
A10	Insufficient Logging & Monitoring

4.2 CWE/SANS Top 25 Most Dangerous Software Errors

The CWE (Common Weakness Enumeration) is a project sponsored by the US Department of Homeland Security's National Cyber Security Division to classify security errors into a formal list that serves as a common language for describing software security weaknesses in architecture, design, or coding, providing a standard for efforts to identify, mitigate and prevent such weaknesses [17].

A collaboration between the SANS/MITER institute and several experts from the United States and Europe has defined the "2011 CWE/SANS Top 25 Most Dangerous Software Errors", which is a result of the selection of 25 most critical errors in software coding list CWE. These errors can lead to serious vulnerabilities in information systems [18]. They are listed in Table 3.

Table 3. CWE/SANS Top 25 most dangerous software errors

Rank	Name	Rank	Name
1	Improper Neutralization of Special Elements used in an SQL Command ('SQL Injection')	14	Download of Code Without Integrity Check
2	Improper Neutralization of Special Elements used in an OS Command ('OS Command Injection')	15	Incorrect Authorization
3	Buffer Copy without Checking Size of Input ('Classic Buffer Overflow')	16	Inclusion of Functionality from Untrusted Control Sphere
4	Improper Neutralization of Input During Web Page Generation ('Cross-site Scripting')	17	Incorrect Permission Assignment for Critical Resource
5	Missing Authentication for Critical Function	18	Use of Potentially Dangerous Function
6	Missing Authorization	19	Use of a Broken or Risky Cryptographic Algorithm
7	Use of Hard-coded Credentials	20	Incorrect Calculation of Buffer Size
8	Missing Encryption of Sensitive Data	21	Improper Restriction of Excessive Authentication Attempts
9	Unrestricted Upload of File with Dangerous Type	22	URL Redirection to Untrusted Site ('Open Redirect')
10	Reliance on Untrusted Inputs in a Security Decision	23	Uncontrolled Format String
11	Execution with Unnecessary Privileges	24	Integer Overflow or Wraparound
12	Cross-Site Request Forgery (CSRF)	25	Use of a One-Way Hash without a Salt
13	Improper Limitation of a Pathname to a Restricted Directory ('Path Traversal')		

4.3 Web Applications Security Consortium (WASC) – Threat Classification

The Web Application Security Consortium (WASC) is composed of an international group of experts, industry professionals and representatives of organizations that seek to produce best practices of security standards, facilitating the exchange and dissemination of ideas, techniques, articles, security guidelines, among other useful documentation used by various companies, educational institutions, governments and software developers worldwide [19].

The WASC Threat Classification is a project resulting from a cooperative effort to clarify and organize attacks and weaknesses that can jeopardize a Web information system, its data or its users. It is compound, according to Table 4, by 49 items arranged in alphabetical order, grouped by attacks and weaknesses, with their respective numbering in the priority classification rank [20]:

Table 4. WASC threat classification

(Nr. Rank) Attacks	(Nr. Rank) Attacks	(Nr. Rank) Weaknesses
03 - Integer Overflows	29 - LDAP Injection	01 - Insufficient Authentication
05 - Remote File Inclusion	30 - Mail Command Injection	02 - Insufficient Authorization
06 - Format String	31 - OS Commanding	04 - Insufficient Transport Layer Protection
07 - Buffer Overflow	32 - Routing Detour	13 - Information Leakage
08 - Cross-Site Scripting	33 - Path Traversal	14 - Server Misconfiguration
09 - Cross-Site Request Forgery	34 - Predictable Resource Location	15 - Application Misconfiguration
10 - Denial of Service	35 - SOAP Array Abuse	16 - Directory Indexing
11 - Brute Force	36 - SSI Injection	17 - Improper Filesystem Permissions
12 - Content Spoofing	37 - Session Fixation	20 - Improper Input Handling
18 - Credential/Session Prediction	38 - URL Redirector Abuse	21 - Insufficient Anti-automation
19 - SQL Injection	39 - XPath Injection	22 - Improper Output Handling
23 - XML Injection	41 - XML Attribute Blowup	40 - Insufficient Process Validation
24 - HTTP Request Splitting	42 - Abuse of Functionality	47 - Insufficient Session Expiration
25 - HTTP Response Splitting	43 - XML External Entities	48 - Insecure Indexing
26 - HTTP Request Smuggling	44 - XML Entity Expansion	49 - Insufficient Password Recovery
27 - HTTP Response Smuggling	45 - Fingerprinting	
28 - Null Byte Injection	46 - XQuery Injection	

5 Selection of Vulnerabilities, Risks and Proactive Controls

Exploring the main vulnerabilities in the coding of software for Web environments and the need for awareness and guidance for developers of information systems, the purpose of this study is also to present some existing recommendations (proactive controls) from specialized agencies to avoid or treat occurrences of exploitation of vulnerabilities by malicious agents.

To provide a more comprehensive selection (categories) of the most critical factors to treat vulnerabilities in Web applications, a study of mapping between the results of the three institutions work products were carried out.

In the bibliographic researches [21, 22], it was verified that each institution carried out a mapping of its vulnerabilities/risks/weaknesses and attacks with ranks of other institutions, where, after comparing then, a list of categories was selected.

It should be highlighted that the indication of the categories and the study is limited to vulnerabilities within the ranks or lists from each institution, with focus on vulnerabilities whose main actions must be carried out by the information systems development team.

That is attacks, vulnerabilities or risks whose proactive action for treatment are not treated through safe coding practices, for example, training of employees to avoid social engineering, among other actions, are not object of study of the present work.

5.1 Selection of Proactive Controls (Recommendations)

After selecting the main categories for vulnerability and risk management in Web information systems, it was also necessary to select which proactive controls that have to be implemented by the development team to minimize or avoid the action of malicious agents.

The treatment of a given vulnerability and/or risk is not necessarily unique. So they can have different solutions, and some vulnerabilities may be interrelated. For example, lack of protection or poor handling of a particular vulnerability may provide an opening for another vulnerability, even if the developer has implemented a right protection action.

After the study of proactive controls has been made on the body of knowledge of the institutions [23, 24], possible controls for use and implementation by developers of Web information systems were selected and grouped by the categories in Table 5.

Table 5. Selection of proactive controls

Category	Actions/proactive controls
Injection	Primary Options: - Use of Prepared Statements (Parameterized Queries) - Use of Stored Procedures - Escaping all User Supplied Input Additional Defenses: - Enforce: Least Privilege - Input/Output Validation, Encoding
Broken Authentication (and Session Management)	- Multi-Factor authentication - Secure Password Storage (Use of Criptografy and Salt) - Secure Password Recovery Mechanism - Session Management, Generation and Expiration - Transport Layer Protection, Security Headers
Sensitive Data Exposure	- Transport Layer Protection, Security Headers - Secure Password and Criptografy Storage (Use of Criptografy and Salt)
Cross Site Scriptiong (XSS)	- Input/Output Validation, Encoding, Escaping, Sanitizing
Unvalidated Input/Output	- Input/Output Validation

(continued)

Table 5. (*continued*)

Category	Actions/proactive controls
Security Misconfiguration	- Secure configuration, Security Headers - Remove unnecessary features, components, frameworks, documentation and samples - Disable default accounts and passwords
Broken Access Control	- Disable web server directory listing and Disable Caching - Anti-automation measures - Authentication Validation (User validation in every page)
XML External Entities	- Input/Output Validation - XML validation against local Static DTD/XML Schema/XSD - Disable XML external entity and DTD processing
Cross Site Request Forgery (CSRF)	- Use of Synchronizer Token
Unvalidated Redirects and Forwards	- Input/Output Validation and Sanitizing
Information Leakage and Improper Error Handling	- Input/Output Validation; Secure configuration - Create a default error handler with sanitized error message
Malicious File Execution	- Input/Output Validation, Encoding, Escaping - Secure configuration
Insufficient Logging and Monitoring	- Input/Output Validation, Encoding, Escaping and Sanitizing
Denial of Service	- Input/Output Validation, Encoding, Escaping and Sanitizing - Anti-automation measures
Buffer Overflows	- Use components from official sources - Input/Output Validation
Insecure Deserialization	- Input/Output Validation
Using Components with known Vulnerabilities	- Remove unnecessary features, components, frameworks, documentation and files - Use components from official sources
HTTP Request/Response Splitting	- Renew the Session ID After Any Privilege Level Change - Input/Output Validation (Heders), Escaping and Sanitizing

As can be seen in Table 5, several actions deal with not only a particular category, but also impact on the treatment of other vulnerabilities, such as Input/Output Validation.

However, for example, although the action of validation serves to treat several categories (as well as other cases), the way the treatment will be implemented will depend on which category is being treated. For instance, the validation of information may have different sources such as HTTP headers, URLs, application forms, and so on, which have led to different types of implementations.

6 Conclusion

According to [25] it is important to emphasize that it is not possible to obtain absolute security, and it is only possible to seek and adopt security policies and practices with the aim of minimizing risks, since, at all times, new vulnerabilities and threats arise. Besides, it is essential, in the case of software, that security actions be applied since the beginning of the development.

This work presented a contribution to information security and software engineering area, related to the treatment of vulnerabilities in Web information systems, in order to minimize the performance of malicious agents.

The brief list examples of security breaches impacts, also showed the importance of the subject and the possible consequences to companies, clients and the society.

References

1. Rexha, B., Halili, A., Rrmoku, K., Imeraj, D.: Impact of secure programming on web application vulnerabilities. In: 2015 IEEE International Conference on Computer Graphics, Vision and Information Security (CGVIS), Bhubaneswar, pp. 61–66 (2015). https://doi.org/10.1109/cgvis.2015.7449894
2. Symantec Corporation: Internet security threat report, vol. 21, April 2016. http://www.ijssst.info/info/IEEE-Citation-StyleGuide.pdf. Accessed 20 Nov 2018
3. Verizon: Verizon DBIR 2016: Web Application Attacks are the #1 Source of Data Breaches, June 2016. https://www.verizondigitalmedia.com/blog/2016/06/verizon-dbir-2016-web-application-attacks-are-the-1-source-of-data-breaches/. Accessed 15 Dec 2018
4. OWASP: Segurança na web: Uma janela de oportunidades. Uma mensagem do OWASP Brasil ao Governo Brasileiro, March 2011. https://www.owasp.org/images/1/16/Seguranca_na_web_-_uma_janela_de_oportunidades.pdf. Accessed 5 Nov 2018
5. Holik, F., Neradova, S.: Vulnerabilities of modern web applications. In: 2017 40th International Convention on Information and Communication Technology, Electronics and Microelectronics (MIPRO), Opatija, pp. 1256–1261 (2017)
6. do Espirito Santo, A.F.S.: Segurança da informação. Departamento de Ciência da Computação - Instituto Cuiabano de Educação (ICE), Cuiabá, MT, Brasil (Unpublished). http://www.ice.edu.br/TNX/encontrocomputacao/artigos-internos/aluno_adrielle_fernanda_seguranca_da_informacao.pdf. Accessed 20 Nov 2018
7. TrendMicro: Data Breaches 101: HowTheyHappen, WhatGetsStolen, andWhere It AllGoes, August 2018. https://www.trendmicro.com/vinfo/au/security/news/cyber-attacks/data-breach-101. Accessed 15 Dec 2018
8. Akamai: State of the Internet - Q3 2017 Report. https://www.akamai.com/us/en/multimedia/documents/state-of-the-internet/q3-2017-state-of-the-internet-security-report.pdf. Accessed 15 Dec 2018
9. Ferreira, F.N.F., Araújo, M.T.D.: Política de segurança da informação – Guia prático para elaboração e implementação, 2nd. edn Revisada. Editora Ciência Moderna Ltda, Rio de Janeiro (2008)

10. Prazeres, A.P.: Princípios para o desenvolvimento de software seguro. Monografia (Especialização em Engenharia de Projetos de Software). Universidade do Sul de Santa Catarina, Florianópolis, SC, Brasil (2015)
11. Kumar, R.: Mitigating the authentication vulnerabilities in web applications through security requirements. Inf. Commun. Technol. (WICT) **60**(2), 651–663 (2016). https://doi.org/10.1109/WICT.2011.6141435
12. Holanda, M.T., Fernandes, J.H.C.: Segurança no desenvolvimento de aplicações: GSIC701. Curso de Especialização em Gestão da Segurança da Informação e Comunicações: 2009/2011. Departamento de Ciências da Computação da Universidade de Brasília. Brasília, DF, Brasil, November 2017
13. OWASP: About the open web application security project, 25 November 2017. https://www.owasp.org/index.php/About_The_Open_Web_Application_Security_Project. Accessed 10 Nov 2018
14. Mon'teverde, W.A.: Estudo e análise de vulnerabilidades web. Trabalho de conclusão de curso (Curso superior de Tecnologia em Sistemas para a Internet). Universidade Tecnológica Federal do Paraná, Campo Mourão, PR, Brasil (2014)
15. Atashzar, H., Torkaman, A., Bahrololum, M., Tadayon, M.H.: A survey on web application vulnerabilities and countermeasures. In: 6th International Conference on Computer Sciences and Convergence Information Technology (ICCIT), Seogwipo, pp. 647–652 (2011)
16. OWASP: The ten most critical web application security risks. OWASP Foundation, Final Release, November 2017. https://www.owasp.org/index.php/Top_10-2017_Top_10. Accessed 25 Nov 2018
17. Mitre Corporation: CWE (Common Weakness Enumeration), 3 February 2018 https://cwe.mitre.org/. Accessed 06 Nov 2017
18. Mitre Corporation: CWE/SANS Top 25 most dangerous software errors, 06 June 2011. http://cwe.mitre.org/top25/. Accessed 25 Nov 2017
19. Web Application Security Consortion: WASC – The web application security consortion. http://www.webappsec.org/aboutus.shtml. Accessed 10 Nov 2017
20. Web Application Security Consortion: The WASC Threat Classification, 2nd edn, 1 January 2010. http://projects.webappsec.org/f/WASC-TC-v2_0.pdf. Accessed 10 Nov 2017
21. OWASP: Vulnerability Classification Mappings, 21 Januray 2010. https://www.owasp.org/index.php/Vulnerability_Classification_Mappings. Accessed 10 Nov 2017
22. Web Application Security Consortion: Threat classification taxonomy cross reference view (2013). http://projects.webappsec.org/w/page/13246975/Threat%20Classification%20Taxonomy%20Cross%20Reference%20View. Accessed 10 Nov 2017
23. OWASP: OWASP Top 10 Proactive Controls 2016. 10 critical security areas that web developers must be aware of. https://www.owasp.org/images/5/57/OWASP_Proactive_Controls_2.pdf. Accessed 10 Nov 2017
24. Mitre Corporation: 2011 CWE/SANS Top 25: Monster Mitigations, 23 June 2011. http://cwe.mitre.org/top25/mitigations.html. Accessed 10 Nov 2017
25. Santiago, H.L.P., Lisboa, G.d.S.: Segurança de sistemas da informação – 'O contexto da segurança dos sistemas de informação. Faculdade Atenas, Paracatu, MG, Brasil (2011). http://www.atenas.edu.br/Faculdade/arquivos/NucleoIniciacaoCiencia/REVISTAS/REVIST 2011/6.pdf. Accessed 28 Nov 2017

A Power Efficient Authentication Model for E-Healthcare System Security Concerns in Context to Internet of Things (IoT)

P. H. Latha(✉)

Department of Information Science and Engineering,
Sambhram Institute of Technology, Bangalore, India
phdlatha2017@gmail.com

Abstract. The growing intellectual advancement in the Internet of Things (IoT) has changed the perception of every individual towards the smart way of using internet and cloud for smart objects interaction. The domain of E-health care application is not away from these advancements in IoT. However, the deployment of these application faces some security concerns as severe obstacle. The authentication of the E-healthcare information is much-concerned issues as the confidential data of patients were exchanged among different entities. Thus, a power efficient authentication model which addresses the security issues of E-health application. Using this model, authentication at both the ends of sensors and a base station (S&BS) is verified and achieved secure exchange of health data. This paper considers arbitrary numbers (i.e., nonces) and Hash Key based authentication during cryptographic communication. The outcomes suggest that with this power consumption is reduced and also lay off the session key arrangement among each S&BS. Also, the performance analysis of HKA approach gives that different kinds of security issues are handled properly with low power consumption.

Keywords: E-healthcare · IoT · Power consumption · Security issues · Hash key based authentication (HKA)

1 Introduction

The last decade of researches in the domain of IoT has witnessed rapid growth in the area of wireless communication. These IoT paradigms aim to enhance the ease of communicating the sensors, wearable devices, phones, objects, etc., through internet media. [1, 2]. Also, the use of IoT has become a global technique which enables everything to be readable, addressable and locatable over the Internet [3]. The IoT is mainly facing/vulnerable to various kind of attacks, and these have a risk of eaves-dropping [4, 5].

Also, the IoT components have resource constraints issues like low power and resource computation. Thus, these parameters cannot be considered mainly in complex wireless security mechanisms [6]. The prime security concern in IoT is data integrity. The authentication in IoT is primary parameter that allows verification of every entity [7–9]. However, for implementation of the authentication mechanism, the IoT

R. Silhavy (Ed.): CSOC 2019, AISC 986, pp. 225–233, 2019.
https://doi.org/10.1007/978-3-030-19813-8_23

characteristics must co-operate with them. The E-health care is one of such application which faces mainly authenticity issue in e-health applications which is need to address in research domain [10]. Thus, this paper introduces a power efficient authentication model for E-healthcare applications by using a hash key authentication code (HKA). The paper is organized with sections like related works (Sect. 2), Research Methodology (Sect. 3), Results in the analysis (Sect. 4) and Conclusion (Sect. 5).

2 Related Work

This section discloses the IoT security in different scenario using various techniques and algorithms to improve security performance. Asplund and Tehrani [11] have introduced the IoT security in social facilities to overcome the critical issues. It provides optimistic results with high security, and less risk has found in the process. In Hamadi and Chen [12] have discussed the trust-based decision-making techniques for health IoT system to improve the performance of the security of medical system. That exhibits the feasibility of the system. Alrawais et al. [13] have presented the fog computing method for IoT by depicting the fog computing in the edge of the internet to improve the security and disclosures the privacy threats. It provides security and privacy from the results of the different case study. Bui and Beigne [14] have discussed the multi-level IoT security with low power and minimum energy utilization using AES technique and PRESENT algorithm. It shows data security and authentication with low cost and improved throughput. Frustaci et al. [15] have introduced the IoT system model including three layers (Perception, transportation, and presentation). That provides a future study path for IoT critical issues. Condry [16] have presented the IoT security using smart edge IoT devices for Industrial application with safe and rapid performance. The result shows secure, scalable and resilient in real-time environment. Gao et al. [17] have discussed the security of IoT using segment operand and basic arithmetic and logical operation. By the analysis, it shows enhanced security which provides an application in watermarking, encryption, and fingerprint by hiding the information. Guan et al. [18] introduced the secure and efficient data acquisition technique for smart grid based on IoT. The cipher text-based encryption and decryptions are utilized, and its results show security with reducing time cost. In Ray [19] have presented the object tracking process using a secure tracking protocol for a different kind of application related IoT. Where cryptographic base PUF (Physical Unclonable Function) is offered to improve the security. Its result found that the method exhibits security and less computational cost. Han [20] have addressed the network-based end to end security framework for cloud IoT to improve the business data. Its results improvement at the security level. Similar kind of security aware research for IoT are found in Burg [21], Guo et al. [22], Hu and Wen [23], Karati et al. [24], and da Cruz and Rodriguse [25], Li et al. [26], Kim [27], Kougianos et al. [28], Gai [29], Granjal et al. [30], Kamal and Thariq [31], Li [32], Bhattarai and Wang [33], Mostafa et al. [34], and Radisavljevic [35]. The previous works of Latha and Vasantha have evolved with various security concerns [36] in the wireless system and offered secure key generation protocol [37], efficient security system [38] and data security for WLAN [39].

The survey of the existing system it is observed that many types of research were addressed the security issues and come up with different technologies to handle different security attacks. However, the growing advancement in the hacking methods is posing challenges in the protection of privacy credentials of the individuals which can be of banking, healthcare, social networking, etc., data over the internet. The E-healthcare application is one of such area where security over IoT is necessary. Thus, there is a need for secure authentication system that can address the E-healthcare applications security issues.

3 Research Methodology

In order to enhance the security level for E-healthcare applications, the proposed Hash key-based authentication (HKA) model introduces authentication at both S&BS nodes. The sensors can be placed over and around the human body. The model helps to bring interfaced authentication with minimal power utilization. The following Fig. 1 indicates the network structure of E-healthcare applications which composes of the block of wearable sensor deployment, web-based backend infrastructure, doctors, care takes and a base station. The model helps in the collection of health-related data from wearable sensors placed over the human body (i.e., ECG, pulse sensors and the sensors measuring the surrounding temperature). The collected data from the system can be forwarded to the base station via wireless interfaces, i.e., Wifi or Zigbee or Bluetooth. In order to observe the captured data, smartphones can be used as it offers mobility among the user and smartphones exhibit better power capacity than the sensors. The Base station then sends the collected information to the doctors or caretakers through backend web infrastructures by wireless media (mobile network or Wifi).

In order to deploy the proposed architecture some of the assumptions were made about the devices used.

- Categorization of Objects: The Objects can be categorized into power/computation constrained sensor nodes and better power/computationally efficient base station.
- Each sensor is represented with identity (Id) and those having the property of performing symmetric and minimum one asymmetric encryption. The base station does the public key structure (PKS) to protect the data after wireless sensor network (WSN).

The proposed authentication model offers interfaced authentication among SBS of E-healthcare system. The functioning of the model can be divided into three different parts which includes: Registration, Authentication and Secure key generation. During the registration, stage is important to stage among SBS. With an assumption cipher suited are defined i.e. hash-function $H_f()$, respective secret key (k_i) and a message authentication code (MAC). Later, an offline supplier comes into the base station for that the respective parameter and Id can be registered for each sensor in the binding table. Further, a hash value on the basis of this Id is computed for the sensor. This H_f can be referred to the “covered identity (C_{id}).” Thus, the binding table is composed of Id, the defined cipher suites, k_i and C_{id} (given in Fig. 2). Once the registration is done,

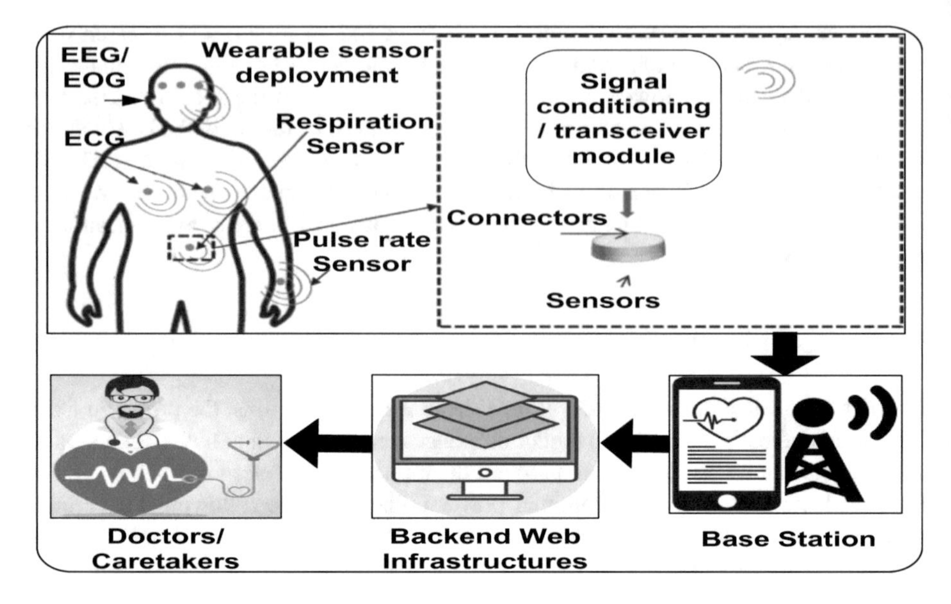

Fig. 1. Network structure of E-healthcare system

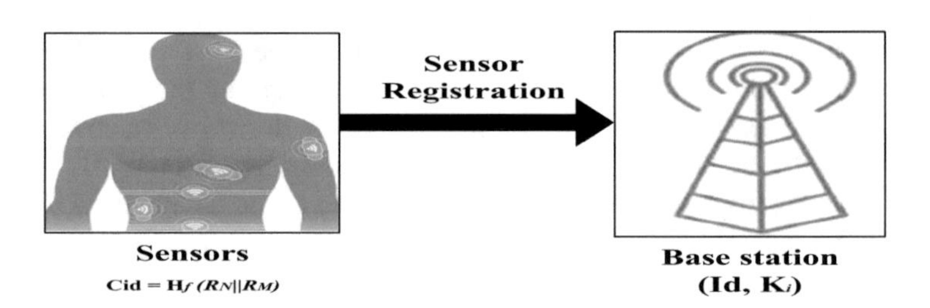

Fig. 2. Processes involved in registration

the entire sensors in base station terminate from registration stage and proceed to the authentication stage.

The authentication stage involved in offering the interfaced authentication at both SBS ends. Every sensor which is needed to be involved in communication with base station must complete this stage. The entire authentication stage is represented in Fig. 3.

i. The sensor node in this stage builds a random value of "R_N" over 8-bytes and forwards a message that involves generated value, its covered identity (C_{id}) and authentication code HKA (C_{id}, Id, R_N) that ensures the message identity.
ii. Computing associated HKA will verify the received message from the base station. If the verification is successful, the base station builds a random value of "R_M" over 8-bytes and resends a message containing received value R_N, HKA (C_{id}, Id, R_N) and R_M.

iii. After receiving the message from base station, the sensor node verifies it with calculations done by HKA. After successful verification, SBS terminates the interfaced authentication.

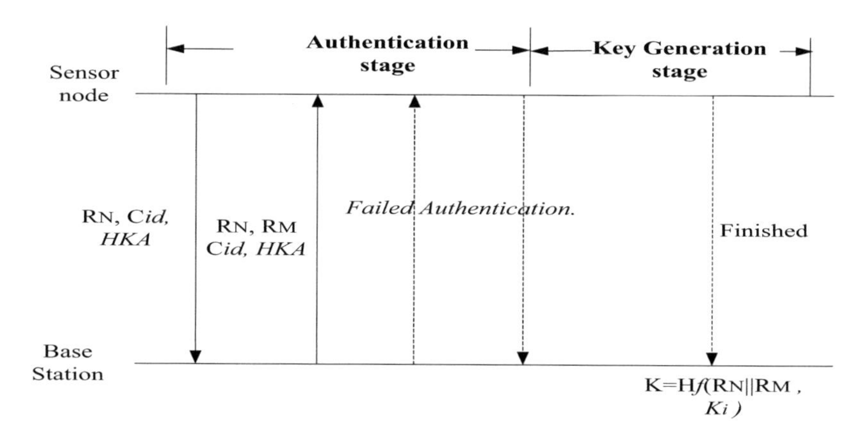

Fig. 3. Steps involved in authentication stage

After the authentication stage, the secret key generation stage will be initiated. In this, as a shared key can be generated to build a secure communication channel. The key can be computed by using the desired function in concatenation with the arbitrary numbers ($R_N \| R_M$) and then the encryption is applied with the k_i of 128 bits. Once the encryption is completed, the "encryption complete" message will be received with k_i. Then, the base station computes, decrypts the messages and performs verification of k_i.

4 Results and Analysis

The outcomes of the proposed authentication models are analyzed by considering the power utilization of the sensor nodes. The power needed for cryptographic process is computed with the power considered for data transmission and receiving. In this, the sensor node was connected with wireless radio which operates on battery cells (total 18.5 kJ). Based on the generated R_N and R_M values the authentication model is evaluated with different techniques and used HAK code for verification. The HAK is computed by using H_f which partitions the messages into blocks of fixed size (128 bits). For comparison, the AES scheme is considered which also protected y HAK code. The following comparative analysis data is obtained with existing systems. The work of Meulenaer et al. [40] has utilized 5.76×10^{-6} J and 6.48×10^{-6} J of power for emission and reception respectively for 1-bytes of data. Similarly, the proposed model consumes 62.15×10^{-6} J for 20-byte of data. Hence the power utilized for authentication of this amount of data is less than [40]. The comparative analysis of energy cost of different exchange techniques is shown in Fig. 4. From this, it is observed that the proposed HKA model can reduce the exchange cost of different messages during authentication phase.

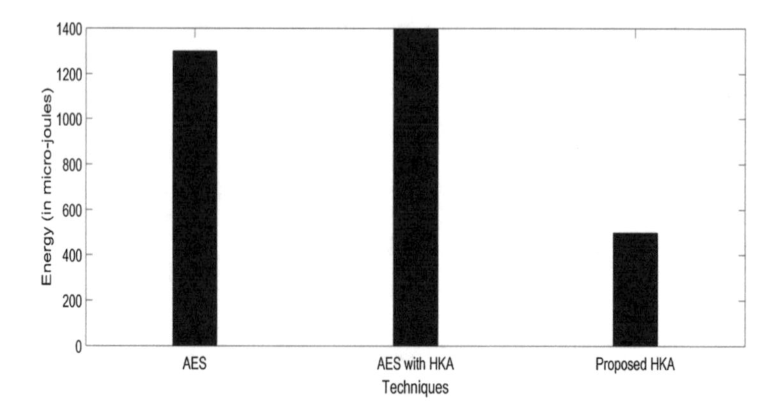

Fig. 4. Energy-cost analysis of the different exchange's techniques

During the authentication stage, the proposed more able to save more energy and also the high-security level at both SBS. The entire cost of the model is computed by considering both the cost of authentication and key generation stage, and it is observed about 832.14×10^{-6}J. This proves that the proposed model is cost-effective against resource constrained IoT environment.

The comparative analysis of the other existing techniques of like SSL mechanism proposed by [41] and Kerberos mechanism introduced by [42] (as shown in Fig. 5). From the figure, it is observed that the proposed scheme can reduce the power cost than existing systems.

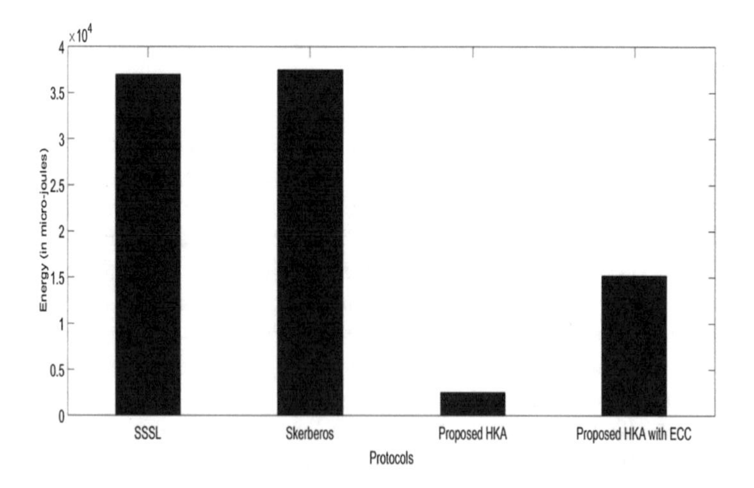

Fig. 5. Comparison with other protocols

5 Conclusion

This paper introduced a power efficient authentication model which addresses the security issues of E-health application by using the generated nonces, covered identity and HAK code at various data exchanges. From outcome analysis, it is observed that the proposed model is cost-effective regarding power consumption. Also, from comparative analysis, it is observed that the proposed authentication has yielded better power consumption with a high level of security than other methods and different techniques introduced in the recent past. With all these significances, it can be said that the proposed model can be implemented in the applications of E-health under resource-constrained environment.

Further, the proposed authentication protocol can be adapted in real time scenarios of e-healthcare applications and others. The futuristic scope of this research can be considered to increase security, reduced memory utilization and optimization of processing time at the higher security level.

References

1. Wu, L., Chen, C.-H., Zhang, Q.: A mobile positioning method based on deep learning techniques. Electronics **8**(1), 59 (2019)
2. Zhang, Z.-K., et al.: IoT security: ongoing challenges and research opportunities. In: 2014 IEEE 7th International Conference on Service-Oriented Computing and Applications (SOCA). IEEE (2014)
3. Zhao, K., Ge, L.: A survey on the Internet of Things security. In: 2013 9th International Conference on Computational Intelligence and Security (CIS). IEEE (2013)
4. Gou, Q., et al.: Construction and strategies in IoT security system. In: IEEE International Conference on and IEEE Cyber, Physical and Social Computing Green Computing and Communications (GreenCom), 2013 IEEE and Internet of Things (iThings/CPSCom). IEEE (2013)
5. Mahmoud, R., et al.: Internet of things (IoT) security: current status, challenges and prospective measures. In: 2015 10th International Conference for Internet Technology and Secured Transactions (ICITST). IEEE (2015)
6. Babar, S., et al.: Proposed embedded security framework for Internet of Things (IoT). In: 2011 2nd International Conference on Wireless Communication, Vehicular Technology, Information Theory and Aerospace & Electronic Systems Technology (Wireless VITAE). IEEE (2011)
7. He, D., Zeadally, S.: An analysis of RFID authentication schemes for Internet of Things in healthcare environment using elliptic curve cryptography. IEEE Internet Things J. **2**(1), 72–83 (2015)
8. Abdolmaleki, B., et al.: Attacks and improvements on two new-found RFID authentication protocols. In: 2014 7th International Symposium on Telecommunications (IST). IEEE (2014)
9. Tewari, A., Gupta, B.B.: Cryptanalysis of a novel ultra-lightweight mutual authentication protocol for IoT devices using RFID tags. J. Supercomput. **73**(3), 1085–1102 (2017)
10. Amin, R., et al.: A light weight authentication protocol for IoT-enabled devices in distributed Cloud Computing environment. Future Gener. Comput. Syst. **78**, 1005–1019 (2018)

11. Asplund, M., Nadjm-Tehrani, S.: Attitudes and perceptions of IoT security in critical societal services. IEEE Access **4**, 2130–2138 (2016)
12. Al-Hamadi, H., Chen, R.: Trust-based decision making for health IoT systems. IEEE Internet Things J. **4**(5), 1408–1419 (2017)
13. Alrawais, A., Alhothaily, A., Chunqiang, H., Cheng, X.: Fog computing for the Internet of Things: security and privacy issues. IEEE Internet Comput. **21**(2), 34–42 (2017)
14. Bui, D.-H., Puschini, D., Bacles-Min, S., Beigné, E., Tran, X.-T.: AES datapath optimization strategies for low-power low-energy multisecurity-level Internet-of-Things applications. IEEE Trans. Very Large Scale Integr. (VLSI) Syst. **25**(12), 3281–3290 (2017)
15. Frustaci, M., Pace, P., Aloi, G., Fortino, G.: Evaluating critical security issues of the IoT world: present and future challenges. IEEE Internet Things J. **5**(4), 2483–2495 (2018)
16. Condry, M.W., Nelson, C.B.: Using smart edge IoT devices for safer, rapid response with industry IoT control operations. In: Proceedings of the IEEE, vol. 104, no. 5, pp. 938–946 (2016)
17. Gao, M., Wang, Q., Arafin, M.T., Lyu, Y., Qu, G.: Approximate computing for low power and security in the Internet of Things. Computer **50**(6) 27–34 (2017)
18. Guan, Z., Li, J., Wu, L., Zhang, Y., Wu, J., Du, X.: Achieving efficient and secure data acquisition for cloud-supported Internet of Things in smart grid. IEEE Internet Things J. **4** (6), 1934–1944 (2017)
19. Ray, B.R., Chowdhury, M.U., Abawajy, J.H.: Secure object tracking protocol for the Internet of Things. IEEE Internet Things J. **3**(4), 544–553 (2016)
20. Han, Z., Li, X., Huang, K., Feng, Z.: A software defined network-based security assessment framework for cloudIoT. IEEE Internet Things J. **5**(3), 1424–1434 (2018)
21. Burg, A., Chattopadhyay, A., Lam, K.-Y.: Wireless communication and security issues for cyber-physical systems and the Internet-of-Things. Proc. IEEE **106**(1), 38–60 (2018)
22. Guo, L., Dong, M., Ota, K., Li, Q., Ye, T., Jun, W., Li, J.: A secure mechanism for big data collection in large scale Internet of vehicle. IEEE Internet Things J. **4**(2), 601–610 (2017)
23. Hu, L., Wen, H., Bin, W., Pan, F., Liao, R.-F., Song, H., Tang, J., Wang, X.: Cooperative jamming for physical layer security enhancement in Internet of Things. IEEE Internet Things J. **5**(1), 219–228 (2018)
24. Karati, A., Islam, S.K.H., Biswas, G.P., Bhuiyan, M.Z.A., Vijayakumar, P., Karuppiah, M.: Provably secure identity-based signcryption scheme for crowdsourced industrial Internet of Things environments. IEEE Internet Things J. **5**(4), 2904–2914 (2018)
25. da Cruz, M.A.A., Rodrigues, J.J.P.C., Al-Muhtadi, J., Korotaev, V.V., de Albuquerque, V.H.C.: A reference model for internet of things middleware. IEEE Internet Things J. **5**(2), 871–883 (2018)
26. Li, X., Niu, J., Bhuiyan, M.Z.A., Fan, W., Karuppiah, M., Kumari, S.: A robust ECC-based provable secure authentication protocol with privacy preserving for industrial Internet of Things. IEEE Trans. Ind. Inf. **14**(8), 3599–3609 (2018)
27. Kim, J.Y., Hu, W., Shafagh, H., Jha, S.: SEDA: secure over-the-air code dissemination protocol for the Internet of Things. IEEE Trans. Dependable Secure Comput. (2016)
28. Kougianos, E., Mohanty, S.P., Coelho, G., Albalawi, U., Sundaravadivel, P.: Design of a high-performance system for secure image communication in the Internet of Things. IEEE Access **4**, 1222–1242 (2016)
29. Gai, K., Choo, K.-K.R., Qiu, M., Zhu, L.: Privacy-preserving content-oriented wireless communication in Internet-of-Things. IEEE Internet Things J. **5**(4), 3059–3067 (2018)
30. Granjal, J., Monteiro, E., Silva, J.S.: Security for the Internet of Things: a survey of existing protocols and open research issues. IEEE Commun. Surv. Tutorials **17**(3), 1294–1312 (2015)

31. Kamal, M.: Light-weight security and data provenance for multi-hop Internet of Things. IEEE Access **6**, 34439–34448 (2018)
32. Li, W., Song, H., Zeng, F.: Policy-based secure and trustworthy sensing for Internet of Things in smart cities. IEEE Internet Things J. **5**(2), 716–723 (2018)
33. Bhattarai, S., Wang, Y.: End-to-end trust and security for Internet of Things applications. Computer **51**(4), 20–27 (2018)
34. Mostafa, B., Benslimane, A., Saleh, M., Kassem, S., Molnar, M.: An energy-efficient multi-objective scheduling model for monitoring in Internet of Things. IEEE Internet Things J. **5** (3), 1727–1738 (2018)
35. Radisavljevic-Gajic, V., Park, S., Chasaki, D.: Vulnerabilities of control systems in Internet of Things applications. IEEE Internet Things J. **5**(2), 1023–1032 (2018)
36. Latha, P.H., Vasantha, R.: Review of existing security protocols techniques and their performance analysis in WLAN. Int. J. Emerg. Technol. Comput. Appl. Sci. (IJETCAS) **7**, 162–171 (2014)
37. Latha, P.H., Vasantha, R.: SAKGP: secure authentication key generation protocol in WLAN. Int. J. Comput. Appl. **96**(7), 25–33 (2014)
38. Latha, P.H., Vasantha, R.: MDS-WLAN: maximal data security in WLAN for resisting potential threats. Int. J. Electr. Comput. Eng. (IJECE) **5**(4), 859–868 (2015)
39. Latha, P.H., Vasantha, R.: An efficient security system in wireless local area network (WLAN) against network intrusion. In: Computer Science On-line Conference, pp. 12–19. Springer, Cham (2018)
40. De Meulenaer, G., Gosset, F., Standaert, F.X., Pereira, O.: On the energy cost of communication and cryptography in wireless sensor networks. In: 2008 IEEE International Conference on Wireless and Mobile Computing Networking and Communications, WIMOB 2008, pp. 580–585. IEEE (2008)
41. Großschädl, J., Szekely, A., Tillich, S.: The energy cost of cryptographic key establishment in wireless sensor networks. In: Proceedings of the 2nd ACM symposium on Information, Computer and Communications Security, pp. 380–382. ACM (2007)
42. Landstra, T., Zawodniok, M., Jagannathan, S.: Energy efficient hybrid key management protocol for wireless sensor networks. In: 2007 32nd IEEE Conference on Local Computer Networks, LCN 2007, pp. 1009–1016. IEEE (2007)

Intelligent Technologies and Methods of Tundra Vegetation Properties Detection Using Satellite Multispectral Imagery

Viktor F. Mochalov[1](✉), Olga V. Grigorieva[1], Viacheslav A. Zelentsov[2], Andrey V. Markov[1], and Maksim O. Ivanets[1]

[1] Mozhaisky Aerospace Academy, Saint Petersburg, Russia
vicavia@yandex.ru

[2] Saint Petersburg Institute of Informatics and Automation (SPIIRAS), Russian Academy of Science, Saint Petersburg, Russia

Abstract. The aim of the study is to develop a script and intelligent technology for collection and processing of heterogeneous data of field measurements and multispectral space imagery in order to identify and assess the state of natural objects. The core principle of the method is adaptive modeling and adjustment of an identification and evaluation technology of the properties of natural objects based on processing of multispectral space imagery, a geobotanical description and field measurements. As an example, we consider the problem of identification and evaluation of tundra vegetation using imagery from Sentinel-2 or Resource-P satellites. The results of the identification and assessment of the state of tundra vegetation are presented by a part of a vegetation map matching territories suitable for grazing deer in different seasons. The identification and assessment are focused on plant community types and possibility to feed reindeer in different seasons of the year. The testing results demonstrate effectiveness of the proposed technology for collection and processing of heterogeneous data.

Keywords: Scenario of big (heterogeneous) data processing · Multispectral remote sensing data · Identification of plant community types · Indicators of assessment · Spectral characteristics (signatures) of vegetation · State of vegetation

1 Introduction

Vegetation of tundra is represented by a wide variety of different communities. Plants over a period of short summer (several weeks) go through all phenological phases of annual development. Phenological phase, moisture content and biomass significantly affect the spectral reflective characteristics of vegetation. Identification of the species composition and assessment of the state of tundra vegetation are important practical tasks. The significance of the problem is determined by the fact that tundra vegetation in the conditions of the Extreme North changes its properties and areola of growth under the influence of global climate changes. Besides that, vegetation is exposed to

R. Silhavy (Ed.): CSOC 2019, AISC 986, pp. 234–243, 2019.
https://doi.org/10.1007/978-3-030-19813-8_24

human impact in the form of vigorous economic activity and intensive reindeer herding. The use of tundra plants as fodder for the reindeer emerged a necessity to identify several classes of pastures and to determine the possibility of their use at different times of the year taking into account the ongoing dynamic changes. Accordingly, assessment of the state of vegetation requires not only determining the total biomass stock, but also the species composition of plants in order to estimate potential resources of fodder in summer, winter and demi-season conditions.

In addition to satellite images, geobotanical descriptions and spectral reflectance factors for different types of plant communities are used as input data. Moreover, spectral reflectance can be measured in the field (in situ) or during imagery processing at test sites.

The traditional technology of multi- and hyperspectral space imagery processing involves test training sampling driven by field spectrometric measurements [1]. At present, assessment methodologies for the state of properties of natural objects, based on processing of hyperspectral data of space imagery, are improving.

Identification of plant community types by processing of multi- and hyperspectral satellite imagery can be performed using a number of familiar methods, for example [2, 3].

However, at present, multi- and hyperspectral data of the required quality can be obtained fragmentarily one or two times during the summer growing season at the selected test sites of the Extreme North. At the same time, it is very difficult to carry out synchronous spectrometric calibration measurements in the field. The systematic field measurements of the spectral reflective signatures of the North vegetation currently cover only certain types of plant communities. In this regard, studies of the state of tundra plant cover often rely only on the analysis of the vegetative index NDVI and other derived indices [4, 5]. The identification of plant community types is extremely rare.

Under the Arctic program implemented by NASA (https://arctic.noaa.gov/) in November 2018 the results of assessment of global changes in the state of tundra vegetation were presented [6]. As initial data the scholars took space imagery from various satellites since 1982. The state of Arctic vegetation was evaluated by the traditional methods of the NDVI index calculation. It was demonstrated that there is a relationship between this indicator and climatic conditions.

The results of similar studies of the state of vegetation in the Arctic are presented in the article by the Swedish scholars [7]. They show regression dependencies between biomass stock and the NDVI index. In other words, species composition of vegetation is not identified.

Under these circumstances, the purpose of this work is to present a technology that provides for identification and assessment of the state of tundra vegetation based on processing of heterogeneous data and multispectral satellite imagery. In order to achieve this goal, it is necessary to form a system of operations (a script) implementing this technology. The operations include: updating methodology and conducting a geobotanical description of test sites; collection of plant spectral signatures; building a list of indicators and pattering spectral features for identification and assessment of the state of vegetation; justifying requirements and processing of multispectral space imagery; verification of identification results and evaluation of the state of vegetation.

The technology should ensure a consistent achievement of the goal in the absence of regularly available data of the required quality from satellites and the required volume of the field spectrometric measurements.

2 Methods

The technology includes a number of operations in a cyclic mode where we can distinguish the main ones: a geobotanical description of test sites, field spectrometric measurements of vegetation, multispectral satellite data collection and processing, tundra vegetation identification and assessment, and result verification.

The technology consists of regular systematic measurements of a controlled object at large hard-to-reach area. We apply methods of adaptive modeling and processing of heterogeneous data, as a rule, in an automated mode. There are some objective constraints due to the complexity of obtaining space imagery and the field work in northern latitudes that are taken into account.

The technological scheme is shown in Fig. 1. The process is carried out accordingly a sequential execution of the specified operations.

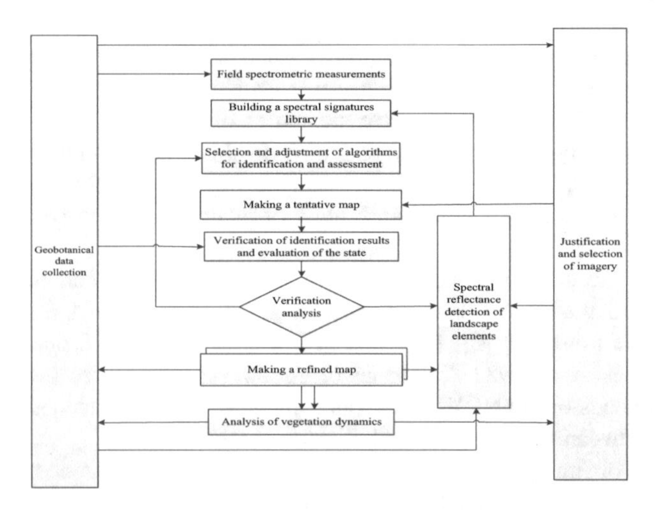

Fig. 1. The technological assessment scheme of the state of tundra vegetation.

In the studies we used the image data produced by Sentinel 2 (S2A_MSIL1C_20170710T082011_N0205_R121_T39WXR_20170710T082008). It was processed with the help of SNAP (ESA) software and sen2cor (ESA) program [8] which performed the atmospheric correction of data. As a result we acquired the spectral reflectance factors of landscape elements. Figure 2 shows a fragment of multispectral image made by Sentinel-2.

Fig. 2. A fragment of space image.

The results of geobotanical descriptions of tundra vegetation, taken during expeditions of 2010–2016, were set up as initial data. Figure 3 shows a map of the field work points.

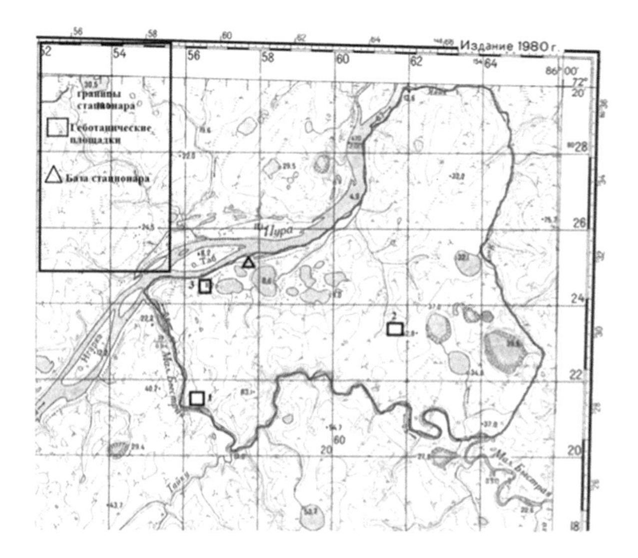

Fig. 3. The field work points.

In the course of the field work in coordination with the geobotanical description of vegetation, the spectrometric measurements of the reflective properties of landscape elements in the visible and near infrared spectral ranges are also carried out. In this case, it is necessary to take into consideration not only the seasonal variability, but also an indicatrix of reflection that depends on the angles of incidence and the angles of sighting at the moment of image creation. However, such measurements for tundra vegetation are not systematized.

The identification of the particular types of vegetation was done on the basis of the results of spectrometric measurements of tundra vegetation [18]. Table 1 shows the reflectance factors of moss, lichen and dwarf birch.

Table 1. The Reflectance factors of tundra vegetation, dimensionless quantity, (0; 1).

Wavelength, nm	Dwarf birch	Moss	Lichen
440	0,022	0,013	0,055
450	-	-	-
460	0,025	0,014	0,08
470	0,026	0,016	0,086
480	0,027	0,018	0,085
490	0,027	0,019	0,082
500	0,029	0,02	0,078
510	0,034	0,025	0,076
520	0,04	0,029	0,082
530	-	-	-
540	-	-	-
550	0,07	0,035	0,093
560	0,067	0,037	0,092
570	0,053	0,039	0,09

The information from Table 1 was used to identify vegetation types in processing of Sentinel-2 space imagery by SNAP software. The identification results of vegetation species were verified by the data of geobotanical description.

The state of the environment is estimated by such parameters as: scattering, absorption and re-reflection of light for different wavelengths of the electromagnetic spectrum of the optical range. Formally, these processes are described in radiative transfer models in different natural environments [9–12], but the practical use of these models supposes a significant number of parameters.

It is difficult to determine the values of these parameters in most cases. Therefore, at present, a method of direct assessment is more often used to evaluate parameters of the state of natural objects. The method is based on the construction of regression models. The models considered a relationship between the reflective characteristics of an object and its estimated properties [13–17].

There should be information on the composition and properties of vegetation collected at the reference points (in situ) as a part of a geobotanical description in order to build a regression model. The research at the reference points, as a rule, does not coincide in time with image collection.

The proposed combined approach assumes that field spectrometric measurements as well as geobotanical descriptions are accepted as the primary materials for the types and properties of plant communities. Measuring of the reflective properties and selected indicators of familiar plant communities and their assessment is based on image processing.

A more detailed assessment of the state of vegetation can be achieved by systematic spectrometric studies of the reflective properties of plant communities in various qualitative states.

The analysis of change dynamics in the state of vegetation is carried out on the basis of space imagery taken in different periods of time. The changes in the state of plant communities are assessed by two main characteristics: changing in boundaries and quantity (biomass, projective cover, etc.). The results of the studies are presented as an analytical map of the vegetation dynamics using geographic information system.

The originality of the technology lies in the fact of cyclic performance of the following operations: use of primary data of field spectrometric measurements for the selected species of plant communities; mask construction (selection and adjustment of algorithms) for automatic identification by multispectral space imagery the areas of certain plant community types; selection and dynamic adaptation of regression model, reflecting the dependence of estimated parameters of vegetation on its spectral reflective characteristics; a comparative analysis of the spatial boundaries of species with the data of the field geobotanical description; creation of digital maps of the main plant community types; systematic processing of space imagery aiming to get spectral reflectance factors of landscape elements considering hydrometeorological data.

3 Results

In accordance with the technology described in the previous section, some fragments of digital maps were created. They reflect the results of identification and assessment of tundra vegetation. It is assumed that the territories of mapping will constantly expand. The list of identified species and estimated indicators will also be expanded when receiving an additional baseline data. Figure 4 shows a fragment of a map reflecting the results of identification and assessment of the state of tundra vegetation.

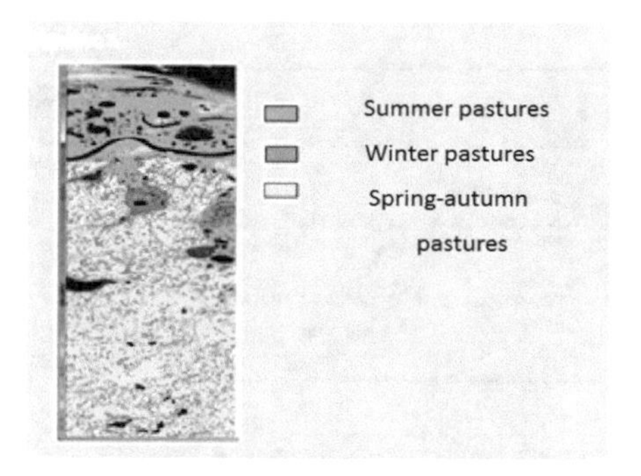

Fig. 4. A fragment of a map reflecting the results of identification and assessment of the state of tundra vegetation used as pastures for the reindeer.

The spectral signatures of some plant communities are shown in Fig. 5. According to the geobotanical data, the vegetation was assessed as a reindeer fodder in different seasons of the year.

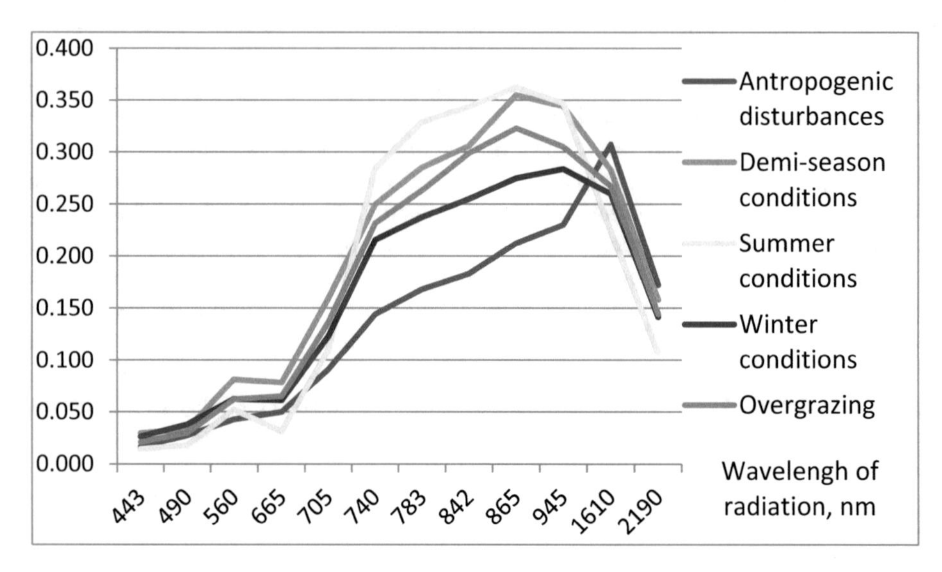

Fig. 5. Spectral signatures (Reflectance factors) of plant communities.

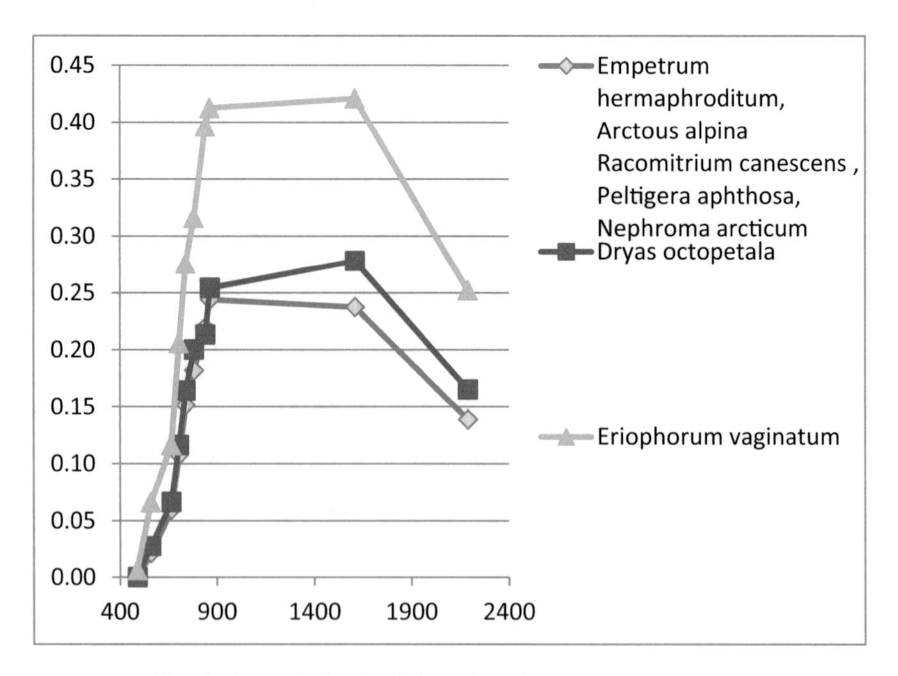

Fig. 6. Example for the filling of a reflectance factors library.

Integrated processing of multi- and hyperspectral satellite imagery for the sites with the field geobotanical descriptions provide the ability to create, adapt and regularly enrich a library of spectral signatures for certain types of plant communities (see Fig. 6) with the abovementioned limitations of the presented technology.

The verification analysis of the results shows a high accuracy in identification of the vegetation types of the North.

4 Discussions

The proposed technology, in comparison with the traditional ones, provides a gradual filling of a spectral signature library and improves the quality of identification and evaluation of tundra vegetation in the absence of a complete set of initial data of field spectrometric measurements.

In general, application of the proposed technology for production and processing of multi- and hyperspectral image data as well as creation of multispectral systems in the visible and near-infrared spectral ranges, open up prospects for the assessment of qualities of environment. For example, an assessment can touch biomass, plant species composition and their qualitative state. The multispectral equipment installed in satellite Sentinel2 and the hyperspectral equipment of satellite Resurs-P serve as sources of initial data for the survey.

However, there are some limitations. Sentinel-2 space imagery has a relatively low resolution (from 10 to 60 m in various spectral channels). In this regard, confident identification of the species composition of homogeneous vegetation is provided at sites with a conventional size of at least 180 $\times$ 180 m. In addition, data must correspond in time (of the phenological phase of plant development) with the values of plant spectral signatures available in the library.

The practical significance of the proposed technology lies in the fact that it will ensure the annual planning of reindeer herding, provide data on the impact of changing climatic conditions on the state of tundra vegetation and timely identify possible adverse effects of anthropogenic loading.

The technology can be applied not only in Russia, but also in other regions of the world with inaccessible terrain (steppes, deserts, emergency zones, conflicts).

In the further research it is planned to substantiate the requirements for the volume of field spectrometric measurements and the amount of points to obtain the estimated parameters with a given accuracy.

Acknowledgements. The research described in this paper is partially supported by the Russian Foundation for Basic Research (grants 16-08-00510, 17-06-00108, state research 0073–2018–0003, International project ERASMUS+, Capacity building in higher education, №. 73751-EPP-1-2016-1-DE-EPPKA2-CBHE-JP, Innovative teaching and learning strategies in open modelling and simulation environment for student-centered engineering education.

References

1. Millard, K., Richardson, M.: On the importance of training data sample selection in random forest image classification: a case study in peatland ecosystem mapping. Remote Sens. **7**, 8489–8515 (2015)
2. Grigoreva, O., Mochalov, V., Zelentsov, V.: Hyperspectral data processing and adaptive modelling for the natural objects properties detection. In: The 6th International Workshop on Simulation for Energy, Sustainable Development & Environment, pp. 7–14 (2018)
3. Manolakis, D., Marden, D., Shaw, G.: Hyperspectral image processing for automatic target detection applications. Lincoln Lab. J. **14**, 79–116 (2003). https://pdfs.semanticscholar.org/00b4/2a7649ac10328fef1d45223484bf9653d995.pdf
4. Elsakov, V.: Spatial and interannual heterogeneity of changes in the vegetation cover of Eurasian tundra: analysis of 2000–2016 MODIS data. Issledovanie Zemli iz kosmosa **14**(6), 56–72 (2017). http://d33.infospace.ru/d33_conf/sb2017t6/56-72.pdf
5. Lavrinenko, I.: Map of technogenic disturbance of Nenets Autonomous District. Issledovanie Zemli iz kosmosa **15**(2), 128–136 (2018). http://d33.infospace.ru/d33_conf/sb2018t2/128-136.pdf
6. Epstein, H., Bhatt, U., et al.: Tundra-Greenness (2018). https://arctic.noaa.gov/Report-Card/Report-Card-2018/ArtMID/7878/ArticleID/777/Tundra-Greenness
7. Becher, M., Olofsson, J., Berglund, L., Klaminder, J.: Decreased cryogenic disturbance: one of the potential mechanisms behind the vegetation change in the Arctic (2018). https://link.springer.com/article/10.1007/s00300-017-2173-5
8. Louis, J., Debaecker, V., Pflug, B., Main-Knorn, M., Bieniarz, J., Mueller-Wilm, U., Eadau, E., Gascon, F.: SENTINEL-2 SEN2COR: L2A processor for users (2018)
9. Brovkina, O., Novotnya, J., Cienciala, E., Zemeka, F., Russ, R.: Mapping forest aboveground biomass using airborne hyperspectral and LiDAR data in the mountainous conditions of Central Europe. Ecol. Eng. **100**, 219–230 (2017). www.elsevier.com/locate/ecoleng
10. Hagan, M., Demuth, H., Beale, M.: Orlando De Jesus Neural Network Design, 2nd edn, 1012 p. PWS Publishing, Boston (1996). ISBN-10:0-9717321-1-6, ISBN-13:978-0-9717321-1-7
11. Kay, S., Hedley, J., Lavender, S.: Sun glint correction of high and low spatial resolution images of aquatic scenes: a review of methods for visible and near-infrared wavelengths. Remote Sens. **1**(4), 697–730 (2009)
12. Grigorieva, O., Markov, A., Zhukov, D., Mochalov, V., Nikolenko, A.: Possibility of use visible and near infrared multispectral and hyperspectral sensors for the bottom classification of shallow seas. Trudy Mozhaisky Aerosp. Acad. **653**, 111–116 (2016)
13. Hagan, M., Menhaj, M.: Training feedforward network with the Marquardt algorithm. IEEE Trans. Neural Netw. **5**(6), 989–993 (1994)
14. Lee, Z., Carder, K., Hawes, S., Steward, R., Peacock, T., Davis, C.: A model for interpretation of hyperspectral remote-sensing reflectance. Appl. Opt. **33**, 5721–5732 (1994)
15. Grigorieva, O., Zhukov, D., Markov, A., Mochalov, V.: The assessment of the coastal waters. Optika atmosery i okeana **29**(7), 1–7 (2016)
16. Zelentsov, V., Potryasaev, S., Pimanov, I., Mochalov, V.: Software suite for creating downstream applications and thematic services on the base of remote sensing data processing and integrated modelling. In: Proceedings of the International Geoscience and Remote Sensing Systems Symposium (IGARSS), Valencia, Spain, pp. 3477–3480 (2018)

17. Zelentsov, V., Potriasaev, S.: Architecture and examples of implementing the informational platform for creation and provision of thematic services using earth remote sensing data. SPIIRAS Proc. **6**(55), 86–113 (2017)
18. Krinov, E.: Spektralnaya otrazatelnaya sposobnost priridnyh obrazovaniy, Moscow, pp. 122–185 (1947)

Sacrifice as Paradigm of Robot Behavior in Group

Viacheslav Abrosimov(✉) and Alexander Mochalkin

Software Engineering Company "Smart Solutions", Samara, Russia
avk787@gmail.com, mochalkin@smartsolutions-123.ru

Abstract. When a group of robots fulfills collective tasks in a conflict environment, it is often necessary to sacrifice one or several robots for the efficient functioning of the group. Sacrifice choice is not a trivial problem. This paper develops a sacrifice role allocation algorithm for such situations based on robots' ranks. We demonstrate that the crucial role here is played by the skill in sacrifice acquired by an robot, its value for the group and the availability of the resources and functionality required for executing the sacrifice role. The efficiency of the suggested approach is illustrated by allocating the sacrifice role in the group penetration through a defense system (the "diffuse bomb" problem).

Keywords: Robot · Group · Role · Sacrifice · Rank · Choice · Fuzzy numbers

1 Introduction

In theory of multirobot systems [1] a basic property of robots concerns their rationality. The rational behavior of robots is connected with situation assessment, decision-making and optimal actions for best result. At the same time, in practice rationality has a negative effect due to its predictability. Really, if side A assumes the rational behavior of side B in a conflict situation, then the former may construct behavioral models for the opponent, simulating and predicting the actions of side B as a rational player. This aspect introduces uncertainty in decision-making in applications.

Theory of multi-robot systems has not thoroughly explored the property of "sacrifice" so far. It arises in joint problem solving by a group of robots: for the sake of a collective task, a natural or ad hoc critical situation (problem) must be solved at the cost of terminating the robot's active functioning. These situations occur while certain tasks are carried out in an unfriendly (antagonistic) environment, e.g., motion within the coverage of an opponent's aerospace defense system, territory surveillance operations in the conditions of radioactive contamination, etc. In such situations, a group has to find or select an robot being able to solve the problem and expressing intention to do it. In other words, an appropriate robot is willing to become a sacrifice, even at the cost of actions inconsistent with its further functioning (i.e., destruction).

As an example, consider group penetration through a defense system [2] for a group of controlled objects. Here the defense system creates an active conflict environment for the attacking objects. As a result, there appear certain zones in the space-time domain where the collective task becomes difficult (or even impossible) to fulfill.

R. Silhavy (Ed.): CSOC 2019, AISC 986, pp. 244–254, 2019.
https://doi.org/10.1007/978-3-030-19813-8_25

These zones are unknown to the attacking objects and the latter may even predict the zones based on the rational behavior of the defense system. A common approach is to construct the optimal avoidance trajectories of the attacking objects for such potentially dangerous zones. However, an alternative solution is to send intentionally one or several attacking objects to the unknown zones in order to reveal the capabilities of the defense system, neutralize its impact and create "safety windows." This approach poses the nontrivial problem of sacrifice choice in a group of robots. Of course, the examples of Japanese kamikaze pilots from World War II are inhuman today. But this problem makes sense for the class of intelligent unmanned aerial vehicles (drones) in the group attack design using rather cheap dummy elements, which are intensively developing now. An especially promising line of research is to study the sacrifice capabilities of robots under the swarm- or flock-type structure of the group. Similar solutions can be used on ground for the group motion of robots in a conflict environment.

In this paper we demonstrate that sacrifice choice is not a trivial problem. We show the crucial role of sacrifice skills. We try to develop a sacrifice role allocation algorithm for such situations based on robots' ranks. And we give the example which illustrate the efficiency of such algorithm for the group penetration through a defense system.

2 Related Works

As a matter of fact, investigators paid limited attention to sacrifice analysis in technical systems. Among related research, consider the works studying different aspects of the robot's emotions in the case of conflicts [3–5] and roles exchange between the objects [6]. So Hill et al. [7] emphasized that cooperative behavior always requires some altruism from the robots. The notion of "categorical altruism" was introduced to describe the robots always preferring altruism, even if their assistance to other robots gains no benefits. On the contrary, an robot demonstrating "conditional altruism" fully implements its interests rather than the interests of other robots. The notion of "situational altruism" proposed in [8] describes the robots balancing between the collective and individual interests. The cited report directly related altruism to negotiations among the robots and their mutual compromises depending on possible conflicts and the available knowledge about the environment. In actions' assessment, a characteristic event is when an robot expresses willingness to perform other tasks, but seeks for obtaining some gains [9–11]. In this case, it is necessary to estimate the robot's losses incurred by different decisions taking into account its individual profit [10]. And in [11] proposed a new class of negotiations among the robots based on risk assessment; an robot expresses willingness to undergo individual losses for collective task fulfillment. This eliminates from negotiations all low-efficiency alternatives, either for an robot or for the whole group.

Sacrifice always calls for reassigning the roles of the robots within the group. Epy article [6] organized the exchange of N roles among N robots by the "one-to-one" principle. The numerical characteristics of Role Exchange Value (REV) and Individual Utility Gain (IUG) were introduced and then involved in an original role exchange algorithm: the main idea is to divide the group of robots into pairs so that role exchange within them guarantees the maximum utility.

The well-known architectures of multi-robot systems (BDI, EBDI and others) do not assist in choosing an appropriate collective behavior strategy in a critical situation for the robots moving in an unfriendly environment. The notion of robot's sacrifice is not directly defined and its measure with respect to the whole group and other robots is not introduced. These observations are also confirmed by the conclusions of the survey [12] covering over 400 papers of the international conferences on robot-oriented technologies and artificial intelligence. Particularly, the authors identified 18 functional requirements (properties) applied to the robots and 11 types of the nonfunctional ones. The property of sacrifice was not mentioned.

Summarizing the outcomes, we may conclude the following: (a) sacrifice as the property of robots is almost not investigated; (b) sacrifice is considered mostly in terms of social but not technical processes; (c) different architectures of multirobot systems (BDI, EBDI) do not yield an appropriate behavioral strategy for a group facing a problem in a conflict environment.

3 Methods

Definition of Essential Attributes for Sacrifice. Assume that a group of robots faces problem $s \in S$. Solution of this problem is possible under definite requirements. An robot allocated to the sacrifice role must satisfy them. A typical robot that fulfills a collective task within a group can be described by the following essential attributes for executing the sacrifice role.

The robot's skill in sacrifice. The paper [13] introduced and further studied the notion of robot's sacrifice. In particular, the willingness (readiness) of robot $i \in N$ to perform "self-sacrifice" was defined as the fuzzy number

$$\mu_i^s = \mu_S(s_i) = \cup_{\alpha+\beta+\gamma}\{\mu_A(\alpha) \cap \mu_B(\beta) \cap \mu_C(\gamma)\}, \tag{1}$$

where

$\mu_A(\alpha) \in [0, 1]$ reflects the robot's commitment to solve problem s;
$\mu_B(\beta) \in [0, 1]$ describes the robot's belief that its capabilities are enough to solve problem s;
$\mu_C(\gamma) \in [0, 1]$ corresponds to the robot's intention to solve problem s.

Here $\mu_A(\alpha), \mu_B(\beta), \mu_C(\gamma)$ are the membership functions of the fuzzy numbers associated with these properties.

A current situation is characterized by the necessary and sufficient conditions of sacrifice: robot i has an appropriate sacrifice paradigm ($\mu_i^s \in [0, 1]$) acquired by learning, as well as resources R_i^s and functionality F_i^s that describes its capabilities to solve problem s.

The robot can be learned to one of the following paradigms:

- the self-sacrifice paradigm (these are robots-"altruists" described by $\mu_i^s = 1$ that sacrifice themselves in any situations and renounce the individual tasks);

- the selfish paradigm (these are robots-"egoists" described by $\mu_i^s = 0$ that always escape sacrifice and continue their tasks within a collective task);
- the sacrifice on demand paradigm (these are robots-"pragmatists" described by $0 < \mu_i^s < 1_i^s$ that sacrifice themselves only under some conditions).

We emphasize, however, that it is important to avoid analogies between the sacrificial behavior of animate beings and technical objects. The sacrifice of controlled objects represented as robots is just a property embedded in their control systems, as a rule, by learning in advance. This property is implemented as the necessary behavior of an robot in an appropriate situation.

1. The robot's value. An robot as a technical element has some value, e.g., production cost. Let $\mu_i^v = \mu_V(v_i) \in [0, 1]$ denote the membership function of robot i to the set V of valuable robots. Obviously, it is unreasonable to sacrifice a valuable robot and hence $\min_N \mu_i^v$ must be achieved by sacrifice choice.
2. The robot's contribution to collective task. Each robot i makes some contribution $w_i(t)$ to the fulfillment of a collective task [13]. For each time moment, the collective objective function forms the algebraic sum of the contributions made by N robots, i.e., $W^\Sigma(t) = \sum_{i=1}^N w_i(t)$. If an robot performs self-sacrifice, some of its tasks within a collective task are not fulfilled. We will associate the loss of an robot with its missed contribution $w_i^-(t)$ to $W^\Sigma(t)$. In practice, such contribution can be described by a linguistic or fuzzy variable for convenience. In the sequel, $\mu_i^w = \mu_W(w_i)$ denotes the membership function of the missed contribution.

The Robot's Ability to Execute the Sacrifice Role. This ability depends on two factors, namely, (a) the resources available to the robot for reaching a dangerous zone and (b) the robot's capabilities for fulfilling a task in the dangerous zone, i.e., the required functionality (e.g., electronic warfare means to neutralize an air defense system). For robot i, we will define the availability of resources by the membership function $\mu_i^r = \mu_R(r_i)$ with respect to the sufficient resources set R and the robot's capabilities by the membership function $\mu_i^p = \mu_F(p_i)$ with respect to the set P of the required functionality for solving problem s.

Therefore, the essential attributes of robot i for the sacrifice role make up the aggregate of the membership functions

$$A := < \mu_i^s, \mu_i^v, \mu_i^r, \mu_i^p, \mu_i^w > \quad (2)$$

that describe its skill in sacrifice, value for the group, resources for the sacrifice role, necessary functionality and missed contribution to the collective task.

All elements of this aggregate are fuzzy numbers with corresponding membership functions [14].

Next, define the capability of the conflict environment to counteract the functioning of i-robot in a zone H as a fuzzy variable with the membership function p_i^H (here $p_i^H = 1$ means the destruction of robot i in the zone H). If the trajectories of $Q < N$ controlled objects pass through the domain H and these objects appear unable to fulfill their tasks, then the losses of the fleet constitute $W^q(t^+) = \sum_{q=1}^{Q} w_q(t^+)$ (in terms of the objective function) and $C^q(t^+) = \sum_{q=1}^{Q} \mu_q^v$ (in terms of the value of the lost objects). In this case, it seems natural to achieve the following effect from sacrifice:

$$E : \min_{Q} W^q(t^+) \text{ and } \min_{Q} C^q(t^+). \tag{3}$$

Ranking of Controlled Objects by Their Priority for Sacrifice Role as Methodology. Clearly, in a group of heterogeneous robots all members have different contributions to the collective objective function, different skills in sacrifice, different values for the group, as well as different problem-solving capabilities in a dangerous zone. To choose an robot-sacrifice, it is necessary to compare all robots in terms of criteria (2). The problem can be stated as follows: given the set of robots G, find an robot $g \in G$ having the highest adequacy to the sacrifice requirements and most fitting this act. In terms of decision-making, it is necessary to develop a system of preferences and find the nondominated alternative g^{nd} among the existing alternative (the set G).

Here a reasonable approach is to involve fuzzy preference relations [14] as the initial data and notions (value, sacrifice, etc.) incorporate natural fuzziness. Let PF^s define a fuzzy preference relation on the set of robots G. Then for any pair of robots $(g^x, g^y) \in PF^s$ the dominance degree of robot g^x over robot g^y $(g^x \succ g^y)$ in terms of a criterion f is described by the membership function $\mu_f(g^x, g^y)$ of the form

$$\mu_f(g^x, g^y) = \begin{array}{l} \{\mu_f(g^x, g^y) - \mu_f(g^y, g^x)\}, \text{if } \mu_f(g^x, g^y) \succcurlyeq \mu_f(g^y, g^x), \\ 0, \text{ if } \mu_f(g^x, g^y) \prec \mu_f(g^y, g^x). \end{array} \tag{4}$$

The following associations can be easily established in our case. The robot's membership function in terms of each criterion f reflects how much this robot corresponds to the sacrifice role. Hence, the fuzzy preference relations in the form of the differences

$$\mu_f(g^x, g^y) = \mu_F(f_x) - \mu_F(f_y) \tag{5}$$

where $F := \{S, V, R, P, W\}$ and accordingly $f := \{s, v, r, p, w\}$, assess the dominance degree of robot g^x over robot g^y in terms of each criterion f.

The main problem of sacrifice role allocation is that an appropriate candidate must satisfy several requirements simultaneously. This feature makes sacrifice choice a multicriteria problem.

Let us accept the following hypothesis on the preferability of the robots. Robot g^x is preferable to robot g^y in terms of sacrifice allocation under the four conditions below:

- the contribution of robot g^x to the collective objective function is smaller than that of robot g^y;
- the willingness of robot g^x to perform self-sacrifice is higher than that of robot g^y;
- robot g^x has more resources for tasks fulfillment than robot g^y;
- the ability of robot g^x to execute the sacrifice role is higher than that of robot g^y;
- the value of robot g^x for the group is smaller than that of robot g^y.

Note that these conditions must be satisfied all together. The integral preferability condition has the form

$$\pi(g^x, g^y) : \left\{ \left[\mu_x^w \le \mu_y^w\right] \cap \left[\mu_x^s \ge \mu_y^s\right] \cap \left[\mu_x^r \ge \mu_y^r\right] \cap \left[\mu_x^p \ge \mu_y^p\right] \cap \left[\mu_x^v \le \mu_y^v\right] \right\} \quad (6)$$

According to the preferability condition (7), the most important thing is that one characteristic exceeds the other, while differences (5) of the membership functions make less sense. In such conditions it is useful to adopt ranks for preference relations.

Consider ranks for separate attributes of an robot. To this effect, compare the characteristics of the robots over all attributes and rank the robots by different components of condition (6). This procedure yields the following table of ranks.

Table 1. The ranks of robots by different attributes

	Robots				
F	1	2	3	j	N
W	μ_1^w	μ_2^w	μ_3^w	μ_j^w	μ_N^w
Rank W	E_{1W}	E_{2W}	E_{3W}	E_{jW}	E_{NW}
S	μ_1^s	μ_2^s	μ_3^s	μ_j^s	μ_N^s
Rank S	E_{1S}	E_{2S}	E_{3S}	E_{jS}	E_{NS}
R	μ_1^r	μ_2^r	μ_3^r	μ_j^r	μ_N^r
Rank R	E_{1R}	E_{2R}	E_{3R}	E_{jR}	E_{NR}
P	μ_1^p	μ_2^p	μ_3^p	μ_j^p	μ_N^p
Rank P	E_{1P}	E_{2P}	E_{3P}	E_{jP}	E_{NP}
V	μ_1^v	μ_2^v	μ_3^v	μ_j^v	μ_N^v
Rank V	E_{1V}	E_{2V}	E_{3V}	E_{jV}	E_{NV}
Total rank	G_1	G_2	G_3	G_j	G_N

The total rank G_i over a column defines the priority of an robot in terms of all attributes (criteria). The specifics of functioning in a conflict environment often require introducing priorities ξ_j for different attributes. Here numerous cases are possible. For instance, the group may value a certain robot and then higher priorities are assigned to the attributes of the robot's value V and its contribution W to the collective objective function. However, if sacrifice is unavoidable, higher priorities are given to the available resources R and the functionality P for the more efficient solution of the collective task. Therefore, in practice the robot having the highest willingness to perform self-sacrifice is not surely the chosen sacrifice; this robot may have insufficient resources or low functionality for problem solving.

In all cases, the robot representing the best candidate for the sacrifice role is defined by the convolution of several criteria, i.e.,

$$A_s = \min_N G_n = \min_N \left\{ \sum\nolimits_F \left(\xi_f * E_{Nf} \right) \right\}. \tag{7}$$

4 Results

Let us set the task to find a solution for sacrifice choice in group penetration through a defense system problem. Let us consider a group of five heterogeneous robots that jointly help to solve the problem of penetration any active objects through a defense system [2]. Their common approach is to make the free areas for the movement of attack by means of their own victims. Let us use for this purpose unmanned aerial vehicles (drones) that are learned in an appropriate paradigm and equipped with necessary equipment. The presence of such zones substantially simplifies the problem constructing the optimal trajectories of the attacking objects.

In our assumptions the drones have different characteristics and skill in sacrifice. We will study three scenarios. In scenario A, the sacrifice is chosen without any priorities to the attributes of the robots. In scenario B, the group needs sacrifice and high priorities are assigned to the characteristics of the resources availability and the functionality of the robots. In scenario C, the group allocates the sacrifice with the minimum value (e.g., cost) and losses in the collective objective function.

The group of robots has the following crisp distribution of the roles (Table 2).

Table 2. The attributes and ranks of drones belonging to a heterogeneous group

The attributes of drones	Notation	The priorities of attributes			The values of attributes the ranks of drones in scenarios A-B-C				
		ξ_f			Drone 1	Drone 2	Drone 3	Drone 4	Drone 5
		A	B	C	The roles of drones in the group				
					False dummy	Intelligence drone	Transport drone	Combat drone	Multipurpose drone
Skill in sacrifice	μ_i^s	1	3	5	1	0.5	0.8	0	0.3
					1, 3, 5	3, 9, 15	2, 6, 10	5, 15, 25	4, 12, 20
Resource availability	μ_i^r	1	2	4	0.6	0.7	0.9	0.4	0.6
					3, 6, 12	2, 4, 8	1, 2, 4	4, 8, 16	3, 6, 12
Functionality	μ_i^p	1	1	3	0.5	0.8	0.6	0.5	0.9
					4, 4, 12	2, 2, 6	3, 3, 9	4, 4, 12	1, 1, 9
Value for the group	μ_i^v	1	5	1	0.7	0.8	0.4	1	1
					2, 10, 2	3, 15, 3	1, 5, 1	4, 20, 4	4, 20, 4
Contribution to the collective objective function	μ_i^w	1	4	2	0.6	0.7	0.3	0.9	0.1
					3, 12, 6	4, 16, 8	2, 8, 4	5, 20, 10	1, 4, 2
Total rank (A-B-C)					13, 35, 37	16, 46, 40	9, 24, 28	22, 67, 67	13, 43, 47

(*continued*)

Table 2. (*continued*)

The attributes of drones	Notation	The priorities of attributes			The values of attributes the ranks of drones in scenarios A-B-C				
		ξ_f			Drone 1	Drone 2	Drone 3	Drone 4	Drone 5
		A	B	C	The roles of drones in the group				
					False dummy	Intelligence drone	Transport drone	Combat drone	Multipurpose drone
Ranks of candidates for sacrifice	A. Without priorities to attributes				2	3	1	4	2
	B. Priority to resources availability and functionality				2	4	1	5	3
	C. With priority to the minimum value and losses in the collective objective function				2	3	1	5	4

Now, consider a group of homogeneous robots with role allocation based on negotiations among them. Such a situation occurs when robots jointly monitor vast territories, exchanging information with each other and redistributing all tasks not performed by an robot among the others [15]. Suppose that the robots have equal value for the group and modify the priorities of attributes, making the collective task most significant.

The initial data and results are presented by Table 3.

Table 3. The attributes and ranks of drones belonging to a homogeneous group

The attributes of drones	Notation	The priorities of attributes			The values of attributes the ranks of drones in scenarios A-B-C				
		ξ_f			Drone 1	Drone 2	Drone 3	Drone 4	Drone 5
		A	B	C	The roles of drones in the group				
					Reconnaissance drone	Reconnaissance drone	Reconnaissance drone	Reconnaissance drone	Reconnaissance drone
Skill in sacrifice	μ_i^s	1	1	3	0.1	0.3	0.6	0.5	0.8
					5, 5, 15	4, 4, 12	2, 2, 6	3, 3, 9	1, 1, 3
Resource availability	μ_i^r	1	2	2	0.4	0.8	0.9	0.6	0.7
					5, 10, 10	2, 4, 4	1, 2, 2	4, 8, 8	3, 6, 6
Functionality	μ_i^p	1	3	1	0.3	0.8	0.6	0.7	0.8
					4, 12, 4	1, 3, 1	3, 9, 3	2, 6, 2	1, 3, 1
Value for the group	μ_i^v	1	3	4	0.7	0.7	0.7	0.7	0.7
					1, 3, 4	1, 3, 4	1, 3, 4	1, 3, 4	1, 3, 4
Contribution to the collective objective function	μ_i^w	1	3	4	0.3	0.9	0.8	0.1	0.7
					2, 6, 8	5, 15, 20	4, 12, 16	1, 3, 4	3, 9, 16
Total rank (A-B-C)					17, 36, 41	13, 29, 41	11, 28, 31	11, 23, 27	9, 22, 29

(*continued*)

Table 3. (*continued*)

The attributes of drones	Notation	The priorities of attributes			The values of attributes the ranks of drones in scenarios A-B-C				
		ξ_f			Drone 1	Drone 2	Drone 3	Drone 4	Drone 5
		A	B	C	The roles of drones in the group				
					Reconnaissance drone	Reconnaissance drone	Reconnaissance drone	Reconnaissance drone	Reconnaissance drone
Ranks of candidates for sacrifice	A. Without priorities to attributes				4	3	2	2	1
	B. Priority to resources availability and functionality				5	4	3	2	1
	C. Without to the minimum value and losses in the collective function				4	4	3	1	2

4.1 Discussions

Let us analyze the results.

In all scenarios (see Table 1) Drone 3 is chosen as sacrifice, i.e., the robot with rather high skill in sacrifice and comparatively high characteristics of the attributes required for executing the sacrifice role (resources availability and functionality). In addition, this drone has the smallest value for the group among the other drones and the corresponding losses in the collective objective function are minor. Despite the highest skill in sacrifice, Drone 1 is rather valuable for the group and also has modest resources and functionality for executing the sacrifice role. In scenario A, it shares rank 2 with Drone 5 that has almost fulfilled its part of the collective task (the contribution to the collective objective function is minimal). Quite expectedly, in all scenarios Drone 4 (the combat drone) is not chosen as sacrifice due to its high value for the group.

The situation slightly changes as we introduce priorities for different attributes: ranks 1 and 2 are still assigned to Drones 3 and 1, respectively.

Drone 5 is chosen as sacrifice in the group of homogeneous robots without initial role allocation (Table 2). It has the maximum skill in sacrifice in combination with considerable resources and functionality. Rank 2 is assigned to Drone 4 due to the smallest contribution to the collective objective function. However, the ranks may change as we introduce priorities for attributes (scenarios B and C). In scenario B the ranks are redistributed for Drones 1, 2, and 3, which can be explained by the high priority of skill in sacrifice and, also, by the impact of the resources availability and functionality (the lowest rank is given to Drone 5 having the smallest functionality). In scenario C the skill in sacrifice is not a crucial factor, as Drone 5 receives rank 2.

5 Conclusion

In this paper we have demonstrated that sacrifice choice is not a trivial problem We have concluded the following: (a) sacrifice as the property of robots is almost not investigated; (b) sacrifice is considered mostly in terms of social but not technical processes; (c) different architectures of multirobot systems do not yield an appropriate behavioral strategy for a group facing a problem in a conflict environment.

So we have developed the sacrifice role allocation algorithm for situations based on robots' ranks. We say that the crucial role here is played by the skill in sacrifice acquired by robot, its value for the group and the availability of the resources and functionality required for executing the sacrifice role.

The efficiency of the suggested approach is illustrated by allocating the sacrifice role in the group penetration through a defense system (the "diffuse bomb" problem).

References

1. Weiss, G. (ed.): Multirobot Systems, p. 920. MIT Press, Cambridge (2013)
2. Korepanov, V.O., Novikov, D.A.: The diffuse bomb problem. Autom. Remote Control **74** (5), 863–874 (2013)
3. Jiang, H., Vidal, J.M., Huhns, M.N.: EBDI. An architecture for emotional robot. In: Proceedings of the AAMAS 2007, Honolulu, Hawai'i, USA (2007). http://jmvidalcse.sc.edu/papers/jiang07a.pdf/
4. Czaplicka, A., Chmieland, A., Hołyst, J.A.: Emotional robots at the square lattice. In: Proceedings of the 4th Polish Symposium on Econo- and Sociophysics, Rzeszów, Poland, vol. 117, no. 4, pp. 693–694 (2010)
5. Pereira, D., Oliveira, E., Moreira, N., Sarmento, L.: Towards an architecture for emotional BDI robots. In: Proceedings of 12th Portuguese Conference on Artificial Intelligence, pp. 40–46. Universidade da Beira Interior (2005)
6. Zhang, X., Hexmoor, H.: Utility-based role exchange. From theory to practice in multi-robot systems. In: 2nd International Workshop of Central and Eastern Europe on Multi-Robot Systems, CEEMAS 2001, Cracow, Poland (2001)
7. Hill, J.C., Johnson, F.R., Archibald, J.K., Frost, R.L., Stirling, W.C.: A cooperative multi-robot approach to free flight. In: Proceedings of AAMAS, pp. 1083–1090 (2005)
8. Hill, J.C., Archibald, J.K., Stirling, W.C., Frost, R.L.: A multi-robot system architecture for distributed air traffic control, pp. 1–11. Brigham Young University, Provo, Utah 84602-4099, USA (2005)
9. Azhana, A., Moamin, A., Mohd, Z., Mohd, Y., Mohd, S.A., Aida, M.: Resolving conflicts between personal and normative goal. In: Proceedings of 7th International Conference on EEEE. Normative Robot Systems Information Technology in Asia, CITA 2011, pp. 1–6 (2011)
10. Buchholz, W., Peters, W.: Equal sacrifice and fair burden-sharing in a public good economy. University of Regensburg (2007). http://www.wiwi.europa-uni.de/de/lehrstuhl/fine/fiwi/team/peters/BuchholzPeters_2007b.pdf
11. Zivan, R, Grubshteiny, A., Friedman, M., Meiselsy, A.: Partial cooperation in multi-robot search (extended abstract). In: Proceedings of the 11th International Conference on Autonomous Robots and Multirobot Systems (AAMAS). Ben Gurion University of the Negev, Beer-Sheva, Israel (2012)

12. Vafadar, S., Barfourosh, A.A.: Requirements engineering in robot-oriented software engineering. Amirkabir University of Technology (Tehran Polytechnic), no. 424, Hafez Ave, Tehran, Iran (2008)
13. Abrosimov, V.: The property of robots sacrifice: definition, measure, effect and applications. Int. J. Reasoning-Based Intell. Syst. **8**(1/2), 76–83 (2016)
14. Piegat, A.: Fuzzy Modeling and Control. Physica-Verlag Heidelberg, New York (2011)
15. Abrosimov, V.: Group control strategy for a fleet of intelligent vehicles-robot performing monitoring. In: Jezic, G., Howlett, R.J., Jain, L.C. (eds.) Proceedings of 9th KES International Conference "Robot and Multi-Robot Systems: Technologies and Applications" (KES-AMSTA 2015), Sorrento, Italy, June 2015, vol. 38, pp. 135–144 (2015)

Value Sets of Ellipsoidal Polynomial Families with Affine Linear Uncertainty Structure

Radek Matušů(✉)

Centre for Security, Information and Advanced Technologies (CEBIA–Tech),
Faculty of Applied Informatics, Tomas Bata University in Zlín,
nám. T. G. Masaryka 5555, 760 01 Zlín, Czech Republic
rmatusu@utb.cz

Abstract. The contribution focuses on the value sets of the ellipsoidal polynomial families with affine linear uncertainty structure. First, it recalls the fundamental terms from the area of robustness under parametric uncertainty, such as uncertainty structure, uncertainty bounding set, family, and value set, with emphasis to the ellipsoidal polynomial families. Then, the illustrative example is elaborated, in which the value sets of the ellipsoidal polynomial family with affine linear uncertainty structure are plotted, including randomly chosen internal points, and compared with the value sets of the classical "box" version of the polynomial family.

Keywords: Value set · Ellipsoidal uncertainty · Spherical uncertainty · Family of polynomials · Affine linear uncertainty

1 Introduction

Parametric uncertainty represents the important, effective, but also relatively simple tool for mathematical description of real-life systems with potentially complex behavior (including nonlinearities, fast dynamics or changes in physical parameters) by means of Linear Time-Invariant (LTI) models.

The systems with parametric uncertainty are frequently given by so-called families of plants or polynomials (e.g. the family of closed-loop characteristic polynomials). Thus, these families and their robustness have been common objects of research interest for several decades [1, 2]. Such family is determined by two main factors – the structure of uncertainty, and the uncertainty bounding set.

The possible structures of uncertainty include (with increasing generality) the independent uncertainty structure, the affine linear uncertainty structure, the multilinear uncertainty structure, and the non-linear uncertainty structure (e.g. the polynomial or completely general uncertainty structures). Furthermore, the special cases are represented e.g. by the single parameter uncertainty or by the uncertain quasipolynomials [1, 3, 4]. This contribution deals with the case of the affine linear uncertainty structure.

The uncertainty bounding set is supposed to be defined by a ball in some appropriate norm [1]. The far most common case utilizes L_∞ norm (a box). The other possible approaches use either L_2 norm (a sphere or an ellipsoid) or L_1 norm (a diamond). This contribution focuses on the uncertainty bounding set given by using L_2

R. Silhavy (Ed.): CSOC 2019, AISC 986, pp. 255–263, 2019.
https://doi.org/10.1007/978-3-030-19813-8_26

norm. The corresponding families of polynomials can be called as the ellipsoidal (or spherical) polynomial families. Strictly speaking, some works, e.g. [1], use the term "spherical polynomial family" solely for a family with independent uncertainty structure and uncertainty bounding set in the shape of an ellipsoid. Thus, it is considered as the analogy to the standard interval polynomial. However, the term "ellipsoidal (or spherical) polynomial family" is used more generally, that is for any uncertainty structure, in this contribution. Despite the fact that the spherical/ellipsoidal uncertainty is not studied so often than the classical "box" uncertainty, there are still several works dealing with the spherical uncertainty and related problems [6–14], including [15] that brought the result applicable to affine linear uncertainty structure. Some more recent results on systems with ellipsoidal uncertainty was published e.g. in [16–18].

Since the most crucial property of all control loops is their stability, the common object of interest for control systems under uncertainty is their robust stability, which means the closed-loop systems are requested to remain stable for all possible members from the assumed family.

There are many tools for robust stability analysis of Single-Input Single-Output (SISO) systems available [1, 2]. The choice of a suitable one depends mainly on the uncertainty structure and the uncertainty bounding set [1, 3, 4]. A universal method is based on the value set concept [1]. In combination with the zero exclusion condition, it represents the efficient approach that is applicable to many cases, and that can be especially useful for complex uncertainty structures with the lack of alternative techniques for robust stability analysis [4, 5].

The contribution intends to present the value sets of the ellipsoidal polynomial families with affine linear uncertainty structure. Some previous results have been already presented e.g. in [6, 7], but these works were focused only on the families with the independent uncertainty structure, i.e. the "spherical version" of the interval polynomials, and they used the Polynomial Toolbox for Matlab for plotting the value sets [8]. This contribution, however, deals with a bit more complicated case represented by the ellipsoidal polynomial families with affine linear uncertainty structure.

2 Uncertainty Structure

As it has been already mentioned in the introductory section, the families of systems with parametric uncertainty are defined by the combination of their structure (especially the uncertainty structure) and the uncertainty bounding set. In this contribution, the following continuous-time uncertain polynomial is supposed:

$$p(s,q) = \sum_{i=0}^{n} \rho_i(q)s^i \tag{1}$$

where q is a vector of real uncertain parameters (or just uncertainty) and $\rho_i(q)$ are coefficient functions. The form of these coefficient functions is very important because it determines the uncertainty structure. Their short classification can be found above in the Introduction. This contribution is focused on the affine linear uncertainty structure, which means that more than just one uncertain parameters can enter into the same coefficient function (i.e. the structure is not independent anymore), and all coefficient functions must have the form of affine linear function [1, 3, 4].

3 Uncertainty Bounding Set

The uncertainty bounding set Q is assumed as a ball in a norm. The typical shapes are a box (for L_∞ norm), a sphere or more generally an ellipsoid (for L_2 norm or weighted L_2 norm), and a diamond (for L_1 norm). The L_∞ norm case is:

$$\|q\|_\infty = \max_i |q_i| \tag{2}$$

and the L_2 norm scenario is based on:

$$\|q\|_2 = \sqrt{\sum_{i=1}^{n} q_i^2} \tag{3}$$

or:

$$\|q\|_{2,W} = \sqrt{q^T W q} \tag{4}$$

where $q \in \mathbb{R}^k$ and $W = diag\left(w_1^2, w_2^2, \ldots, w_k^2\right)$ is a positive definite symmetric matrix (weighting matrix) of size $k \times k$. Under assumption of $r \geq 0$ and $q^0 \in \mathbb{R}^k$, the ellipsoid (in $\mathbb{R}^k$) which is centered at q^0 can be expressed by means of one of these two equivalent inequalities:

$$\begin{aligned} &(q - q^0)^T W (q - q^0) \leq r^2 \\ &\left\|q - q^0\right\|_{2,W} \leq r \end{aligned} \tag{5}$$

For $k = 2$ (two-dimensional space), the ellipse of uncertain parameters can be easily visualized [1, 6, 7].

4 Ellipsoidal Polynomial Family

Generally, the family of continuous-time polynomials can be denoted as [1, 3, 4]:

$$P = \{p(s, q) : q \in Q\} \tag{6}$$

where $p(s, q)$ was defined in (1) and Q is the uncertainty bounding set.

In this contribution, $p(s, q)$ is assumed to have the affine linear uncertainty structure and Q is supposed to be an ellipsoid. Thus, the resulting family of polynomials (6) is called the ellipsoidal polynomial family here.

Formally, there are two basic representations of spherical polynomial families with independent uncertainty structure available. The first one with the zero-centered polynomial $p(s, q)$, and the second one with the zero-centered uncertainty bounding set Q [1, 6, 7].

5 Value Sets

As adumbrated above, the method that combines the value set concept with the zero exclusion condition represents universal and widely applicable technique for analysis of robust stability [1, 3, 4].

Suppose a family of polynomials (6). Its value set at frequency $\omega \in \mathbb{R}$ is given by [1]:

$$p(j\omega, Q) = \{p(j\omega, q) : q \in Q\} \tag{7}$$

and so $p(j\omega, Q)$ is the image of Q under $p(j\omega, \cdot)$. That is the value set can be practically constructed by substituting s for $j\omega$ and letting q range over Q.

The typical shapes of the value sets for the polynomial families with Q in the shape of a box and various uncertainty structures can be found e.g. in [3, 4].

Once the value sets are obtained for selected samples of non-negative frequencies, the robust stability can be verified by means of the zero exclusion condition [1]. For more information on the parametric uncertainty, the value sets, and the robust stability testing see e.g. [1, 3, 4].

From the viewpoint of spherical uncertainty, the formulas for computation of (ellipsoidal) value sets of a spherical polynomial family with independent uncertainty structure were derived in [1] on the basis of the proof of the Soh-Berger-Dabke theorem [12].

In this contribution, the value sets for an ellipsoidal polynomial family with affine linear uncertainty structure are numerically calculated and plotted.

6 Illustrative Example

Assume a polynomial family with affine linear uncertainty structure defined by the combination of the uncertain polynomial:

$$p(s, q) = s^3 + (2q_1 + q_2 + 2)s^2 + (q_1 + 2q_2 + 1)s + (q_1 + q_2) \tag{8}$$

and two various uncertainty bounding sets:

(a) L_∞ norm (box – rectangle):

$$\begin{aligned} q_1 &\in [1, 3] \\ q_2 &\in [1.5, 2.5] \end{aligned} \tag{9}$$

(b) Weighted L_2 norm (ellipsoid – ellipse):

$$\begin{aligned} &\left\|q - q^0\right\|_{2,W} \leq 1 \\ &q^0 = [2, 2] \\ &W = \begin{bmatrix} w_1^2 & 0 \\ 0 & w_2^2 \end{bmatrix} = \begin{bmatrix} 1 & 0 \\ 0 & 2 \end{bmatrix} \end{aligned} \tag{10}$$

that is:

$$\left(q_1 - q_1^0\right)^2 + 2\left(q_2 - q_2^0\right)^2 \leq 1 \tag{11}$$

The elliptical uncertainty bounding set (10) and 1000 randomly chosen internal points are visualized in Fig. 1. Moreover, the rectangle formed by the displayed axes in Fig. 1 concurs with the rectangular uncertainty bounding set (9).

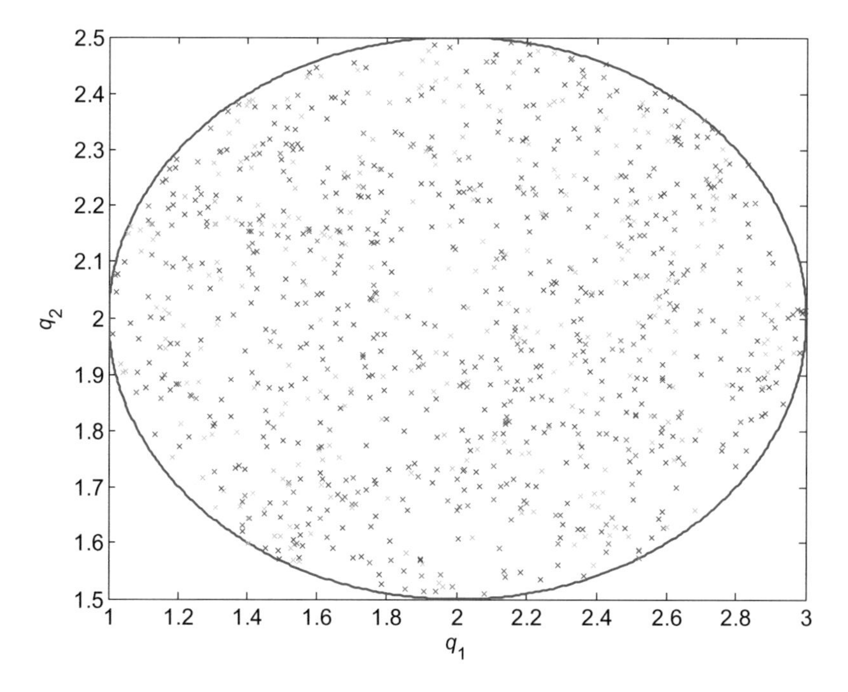

Fig. 1. Elliptical uncertainty bounding set (10) with 1000 randomly chosen internal points.

The value set of the elliptical polynomial family with affine linear uncertainty structure given by (8) and (10) for the frequency $\omega = 1$ is shown in Fig. 2. The plotted value set consists of its boundaries (related to the boundaries of the elliptical uncertainty bounding set from Fig. 1) and the images of 1000 randomly selected points within the elliptical uncertainty bounding set. As can be seen, all internal points from the parameter space remain inside the value set in the complex plane as well. The other way around, the boundaries of the value set are mapped only from the boundaries of the uncertainty bounding set for the studied case of the ellipsoidal polynomial family. However, this is generally true only for one direction as some parts of the boundaries in the (more than two-dimensional) uncertain parameter space can be mapped into the interior of the value set – see example at page 142 in [1] where an edge of an uncertain cuboid is mapped to the inside the value set. The abovementioned results are valid for the affine linear uncertainty structure (and simpler independent one), but they need not to be true for more general and more complex uncertainty structures.

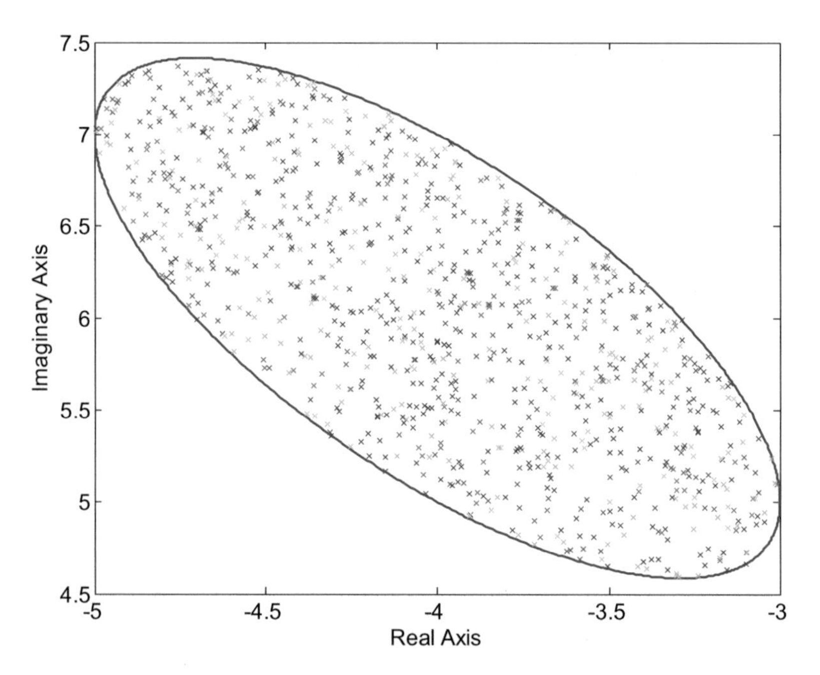

Fig. 2. The value set of the family (8), (10) for $\omega = 1$ with the images of 1000 randomly chosen internal points from the elliptical uncertainty bounding set (Fig. 1).

Furthermore, Fig. 3 compares the value set of the ellipsoidal version of the family (8), (10) (blue solid curve) with the value set of the classical “box” version of the family (8), (9) (black dashed curve) – both for $\omega = 1$. The same comparison but for the value sets in the range of frequencies from 0 to 3 with the step 0.1 is shown in Fig. 4.

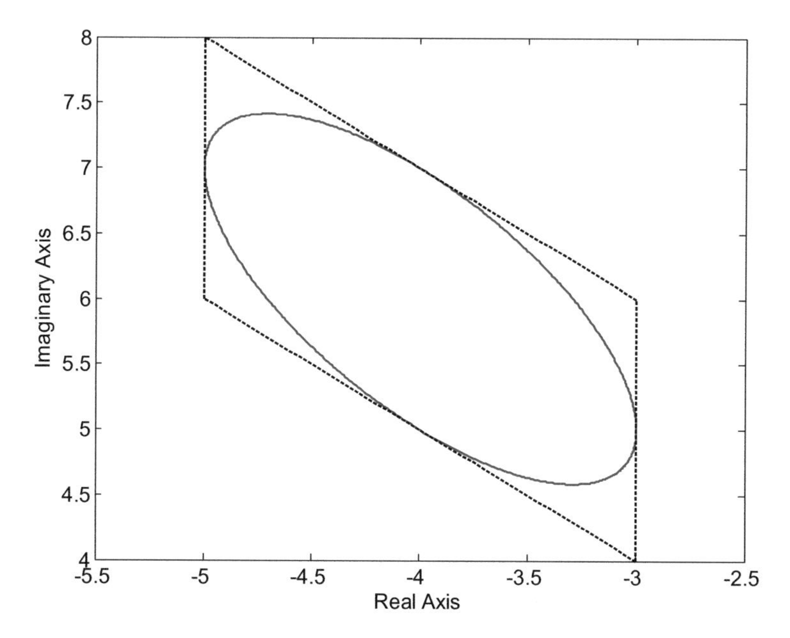

Fig. 3. Comparison of the value sets of the family (8), (10) (blue solid curve) and of the family (8), (9) (black dashed curve) for $\omega = 1$.

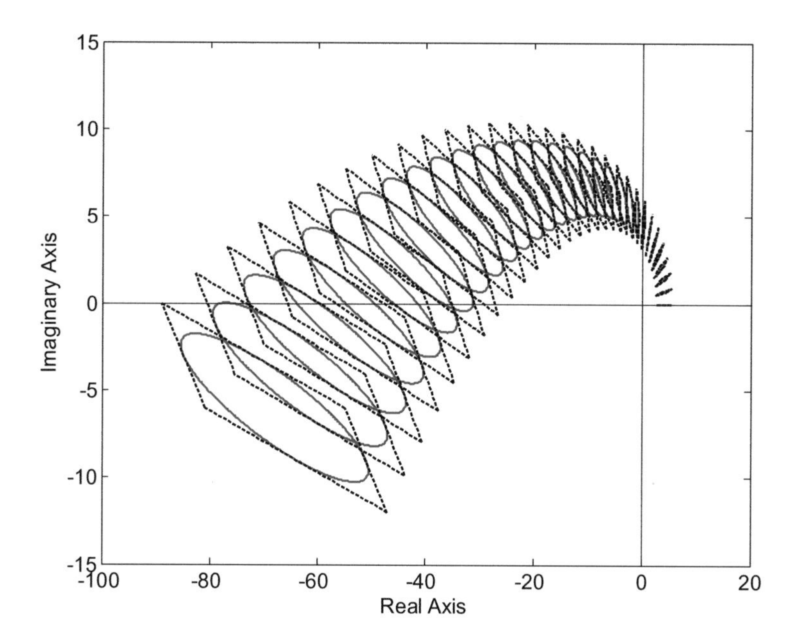

Fig. 4. Comparison of the value sets of the family (8), (10) (blue solid curves) and of the family (8), (9) (black dashed curves) for $\omega = 0 : 0.1 : 3$.

7 Conclusion

This contribution was focused on the ellipsoidal polynomial families with affine linear uncertainty structure and their value sets. After recalling the basic terms, it was shown that, similarly to the case of the box uncertainty bounding sets, the boundaries of the value set come only from the boundaries of the ellipsoidal uncertainty bounding set for the families with affine linear uncertainty structure. For this purpose, the value set of the ellipsoidal polynomial family with affine linear uncertainty structure was plotted including randomly chosen internal points. Furthermore, it was compared with the value set of the classical "box" version of the polynomial family.

Acknowledgments. This work was supported by the Ministry of Education, Youth and Sports of the Czech Republic within the National Sustainability Programme project No. LO1303 (MSMT-7778/2014).

References

1. Barmish, B.R.: New Tools for Robustness of Linear Systems. Macmillan, New York (1994)
2. Bhattacharyya, S.P.: Robust control under parametric uncertainty: an overview and recent results. Annu. Rev. Control **44**, 45–77 (2017)
3. Matušů, R., Prokop, R.: Graphical analysis of robust stability for systems with parametric uncertainty: an overview. Trans. Inst. Meas. Control **33**(2), 274–290 (2011)
4. Matušů, R., Prokop, R.: Robust stability analysis for systems with real parametric uncertainty: implementation of graphical tests in Matlab. Int. J. Circuits Syst. Signal Process. **7**(1), 26–33 (2013)
5. Matušů, R., Pekař, R.: Robust stability of thermal control systems with uncertain parameters: the graphical analysis examples. Appl. Therm. Eng. **125**, 1157–1163 (2017)
6. Matušů, R., Prokop, R.: Robust stability analysis for families of spherical polynomials. In: Intelligent Systems in Cybernetics and Automation Theory. Proceedings of CSOC 2015. Advances in Intelligent Systems and Computing, vol. 348, pp. 57–65. Springer, Cham (2015)
7. Matušů, R.: Spherical families of polynomials: a graphical approach to robust stability analysis. Int. J. Circuits Syst. Signal Process. **10**, 326–332 (2016)
8. Hurák, Z., Šebek, M.: New tools for spherical uncertain systems in polynomial toolbox for Matlab. In: Proceedings of the Technical Computing Prague, Prague, Czech Republic (2000)
9. Tesi, A., Vicino, A., Villoresi, F.: Robust stability of spherical plants with unstructured uncertainty. In: Proceedings of the American Control Conference, Seattle, Washington, USA (1995)
10. Polyak, B.T., Shcherbakov, P.S.: Random spherical uncertainty in estimation and robustness. In: Proceedings of the 39th IEEE Conference on Decision and Control, Sydney, Australia (2000)
11. Chen, J., Niculescu, S.-I., Fu, P.: Robust stability of quasi-polynomials: frequency-sweeping conditions and vertex tests. IEEE Trans. Autom. Control **53**(5), 1219–1234 (2008)
12. Soh, C.B., Berger, C.S., Dabke, K.P.: On the stability properties of polynomials with perturbed coefficients. IEEE Trans. Autom. Control **30**(10), 1033–1036 (1985)
13. Barmish, B.R., Tempo, R.: On the spectral set for a family of polynomials. IEEE Trans. Autom. Control **36**(1), 111–115 (1991)

14. Tsypkin, Y.Z., Polyak, B.T.: Frequency domain criteria for l^p-robust stability of continuous linear systems. IEEE Trans. Autom. Control **36**(12), 1464–1469 (1991)
15. Biernacki, R.M., Hwang, H., Bhattacharyya, S.P.: Robust stability with structured real parameter perturbations. IEEE Trans. Autom. Control **32**(6), 495–506 (1987)
16. Sadeghzadeh, A., Momeni, H.: Fixed-order robust H_∞ control and control-oriented uncertainty set shaping for systems with ellipsoidal parametric uncertainty. Int. J. Robust Nonlinear Control **21**(6), 648–665 (2011)
17. Sadeghzadeh, A., Momeni, H., Karimi, A.: Fixed-order H_∞ controller design for systems with ellipsoidal parametric uncertainty. Int. J. Control **84**(1), 57–65 (2011)
18. Sadeghzadeh, A., Momeni, H.: Robust output feedback control for discrete-time systems with ellipsoidal uncertainty. IMA J. Math. Control Inf. **33**(4), 911–932 (2016)

Optimization of Models of Quantum Computers Using Low-Level Quantum Schemes and Variability of Cores and Nodes

Viktor Potapov(✉), Sergey Gushanskiy, and Maxim Polenov

Department of Computer Engineering, Southern Federal University, Taganrog, Russia
vitya-potapov@rambler.ru,
{smgushanskiy,mypolenov}@sfedu.ru

Abstract. Basic concepts in the field of quantum information such as quantum circuits, logic gates and qubits were identified and analyzed in the present paper. It provides a basis for the development of several concepts (line, node, interaction of qubits and their neighborhood) of the theoretical and practical components of low-level random quantum circuits. Also, various optimizations, regulated by quantitative correlations of cores (one or several cores) and nodes (one or several), are derived. It can be used to implement simulation of a multi-node quantum circuit.

Despite the fact that the theory of quantum computation is not yet ready for a conversion from theory to practice, one can imagine a possible form of the future quantum computer and its interface. In addition, it is proved that most non-trivial tasks cannot be solved without the use of entanglement. Recently there has been a rapid growth of interest in quantum computers, especially after the sale of operating quantum computers. The main advantage of quantum computation in comparison with traditional calculations is that they use the property of quantum particles to be simultaneously in many states, which is called superposition. The use of quantum computers can significantly increase the speed of solving computational problems and, most importantly, exponentially increase the speed of NP-complete problems solving, which can be solved for unacceptable time on classic computers.

Keywords: Qubit · Entanglement · Quantum scheme · Wave function · Quantum gate

1 Introduction

Quantum calculators can solve problems that classic machines cannot do. Among such areas are quantum chemistry, material science, machine learning and cryptography.

Experimental devices containing about 50 qubits [1] will soon be available and will be able to perform well-defined computational tasks [2] that would require the most powerful classical supercomputers. However, in order not to repeat the mistakes of the past in the field of computing systems, it is necessary to solve optimization, theoretical and practical issues facing quantum computing devices and their components at this

R. Silhavy (Ed.): CSOC 2019, AISC 986, pp. 264–273, 2019.
https://doi.org/10.1007/978-3-030-19813-8_27

preliminary stage. Particularly, quantum schemes, optimization of the distribution of gates and qubits, which will reduce the number of links in a multisite setting and the number of k-qubit cores with gates at the level of a single node. A prime advantage will also serve to reduce the clock cycle of the quantum scheme.

2 Low-Level Random Quantum Schemes

A quantum scheme is a sequence of physical transformations from a finite set of basic elementary transformations – gates. At the input of a quantum circuit gets quantum bits. The result of its work is probabilistic. Physically, you can only implement linear, constant-preserving sum of squares of coefficients (unitary) transformations over a small number of quantum bits. Consequently, any transformation is uniquely defined by values on the basis states and the transformation over k quantum bits can be written as a matrix $2^k \times 2^k$.

Although one of the computational problems proposed for demonstrating the computing power of a quantum computer – performing low-level random quantum circuits – is not scientifically useful, its implementation schemes are still very useful for calibrating and testing temporary quantum devices. The low-level random quantum scheme is shown in Fig. 1 and demonstrates quantum superiority.

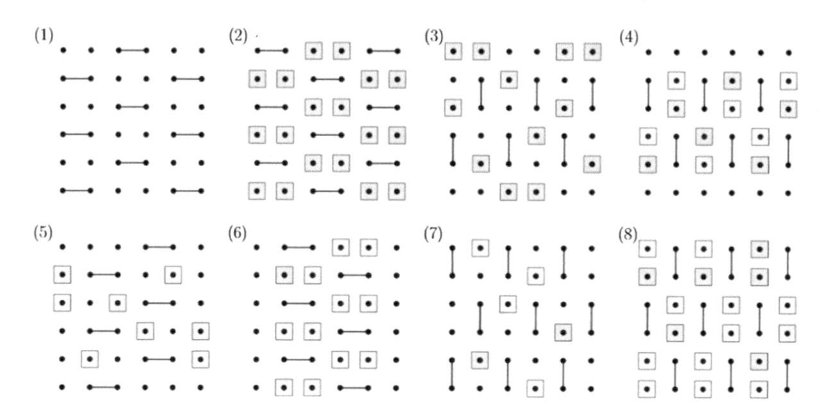

Fig. 1. Low-level random quantum scheme

Points in Fig. 1 represent the quantum bits of information used in the quantum circuit. No quantum scheme can function without a specific set of quantum gates, shown in Fig. 1 in squares. This low-level random quantum scheme is a special case, so there is no fundamental difference in the choice of a specific quantum of operation of the quantum scheme.

Identical schemes were created using the following rules: during cycle 0, Hadamard gate is applied to each qubit. As a result, eight different patterns (6×6 qubits) of controlled Z (CZ) gates are reapplied until the desired depth of the pattern is reached. CZ gates, which are logical two-qubit gates and described by Pauli matrices, are represented by a line between two qubits (we will further identify this line with the notion of a node). Physically, a node is a part of the executable code within the framework of a quantum calculator simulator. The line is converted to a node by adding a specific gate to the scheme. The choice of a gate is limited only to the framework of the applicable set of basic gates. This pattern ensures that all possible qubit interactions in this two-dimensional nearest neighbor architecture are executed every 8 cycles. Single qubit gates are randomly selected.

Therefore, in addition to verifying quantum algorithms and conducting studies of their work in the presence of interference, the quantum circuit model can provide the tools for carrying out calibrations and control measurements, allowing to obtain the most efficient computational structure.

Restrictions in low-level random quantum schemes are similar to classical ones, in which the total number of branches (lines) of a quantum scheme and the number of gates are limited to the scale of the scheme itself.

Quantum simulators are comparable with the tools of structural modeling, allowing to implement the simulation of the classical calculator. The quantum simulator [3], developed earlier, was implemented and optimized by means of a multi-level approach. The main form of the interface built by the quantum computer model is shown in Fig. 2.

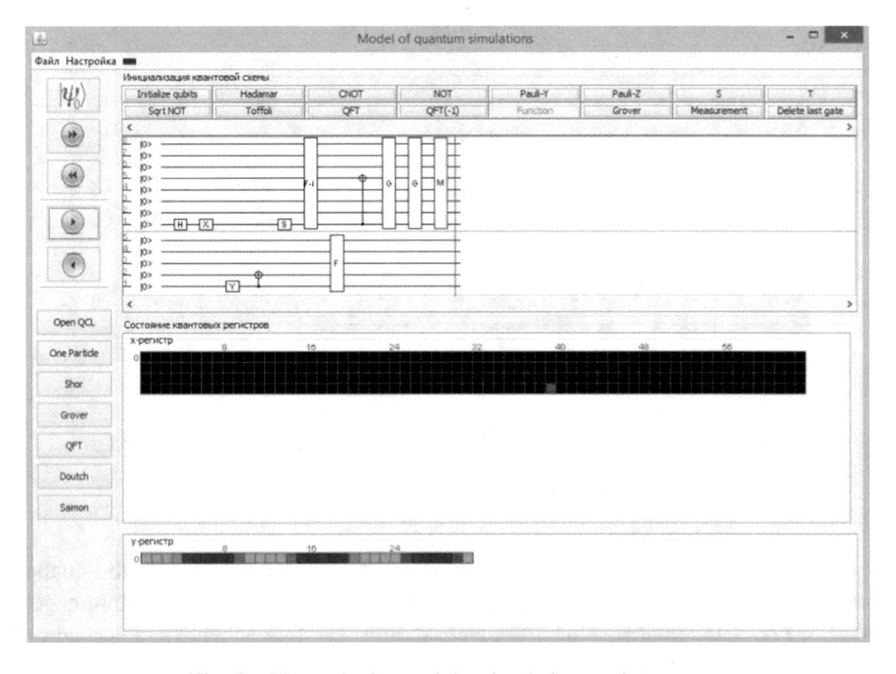

Fig. 2. The main form of the simulation environment

The quantum control keys are located on the left side of the main form. This area of the simulation environment provides to the model the ability to work in an automatic or step-by-step mode; it is also possible to delete the last selected and entered element in the quantum scheme or to completely clean the entire circuit. In the upper part there is a navigation menu that manages and configures the model of quantum computer, the center of the top is a collection of quantum logic gates, and below is the color complex diagram of the quantum states of the registers X and Y.

3 Options of Optimization

For large systems simulation there is no need for storage of a $2^n \times 2^n$ matrix operating on a state vector. You can use its regular structure and apply methods that, considering its state, simulates multiplication by this matrix. The standard implementation has two state vectors (input and output). In dynamic systems, the reaction *y(t)* at time *t* depends not only on the input value *u(t)* at time *t*, but also on the initial condition, i.e. initial state. Let the dynamic system [4] have *r* input variables and *m* output, as well as *n* internal variables. Variables $x_i(t)$, $i = \overline{1, n}$ are state variables, if you set their values $x_i(t_0)$, $i = \overline{1, n}$ at the time t_0 and the law of change of input variables $u_k(t)$, $k = \overline{1, r}$ on the interval $[t_0; t]$, then you can uniquely determine the values $x_i(t)$ for all *i* at time *t*. Variables $x_1(t), \ldots, x_n(t)$ characterize the initial state of the system. Define vectors:

$$x(t) = \begin{bmatrix} x_1(t) \\ \ldots \\ x_n(t) \end{bmatrix} = [x_1(t), \ldots, x_n(t)]^T \text{ – vector of state variables, state vector.}$$

$u(t) = [u_1(t), \ldots, u_r(t)]^T$ – vector of input variables.
$y(t) = [y_1(t), \ldots, y_m(t)]^T$ – vector of output variables.

To define one record of the output vector, two complex products [5] and one complex addition must be performed on two records of the input vector by using a common one-qubit gate. In general, the operation can be represented by the formula

$$2 * (4[\text{mul}] + 2[\text{add}]) + 2[\text{add}] = 14 \text{ FLOP}$$

For one complex double-precision record, 16 bytes of memory are required; the input vector must be unloaded from memory; output vector is written back into memory. Thus, the work intensity is less than 1/2, this is an example of the fact that the memory capacity limits most systems.

Single Core
To reduce memory requirements by a factor of 2, complex sparse matrix-vector multiplication can be performed due to cache access [6]. The more the qubit is included in the gates, the more operations must be implemented, which allows the use of hardware computing power more efficiently. Indeed, the number of operations grows exponentially with an increase in the number of qubits, since the use of a k-qubit gate allows one scalar product [7] of dimension 2^k to be executed.

In order to operate a k-qubit gate (dimension $2^k \times 2^k$) on a state vector of size $2n$, where *n* is the number of qubits, the values corresponding to each of the 2^k indexes of the gate matrix must be represented as a time vector of dimension 2^k, which than being loaded into the state vector, multiplied by the gate matrix. The indices of the records of the state vector, represented in binary code, are bit sequences of the form

$$c_{n-k-1} \cdot x_{i_{k-i}} \ldots c_j \ldots x_{i_1} \ldots c_0,$$

where i_0, i_1, i_{k-1} are the indices of k qubits to which the gate was applied. Extracting and combining the bits x_{ij} of the write index: $x = x_{i_{k-1}} \ldots x_{i_1} x_{i_0}$ gives the index of the record relative to the time vector. All 2^{k} entries that have the same substring index $c = c_{n-k-1} c_{n-k-2} \ldots c_0$ are the part of the product of the matrix by the vector. Once all the entries multiplied by the matrix are obtained and saved as a state vector, you can move on to the next substring of the index with c′ = c + 1. In general, 2^{n-k} complex matrix-vector multiplications of dimension 2^k are realized.

Thus, one matrix is used 2^{n-k} times, and the elements of the matrix can be represented in advance so that the indices of qubits and, as a result, access to memory, are always ordered. Matrix-vector product

$$\upsilon'_l = \sum_{i=0}^{2^k-1} m_{l,i} * \upsilon_i$$

may require all records of the temporary vector υ (loading them from memory). In this case, the calculation is locked, and the block size is determined using automatic code generation or comparative evaluation [8]. For the index of each of the blocks: $b = 0, 1, \ldots, \frac{2^k}{B} - l$ all the indices l of the temporal output vector, before proceeding to the next block, are converted according to

$$\upsilon'_l + = \sum_{j<B} m_{l,i(b,j)} * \upsilon_{i(b,j)}, \text{ where } i(b,j) = b * B + j.$$

To parallelize updates as a means of ordering the values of l, explicit vectorization is used. And, because there are complex values of double precision, vectorization [9], theoretically allows you to speed up the implementation of 2 or even 4 times. a_R and a_I are the real and imaginary parts of a. Multiplying one complex record $\upsilon_l = (\upsilon_R, \upsilon_I)$ of the time vector v by one complex record of the gate matrix $m = (m_R, m_I)$ and adding the result with the time output vector υ' can be written as

$$(\upsilon'_R, \upsilon'_I) + = (\upsilon_R m_R - \upsilon_I m_I, \upsilon_I m_R + \upsilon_R m_I)$$

However, the implementation of this update leads to the depletion of computing resources due to artificial dependencies and additional permutations. However, the following reorganization is possible, contributing to an increase in the maximum achievable performance.

$$(\upsilon'_R, \upsilon'_I) + = (\upsilon_R m_R, \upsilon_I m_R), (\upsilon'_R, \upsilon'_I) + = (\upsilon_I \cdot -1 \cdot m_I, \upsilon_R m_I)$$

It is worth noting that υ_1 can be rearranged once at boot (and then in the register), since it is reused for 2^k complex works. Since the matrix m is used in 2^{n-k} matrix-vector products, the preliminary calculation for constructing these two matrices consisting of (m_R, m_R) *and* $(-1 \cdot m_I, m_I)$ is free.

Single Node

The described above optimization does not detract from the fact that the operational efficiency of using a single-qubit gate [10] is very low, which makes it difficult to fully use the power of multi- and multi-core processors [11]. However, as mentioned earlier, the more qubits a gate contains, the more operations are required, and if their use remains memory-related, multi-qubit gates (for example, Toffoli [12]) and one-qubit gates (Not [13], Hadamard [14]) work out almost equal time. The gain, in addition to increasing the intensity of the work, is that large gates can be used to perform a whole sequence of one- and two-bit gates. In particular, the set of gates acting on k different qubits can be combined into one multi-qubit k-qubit gate.

The value of k is chosen depending on the maximum performance, memory capacity, cache size and scheme. Cache parameters are especially important when applying gates to qubits with a large index. With low associative caching, this causes conflicts if the kernel is small. Since 2^k values must be loaded from the state vector (the minimum size of which is 2^m, m is the smallest qubit index) for each of the 2^{n-k} matrix-vector product, the 2^k-directional cache lines must be correlated with the corresponding location (address), does not matter how big is m. This allows to get direct access to the data in the cache during subsequent operations of matrix-vector multiplication.

Many Nodes

To simulate the properties of more than 30 qubit quantum solvers [15], many nodes are necessary for the state vector to be stored in memory. To ensure the connection of 2^g nodes, each of which has its own state vector of size 2^l (g, l is the number of common and local qubits, respectively). Applying gates to local qubits, i.e. qubits with index $i < l$ does not require communication. Cubits with the index $i \geq l$ – on the contrary, communication is necessary.

There are two basic schemes that can be used to implement the simulation of a multinodal quantum scheme. The first one contains common qubits (a typical example is the neighboring qubits of the quantum scheme in Fig. 1) and uses common gates (the gate community stems from the generality of qubits) using two pair exchanges of half of the state vector. The second circuit swaps common and local qubits, applying gates to local qubits in the usual way and, if necessary, swaps them with common qubits again. The exchange with the global qubit and its subsequent return immediately requires the same connection as the first scheme. Consider the exchange scheme in more detail.

An Example with 1 Qubit. In the case of two digits, the exchange of the highest-order qubit (the highest bit of the local index) with the general qubit (the first bit of the digit number) can be implemented as follows: the first block of the 0-digit remains unchanged, since replacing 0 with 0 is useless. The exchange of 0 (total) to 1 (local) for the second block requires the transfer of the entire block to 1 bit, where these coefficients are associated with a local qubit equal to 0.

An Example with 2 Qubits. In order to change two common qubits with two local qubits of higher order in the case of 4 digits, the i-th quarter of the state vector of each digit is translated into the i-th digit.

To allow arbitrary exchange of local qubits, first optimized kernels are used. With their help, the exchange between qubits with the highest index and those with which the exchange should be carried out. Then a group local exchange is performed on the "all-to-all" principle and, if necessary, other local exchanges (with qubits of smaller indices) in order to increase the locality of data in the cores consisting of k-qubit gates.

4 Optimization Schemes: Distribution of Gates and Qubit

In addition to implementing optimization, circuit optimization is needed, aimed at reducing communication "steps" and using accurate and more efficiently configured cores. There are various options for optimizing the distribution of gates and qubits based on "quantum superiority" schemes. Optimizations are general in nature and can be applied in any quantum chain. In fact, "quantum superiority" schemes can be designed in a way that is least suitable for these types of performance optimization. The construction of such random, low-level quantum circuits is shown in Fig. 1. The circuits are designed for use as part of a quantum computational architecture with bidirectional communication. All possible two-bit gates are used within 8 cycles, which makes the system very confusing. The simulator can skip the Hadamard gate in the 0 cycle to initialize the wave function [16] directly. Moreover, the simulation of the last CZ-gates [17] is not performed, since they only change the phases of the probability amplitudes α_i [18].

5 Distribution of Gates

The most important optimization of quantum schemes is the distribution of gates, since it drastically reduces the number of connections in a multi-node setup and the number of k-qubit cores with gates at the level of a single node. Optimization can be divided into 3 stages:

Minimize the Number of "Steps" of Communication. In the first step, distribution of gates minimizes the number of general local exchanges, which is the most important parameter of the multisite configuration. The execution of each clock cycle of the circuit itself requires at least one "step" of communication on each cycle, characterized by a common off-diagonal gate. However, in accordance with the multi-node principle, the stage turns out to be useful not for the operation of common gates, but rather for the exchange of common and local qubit and the subsequent local testing of gates. For the scheme to be the most useful, the distribution of gates algorithm reorganizes, if possible, the operation of the gates.

*Minimizing the Number of k-*qubit Gates. In the second stage, all the gates are distributed so that it is possible to chain the successive 1- or 2-qubit gates into k-qubit ones and use the k-qubit gate instead of multiple one- or two-qubit ones. It is necessary to increase the number of k qubit per one cluster, while fulfilling the condition $k \leq k_{max}$, where the maximum k_{max} is the value of k, for which the performance of the core of the k-qubit gate will show good performance in the final system.

Local Adjustments for General Exchanges. The last cluster at each stage contains a smaller number of one- and two-qubit gates (Fig. 3).

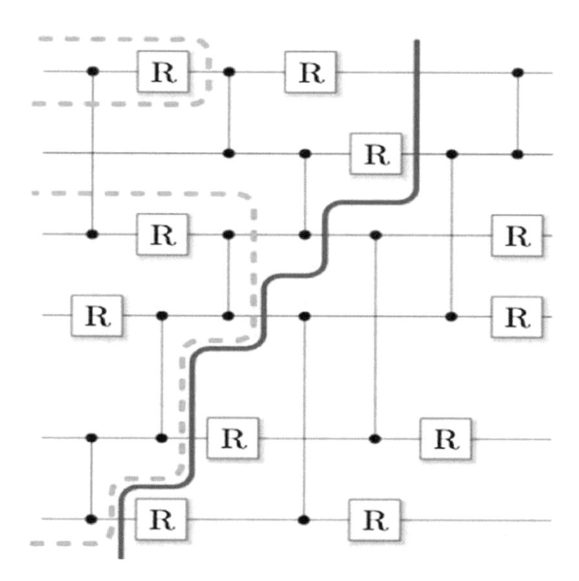

Fig. 3. Example of gates distribution

To increase the average number of gates in each cluster, thereby reducing the total number of clusters in the circuit, you need to remove the last cluster at each stage through general local exchanges before, and, if possible, without increasing their total number.

6 Distribution of Qubits

The bit position in each qubit is optimized according to the number of clusters [19], which is accompanied by a decrease in performance due to the associativity of the last level cache. Since this performance degradation occurs only when gates are applied to high-order bit positions, a reassignment (redistribution) can be applied. The following heuristics allows to reduce the solution time by half: to connect the qubit and bit locations of 0 so that the number of clusters that have a bit location of 0 is maximum. Continue to ignore all the clusters, which act on this qubit. Perform the same actions for the position of bits 1, 2 and 3.

7 Conclusion

Optimization and implementation of quantum computing in the field of quantum calculator models, based on the described approach, allow to:

- Calibrate, verify, and test quantum computing devices;
- Reduce memory requirements by n times and thereby speed up the computational process by performing complex sparse matrix-vector multiplication thanks to cache access;
- Make the most rational choice of one of the existing basic schemes that can be used to implement the simulation of a multinodal quantum scheme;
- Find new ways to apply quantum computing to simulate any parameters in the task being performed.

The basic concepts in the field of quantum information such as quantum circuits, logic gates and qubits were identified and analyzed. This served as the basis for the development of several concepts (line, knot, the interaction between qubits and their neighborhood) to the theoretical and practical components of low-level random quantum schemes. Various combinations of optimization, regulated by quantitative ratios of the nuclei (one core or several) and nodes (one node or several), which can be used to implement the simulation of a multinodal quantum scheme were produced either.

Acknowledgments. This work was carried out within the State Task of the Ministry of Science and Higher Education of the Russian Federation (Project part No. 2.3928.2017/4.6) in Southern Federal University.

References

1. Sukachev, D.D., Sipahigil, A., Lukin, M.D.: Silicon-vacancy spin qubit in diamond: a quantum memory exceeding 10 ms with single-shot state readout. Phys. Rev. Lett. **199** (2017)
2. Lukin, M.D.: Probing many-body dynamics on a 51-atom quantum simulator. Nature **551**, 579 (2013)
3. Potapov, V., Gushansky, S., Guzik, V., Polenov, M.: Architecture and software implementation of a quantum computer model. In: Advances in Intelligent Systems and Computing, vol. 465, pp. 59–68. Springer (2016)
4. Raedt, K.D., Michielsen, K., De Raedt, H., Trieu, B., Arnold, G., Richter, M., Lippert, T., Watanabe, H., Ito, N.: Massively parallel quantum computer simulator. Comput. Phys. Commun. **176**, 121–136 (2007)
5. Boixo, S., Isakov, S.V., Smelyanskiy, V.N., Babbush, R., Ding, N., Jiang, Z., Martinis, J.M., Neven, H.: Characterizing quantum supremacy in near-term devices. arXiv preprint arXiv: 1608.00263 (2016)
6. Stierhoff, G.C., Davis, A.G.: A history of the IBM systems Journal. IEEE Ann. Hist. Comput. **20**(1), 29–35 (1998)
7. Lipschutz, S., Lipson, M.: Linear Algebra (Schaum's Outlines), 4th edn. McGraw Hill, New York (2009)

8. Collier, D.: The comparative method. In: Finifter, A.W. (ed.) Political Sciences: The State of the Discipline II, pp. 105–119. American Science Association, Washington, DC (1993)
9. Vectorization. https://en.wikipedia.org/w/index.php?title=Vectorization&ldid=829988201
10. Williams, C.P.: Explorations in Quantum Computing. Texts in Computer Science, pp. 51–122. Springer, Heidelberg (2011). Chapter 2. Quantum Gates
11. Olukotun, K.: Chip Multiprocessor Architecture – Techniques to Improve Throughput and Latency. Morgan and Claypool Publishers, San Rafael (2007)
12. Potapov, V., Guzik, V., Gushanskiy, S., Polenov, M.: Complexity estimation of quantum algorithms using entanglement properties. In: Informatics, Geoinformatics and Remote Sensing (Proceedings of 16-th International Multidisciplinary Scientific Geoconference, SGEM 2016, Bulgaria), vol. 1, pp. 133–140. STEF92 Technology Ltd. (2016)
13. Inverter (logic gate). https://en.wikipedia.org/w/index.php?title=Inverter_(logic_gate)&oldid = 844691629
14. Lachowicz, P.: Walsh – Hadamard Transform and Tests for Randomness of Financial Return-Series. http://www.quantatrisk.com/2015/04/07/walsh-hadamard-transform-python-tests-for-randomness-of-financial-return-series/ (2015)
15. Potapov, V., Gushanskiy, S., Guzik, V., Polenov, M.: The computational structure of the quantum computer simulator and its performance evaluation. In: Software Engineering Perspectives and Application in Intelligent Systems. Advances in Intelligent Systems and Computing, vol. 763, pp. 198–207. Springer (2019)
16. Zwiebach, B.: A First Course in String Theory. Cambridge University Press, Cambridge (2009)
17. Potapov, V., Gushanskiy, S., Samoylov, A., Polenov, M.: The quantum computer model structure and estimation of the quantum algorithms complexity. In: Advances in Intelligent Systems and Computing, vol. 859, pp. 307–315. Springer (2019)
18. Universe of Light: What is the Amplitude of a Wave? Regents of the University of California. http://cse.ssl.berkeley.edu/light/measure_amp.html
19. Sternberg, R.J., Sternberg, K.: Cognitive Psychology, 6th edn. Cengage Learning, Wadsworth (2012)

Economics of Using Blockchain in the System of Labor Relations

Ruslan Dolzhenko(✉)

Ural State University of Economics, Yekaterinburg, Russia
rad@usue.ru

Abstract. The article considers the prospecting opportunities of using blockchain technology in the system of labor relations from the point of economics. The Transaction Cost Economics has been adopted as a conceptual basis for the argumentation since it deals with the category of "transaction costs" as a mandatory attribute of any economic relations with the exception of ideal socialist relations, where these are minimal. The objective of the article is as follows: based on the use of Transaction Cost Economics assumptions, to show the possibilities and prospects for using the blockchain in the system of labor relations as a transaction chain, which eliminates fraud, theft, breach of the rights, opportunism, i.e. minimizes transaction costs. In addition, the article researches into the need of updating the classical ideas introduced by Oliver E. Williamson on classes of regulatory structures that guarantee execution of labor relations due to the fact that the wide-scope blockchain implementation could prospectively reduce transaction costs so that these can be accompanied by a new form of regulatory structures, which the author proposes to refer to as a "smart contract". The authors describe the justifications and examples of possible directions for the blockchain technology use to formalize specific aspects of labor relations subject to reduction in transaction costs. The article considers the possibilities, prospects, and grounds lacking in the possibilities of using blockchain technology in the labor relations system.

Keywords: Labor · Labor relations · Transaction costs · Blockchain · Smart contract

1 Introduction

The idea of blockchain has occupied a specific place in the public mind: on the one hand, it is considered as a promising technology that can be exploit in almost any sphere of human relations, on the other hand, it has created the basis for the emergence of cryptocurrencies, overheated interest in bitcoins and other similar phenomena, which erode the blockchain image. Moreover, there often appears the impression that the more people are discussing this topic, the less they actually understand in this technology essence since specific conceptual features of its actual application remain unrevealed by the public. Some of the scientists who primarily focus on information technologies specify the scope of possibilities of a distributed database usage, including those in the field of labor relations, however, their statements are going too far beyond

R. Silhavy (Ed.): CSOC 2019, AISC 986, pp. 274–282, 2019.
https://doi.org/10.1007/978-3-030-19813-8_28

the traditional ideas and cause some skepticism, especially when it comes to fundamental possibilities rather than individual formalizations of entries made on transactions using encryption, distributed storage, and other features of blockchain.

Hence, this article focuses on the essence of this technology in order to select those features which affect the possibilities of its use in the labor relations field, specify the already implemented solutions, and attempt to propose its future prospects with the account of views existing in economics, in particular, the Transaction Cost Economics and Contract Theory. However, allow us firstly to give an overview of this technology.

2 The Essence of Blockchain Technology and Its Prospective Use in the Labor Relations System

From a global perspective, blockchain is a network designed to process transactions with a set of rules prescribed ("the Trust Protocol"), following which the participants can have a public ledger of transactions and set a network state at a certain moment. In the long term, a blockchain technology enables input of "the Trust Protocol" in any relations through the use of IT. Bitcoin invented by Nakamoto [8] is one of the forms of a blockchain practical application.

It is worth mentioning that its conceptual grounds were formed back in the 20th century, but there were required a concourse of circumstances, a necessary level of development of specific technologies for the blockchain to come into active use.

Some researchers emphasize that this is just the beginning of an era for this technology, which has all the prospects to become the basis for innovations in people's lives. In the recent few years, the understanding of blockchain capabilities was expanded through the development of the smart contracts idea. Smart contract is a computer protocol employed to automate and facilitate financial transactions [2]. This term was firstly used by Nick Szabo in the end of the 20th century and it took it a decade to gain popularity among field users and researchers into blockchain. Smart contracts have become a key feature of the Ethereum project developed by V. Buterin. The detailed overview of this project can be found in the paper [9, 12].

One of the first scientists to have outlined the prospects of blockchain use in public life and, in particular, in the sphere of labor, was Don Tapscott, an author of a number of popular science books in the field of management. His ideas on the prospects of this technology were presented in the book "How blockchain will change organizations" as well as in the article [10]. It describes the application scope and points out that human civilization has not yet fully realized the capabilities that blockchain offers.

A similar viewpoint is shared by Steve Hamm, who in his article emphasizes that blockchains based on smart contracts can lead to significant changes in specific industries and emergence of new business models. However, in all fairness, foreign scientific environment is poorly aware of the blockchain capabilities and spheres of its application and study. For instance, 15 articles in the Web of Science database discuss the economic aspects of blockchain and only 3 papers disclose it in conjunction with a smart contract.

As of today, Russian scientists have not raised the topic of blockchain in the field of labor relations, there exist only a few works to cover the prospects of this technology use on the Internet of Things, changes in approaches to legal regulations use, in particular, the labor law, and some other works [1–3]. Moreover, all scientific and popular articles on this topic are considering how the function of human resources management in a company will change, i.e. reducing the scope of blockchain only to certain areas of HR management at a micro level. Thus, all other aspects of labor relations at a country level, labor market, regions, network structures and subjects of labor remain beyond the focus of blockchain practitioners and theoreticians of social sciences. In any case, allow us considering the application scope of this technology in personnel management found at the current level of the technology development.

Popular ideas about how the HR management will change due to the fact that companies will start using blockchain in their activities were described in the article "Labor activity under smart contracts: how blockchain will change HR" by V. Sveshnikov, co-founder and head of the Stafory company with its key product called "Robot Vera". In this article the author highlighted the most important changes that can happen in a company (see Fig. 1) when it begins to actively use blockchain in specific directions of its activity.

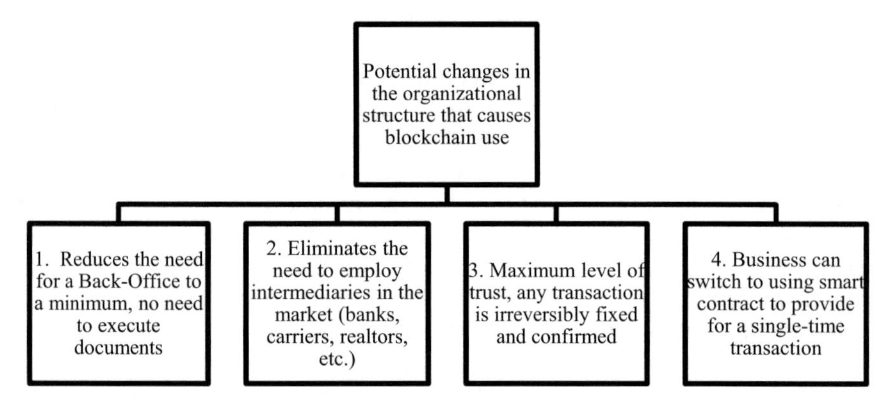

Fig. 1. Potential changes in a company caused by the practical use of blockchain. Source: author's work

There already exist practical examples of building business relations using blockchain. The main ones are described by Christidis and Devetsikiotis [7], some of them are related to the implementation of labor relations between employees and employers. The most famous project in this field is ChronoBank, a company that supports the platform for online payments using blockchain technology, its main activity is the area of HR management and employees recruiting. The company was founded in Australia in 2016, and for the past 2 years it has demonstrated strong growth in the market. However, some single examples of blockchain use in practice prove it has not received proper distribution so far.

Large consulting companies point out that in the next 2–3 years blockchain will start to be actively used in HR management. In 2017, PWC organized and hosted round table discussions on the prospects for blockchain application in the field under consideration to show that experts see the following key directions of introducing this technology in this area (see Fig. 2).

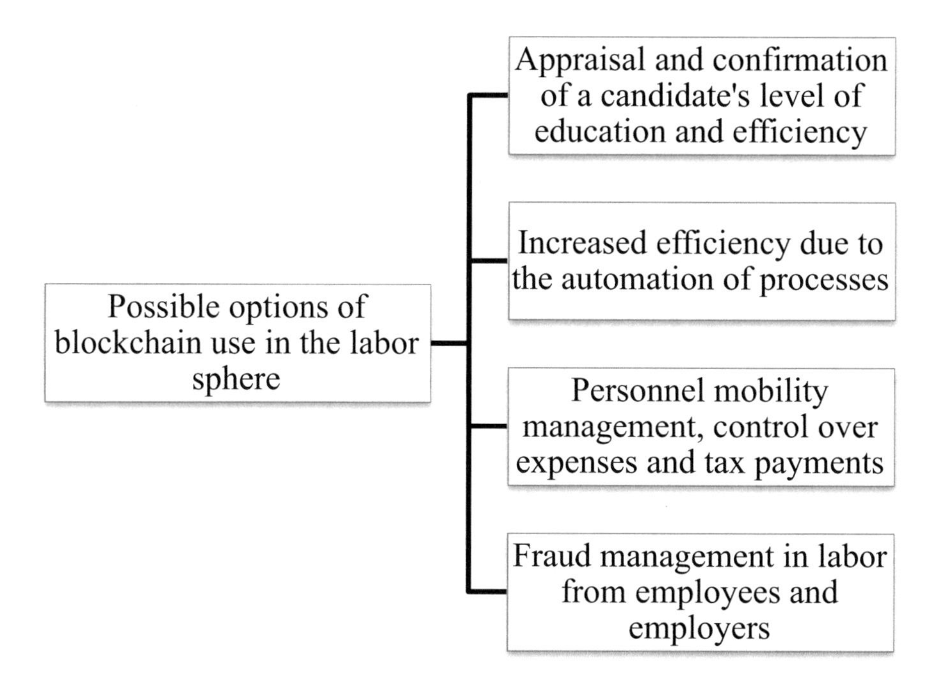

Fig. 2. Key directions of blockchain use in the field of labor. Source: author's work

Thus, the blockchain technology can be used in the field of labor relations, but what in terms of Economics can determine the success of its implementation? The Transaction Cost Economics (TCE) can assist in finding the answer to this question. Allow us referring to the TCE.

3 Conceptual Grounds for the Blockchain Use in the System of Labor Relations from a Position of Economics

What are the economic grounds of using blockchain under present conditions? In order to deal with this issue, we need to determine a basic economic school since the science is currently represented by different trends. The principles of the latter enable to differently assess the grounds for existence and implementation of economic relations. In our opinion, such a ground can be neo-institutionalism, which actively develops and enables to assess changes in the public life frontier due to a number of advantages over other trends.

Special focus is laid on the fact that this viewpoint is shared by Don Tapscott above, who provides the arguments through the prism of this theory in his work. He notes that according to R. Coase, who represents the classics of neo-institutional school of economics, there exist various types of transaction costs associated with conducting relations in a company format: search costs, coordination costs, drafting contract costs, and reputation-building costs [10].

Below we consider in detail the content of this category and attempt to assess the impact that a blockchain could have on them.

Relations within which there is a transfer of ownership can be referred to as transactions; costs associated with their implementation are considered by Economics (and not only) as transaction costs that are one of the most important components of neo-institutional theory of Economics.

In general terms, the concept can be defined as the cost of economic interaction, whatever form it may have. It is a known fact that R. Coase was the first to use this concept in economics in 1937 in his famous article "The Nature of the Firm", a key paper for the modern economic thought, and in 1960 the author expanded it in the article "The Problem of Social Cost" [6]. He believed that in order to execute a transaction in the market it is necessary to make a number of additional steps that ensure confidence that the terms of the economic agreement are executed (to form "islands of consciousness" providing guarantees and saving transaction costs). These may include legal advice, negotiations, contract drafting, etc. These actions increment additional costs for the participants of economic relations.

They are required to formalize the interactions between economic actors, to provide guarantees, to protect against fraudulent activities, but they also form additional costs for the company and common sense makes it focus on minimizing such through various ways. According to Tapscott, blockchain use will help minimize all categories of transaction costs. The fact of withdrawing the cost reduction tool beyond the company contour can potentially fully change existing models of a company. Thus, one of the key possibilities related to the widespread blockchain introduction is the reduction of transaction costs, which led to existing forms of fastening of relations of a certain type. It is clear that the application scope of this technology is limited to the IT sphere, where the Internet is not involved in the relations, a blockchain has no ground for existence. However, this possibility can be differently assessed. Using the economic and mathematical logics, R. Coase attempted to show that "socialism" is the best remedy for transaction costs, as it does not require any market pricing mechanism and, hence, there are no costs of its use. Blockchain built in labor relations chain (iterations), by default reduces transaction costs to a minimum, lowers the uncertainty of transactions, and eliminates the possibility of participants' opportunism. Due to a special IT-Protocol that enables its realization, and an encryption system that protects the protocol and blockchain from modifications, the relationships between actors can automatically formalize in line with their consequences as set forth in the smart contract.

We propose to refer to the fields where the technology can be presented as "blockchain relations", i.e. relations formalized through the use of blockchain technology. Further we consider the economic grounds of their use.

How is the case with transaction costs in relations that do not require any additional actions between actors, except for those related to defining the terms ("the Trust Protocol") of a smart contract before the first transaction and may not be changed except for the cases of applying to the so-called "fork"? "Fork" is the term used to define the process of introducing forced principal changes into the basic blockchain protocol resulting in splitting the block chain into two parts. In fact, the possibility of fork is one of the key sources of transaction costs for participants in the relations since it may lead to establishing fundamentally different sequences of blocks not related to the basic chain.

The important principle of the Transaction Cost Economics is known to be that each class of transactions between the actors of relations corresponds to a specific class of regulatory structures for their execution with minimal transaction costs. In our opinion, the emergence of blockchain technology and its imposition on various classes of public relations requires a revision of the base diagram proposed by Williamson, who included three classic classes of regulatory structures: market, hierarchy (firm) and "hybrids", an intermediate option between the first two [11]. The differences between them are mainly three core characteristics of a relationship: the power of incentives, administrative control, and type of contracts used. The difference in the first two characteristics is obvious:

The market is characterized with highly powerful incentives, as well as significant losses from the possibility of bargaining between the actors of relations.

A firm is distinguished by incentives of low power and executors management through the use of administrative control (orders) that allow to completely eliminate costly two-way "bargaining".

"Hybrid" as a form of regulatory structure is in the intermediate position between the market and the firm: its incentives are weaker than in the market, but stronger than in firms, the possibilities of using administrative control are wider than in the market, but more limited than in firms.

Each form is preferable under certain conditions. The latter are determined by the specificity of the resource about which the relationship is developed, and the amount of the transaction costs incurred by actors. Moreover, the list of forms is not static and is liable to development [5].

The more costs are found, the higher is the demand for a clear focused regulation through the use of the firm. With the view to the above, the amount of the transaction costs will be minimal in the organized chain of relations on the basis of blockchain and use of smart contracts. In addition, specificity of the resource does not play a significant role since blockchain enables formalizing almost any forms of relations.

Thus, blockchain in conjunction with smart contract is the basis for the principal possibility of new forms of regulatory structure to emerge. This is characterized by extremely weak incentives, minimum (zero) losses from bargaining due to the fact that it is not provided for and maximum capabilities of administrative control since the terms set forth in the conditions of "the Trust Protocol" cannot be breached (see Fig. 3).

If a transaction involves the use of non-specific resources a customer does not care who will perform the activity, and it can use the market mechanism of regulation. As the specificity of the resource and the transaction costs amount increase at some point (M-H point in the Fig. 3), the customer will have to switch to a hybrid form of relations, which provides greater guarantees. Further growth of these indicators will cause the necessity to account for the need of guarantees to perform the contract, then the customer will need larger and larger administrative controls to reduce transaction costs, all this leads to execution of such relations in the form of a firm (H-F point in the

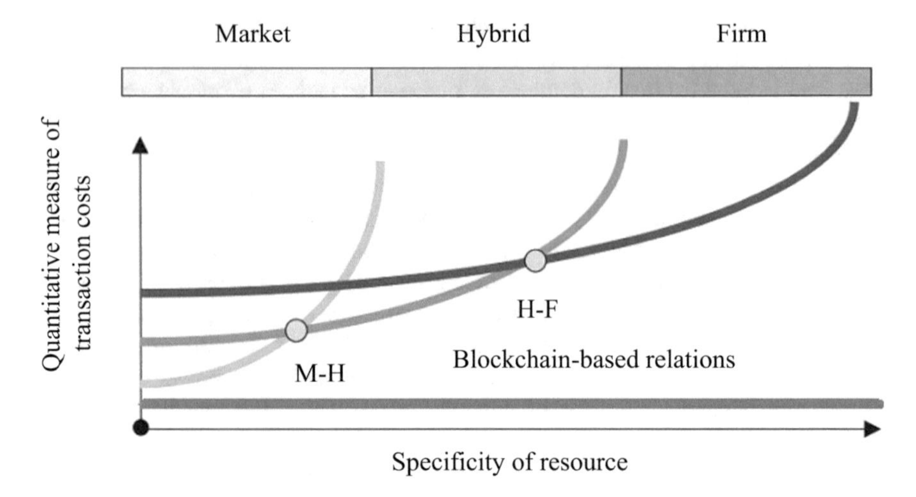

Fig. 3. Relation of the transaction costs amount to the specificity of the resource and forms of regulatory structures. Source: prepared on the basis of [4].

Fig. 3), where relations are implemented by order, and disputes are resolved through the reference to a higher authority in a company.

Oliver E. Williamson deserves credit for his having identified the factors that determine the amount of the transaction costs. According to him, it depends on the following: frequency of transactions, uncertainty and the specificity of the assets involved in the transaction. The choice of the form of the relations management and control type of transaction depends on the amount of transaction costs associated with such relations and, therefore, these three factors.

In the long term, blockchain technology implementation and expansion of its application scope in the relations sphere can lead to a reduction in the proportion of the "firm" regulatory form, the latter in some cases would simply be redundant. Thus, the matrix of selecting the preferred management form and contract type subject to various values of asset specificity, uncertainty and frequency of transactions in case of blockchain application can be displayed as follows (see Table 1).

In compliance with the matrix above, it can be concluded that under almost all conditions the customer of the relations prefers applying to the market regulatory mechanism using blockchain and only in specific cases where the relations with the executor of labor affect highly specific resources (e.g., unique knowledge, experience and competence of an employee) the customer may find reasonable to realize them through a hybrid form of relations in order to address unique challenges on a temporary basis and extremely rarely under high uncertainty and high frequency of transactions the customer may do that through the firm. And even in such a case it is the case due to the fact that no appropriate smart contracts have been developed for such relations and their terms have not been outlined for a particular occasion.

In conclusion, allow us highlighting the key advantage of a blockchain, which bears the possibility of a radical transformation in the labor system. In case it is used, the opportunism of actors of relations and uncertainty level of their transactions will be

Table 1. Matrix of selecting the management type depending on the specificity of labor, the uncertainty of the result and the frequency of labor transactions execution in case of blockchain usage. Source: author's work.

		Specific nature of a resource			
		Non-specific	Highly specific	Non-specific	Highly specific
Frequency of transaction	Accidental	Nonrecurring relation in the anonymous market	Market (standard smart contract)	Market (standard smart contract)	Hybrid (complex smart contract)
	Constant	Market (standard smart contract)	Market (standard smart contract)	Market (standard smart contract)	Firm (complex smart contract)
		Low or medium non-specificity		High non-specificity	

minimal since the trust between them is based on the principles of mathematics, economics and cryptography, and is maximally formalized. Thus, the presence of the third party between the relations participants, which is necessary to confirm eligibility of counterparties interaction, the start, execution, and end of the transaction is no longer required. If the key actors of labor relations are the employee and the employer, and the fact of their observing the terms for the interaction is provided by a complex system of institutions, the availability of specialized executers who organize and formalize the relations (labor market, employment services, human resources department, HR documents workflow, etc.), the widespread use of blockchain in the system of labor relations in the future enables eliminating the need for institutional support system, i.e. replaces the entire complexity and diversity of institutions with a set of rules (a "Protocol"). Everything that does not create value to people, that incurs a large amount of transaction costs could be formalized and automated through the use of blockchain technology and smart contracts. Is not this an allusion to the revolution?

4 Conclusion

The conclusion to be drawn in the result of the consideration over the blockchain in the field of labor relations is unequivocal. The impact of this technology on the prospects for the labor formalization will reveal itself in the coming years and it is not only about the need for the employers to implement blockchain in specific areas of HR management, but also about personnel employed in labor relations. Indeed, blockchain can ensure the implementation of the "talent on demand" principle, when a firm hires key experts on the open market, with minimal transaction costs. Even now there exist attempts to use this technology in the field of HR management, to formalize the documents on education (instead of traditional diplomas), labor relations (cancelling work books), searching and hiring talented employees, etc. However, the overview shows that in terms of economic theory perceptions, the scope of blockchain application can be much larger and more diverse. The analysis completed in this article allows concluding that blockchain is a possible basis for defining a new element in the chain of institutional agreements forms proposed by Oliver E. Williamson.

Funding. The reported study was funded by RFBR according to the research project №. 19-010-00933

References

1. Vlasov, A.I., Karpunin, A.A., Novikov, I.P.: System analysis of data exchange and storage blockchain. Mod. Technol. Syst. anal. Model. **3**(55), 75–83 (2017)
2. Genkin, A.S., Mavrina, L.A.: Blockchain plus "smart" contracts: advantages of application and emerging issues. Econ. Bus. Banks **2**(19), 136–149 (2017)
3. Genkin, A.S., Mikheev, A.A.: Blockchain on the Internet of Things. Insur. Bus. **10**(295), 3–11 (2017)
4. Kiryanov, I.V.: Quantitative assessment of the transaction costs of the organization: general methodical approach. Bull. NSUEF **1**, 78–101 (2015)
5. Kotlyarov, I.D.: The evolution of approaches to understanding the nature of the economic cell. Bull. Moscow Univ. Ser. 6: Econ. **5**, 3–25 (2016)
6. Coase, R.: Nature firms. SPb, School of Economics, pp. 11–32 (1995)
7. Christidis, K., Devetsikiotis, M.: Blockchains and smart contracts for the Internet of Things. IEEE Access **4**, 2292–2303 (2016)
8. Nakamoto, S.: Bitcoin: a peer-to-peer electronic cash system (2008)
9. Smart Contracts on Bitcoin Blockchain BitFury Group Sep 04 (Version 1.1) (2015)
10. Tapscott, D., Tapscott, A.: How blockchain will change organizations. MIT Sloan Manage. Rev. **58**(2), 10–13 (2017)
11. Williamson, O.E.: Transaction cost economics and organization theory. Ind. Corp. Change **2** (2), 113–117 (1993)
12. Wood, G.: Ethereum: a secure decentralized generalized transaction ledger. Ethereum Project Yellow Paper (2014)

Ball & Plate Model on ABB YuMi Robot

Lubos Spacek(✉) and Jiri Vojtesek

Department of Process Control, Faculty of Applied Informatics,
Tomas Bata University in Zlin, Nad Stranemi 4511, 760 05 Zlin, Czech Republic
{lspacek,vojtesek}@utb.cz

Abstract. The purpose of this paper is to present results of control of Ball & Plate model using the collaborative industrial robot YuMi. The paper presents first results after simulations and feasibility were examined in the pilot study. The linear quadratic (LQ) 2DoF controller was used because of its ease of use, robustness, and reliability. Following this study can help with projects that rely on fast synchronization of robots and external sensors such as dynamic scanning or laser cutting.

Keywords: Ball & Plate · YuMi · Robot · LQ control · Discrete-time

1 Introduction

There are many solutions for controlling the Ball & Plate model, such as research projects [1] and [2] or many hobby solutions aimed more for building the hardware than optimal control. However, there are not many solutions where an industrial collaborative robot is used. This paper is supposed to fill the hole by designing the Ball & Plate model for industrial collaborative robot YuMi build by ABB [3]. This paper follows the pilot study done in a similar manner to check the feasibility of the solution [4]. Satisfying results encouraged authors to continue to the next phase presented in this paper. The linear quadratic (LQ) 2DoF controller [5] was used to achieve fast implementation and reliability [6]. Authors have positive experiences with this type of controller for its "user-friendly" nature and satisfying results.

The paper is organized as follows. The model and the robot are described in the first chapter. The identification of the robot's arm dynamics and controller design follow afterward. Results of the real system and discussions close this paper.

1.1 Ball & Plate Model

The Ball & Plate model is an unstable system used mainly in education and hobby projects. The model can be also used for testing of control algorithms for unstable systems. The model in this paper consists of a plate made from a resistive touch screen which also serves as a sensor for obtaining a position of the ball.

R. Silhavy (Ed.): CSOC 2019, AISC 986, pp. 283–291, 2019.
https://doi.org/10.1007/978-3-030-19813-8_29

The 4-wire resistive touch screen is used, connected to a touch screen controller, which is connected to the control unit by USB cable. The control unit is chosen to be Raspberry Pi 3 and the touch screen acts as an event device in its operating system. Raspberry Pi is not the best choice for this kind of project, because it is not a real-time processing unit, it has its own operating system and thus it is not able to provide a precise sampling. Advantages of Raspberry Pi are a quick and easy implementation of the code, it has many communication interfaces and it was assumed (rightfully) that the bottleneck of the whole application will be the robot.

The resistive touch screen is a foil on a glass plate, thus it is used as the plate for the ball itself. The walls are made from flexible 3D printer filament. These walls are there only for testing purposes as the goal is for the ball not to fall over the edge. The plate has dimensions 322×247 mm, which is not a square base, but this does not affect the control strategy.

The ball needs to be heavy enough to work on the resistive touch screen, so a steel bearing ball was used for this purpose. It is made from 100Cr6 steel and with 25 mm diameter weighs about 64 g. The whole setup can be seen in Fig. 1.

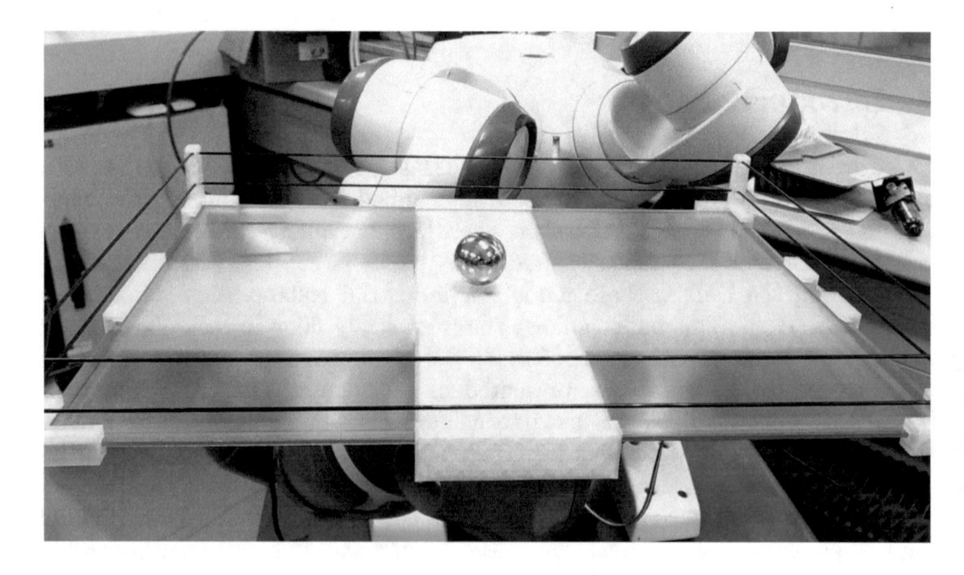

Fig. 1. Ball & Plate model with robot

1.2 YuMi® Collaborative Robot

YuMi is a collaborative, dual-arm industrial robot designed by ABB for small part assembly (Fig. 2) [3]. Each arm has 7 degrees of freedom (7DoF), thus providing better maneuverability. An industrial robot is designed for different type of operations and real-time control needs specific approach in industrial robotics.

Industrial robots follow point-to-point paths and they expect to finish the current path. This could lead to unexpected behavior in terms of non-linear sampling time. This problem can be addressed by using Externally Guided Motion (EGM) [7], which can significantly improve the lag of the robot. However, the EGM is available only for 6-axis robots, so there needs to be more research done in this area. Another solution can be the Robot Operating System (ROS) [8], which provides better control over the manipulator than the controller of the robot itself, but it does not directly solve the problem. The overall view of the robot YuMi holding the plate in simulation environment RobotStudio is shown in Fig. 3.

Fig. 2. ABB IRB 14000 YuMi

2 Methods

2.1 Identification

The identification of the system cannot be supported by a precise mathematical model, because parameters of motors in the robot are not known. It should be also noted that one arm is the 7DoF system of motors connected in series, which complicates the situation even more. This is the reason why the model was experimentally identified and approximated to the appropriate structure of continuous transfer function, which is subsequently discretized.

Measured data were averaged and the best fit for them was (surprisingly, to the complicated nature of the system) a continuous transfer function with 3 integrators, as can be seen in Eq. 1 and its discretized form in Eq. 2. It is better

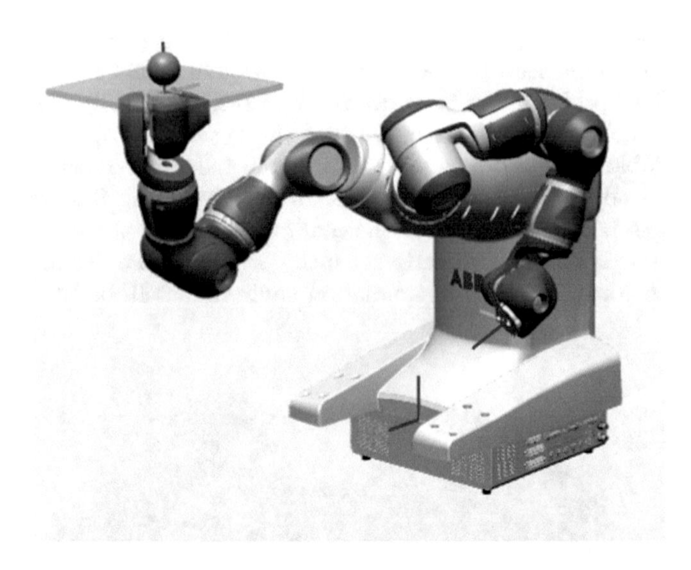

Fig. 3. The overall view of YuMi holding the plate

to identify the system in continuous form, as it requires only one parameter to identify (instead of 3 in the discrete form). The robot is clearly a non-linear system and trying to linearize it with simple transfer function is a bold move, but because there are still certain limits (e.g. slow robot movements), it is not wise to try designing very complicated or optimal solutions if the basic approach will not work.

$$G(s) = \frac{C}{s^3}, \tag{1}$$

where $G(s)$ is continuous transfer function with complex variable s and C is the acceleration gain of the system.

$$G(z^{-1}) = \frac{B(z^{-1})}{A(z^{-1})} = \frac{b_1 z^{-1} + b_2 z^{-2} + b_3 z^{-3}}{1 - 3z^{-1} + 3z^{-2} - z^{-3}}, \tag{2}$$

where $G(z^{-1})$ is discrete transfer function with complex variable z^{-1} and $B(z^{-1})$, $A(z^{-1})$ polynomials obtained from discretization.

Measurements were made for two angles (for x and y coordinates[1]) because the system is not symmetric and it cannot be assumed these movements have the same dynamics. The identification is based on step responses of the system for 5-degree angles in each direction. Averaged measurements and their approximations to transfer function in Eq. 1 for x and y coordinates are shown in Figs. 4 and 5 respectively.

[1] x coordinate angle is angle of the plate that changes only x position of the ball and y coordinate angle is angle of the plate that changes only y position of the ball.

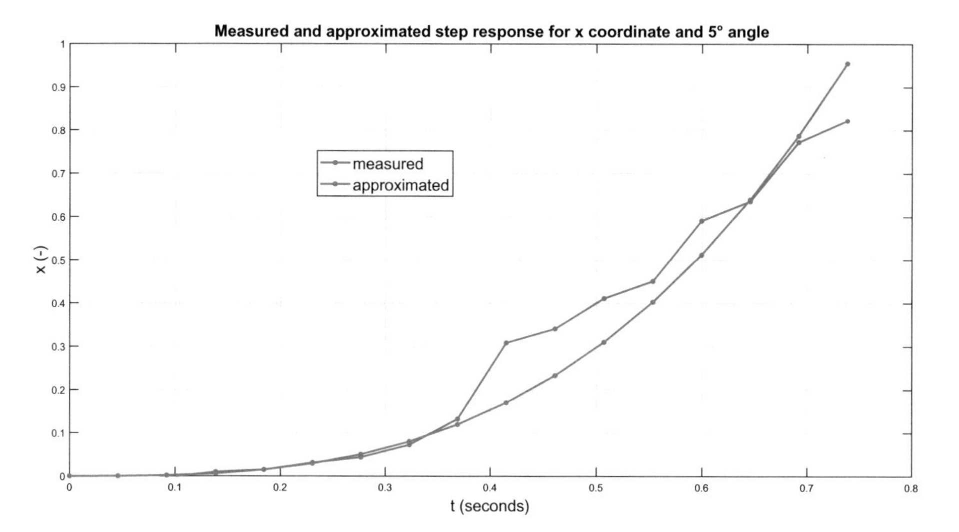

Fig. 4. Step responses of the system for x coordinate

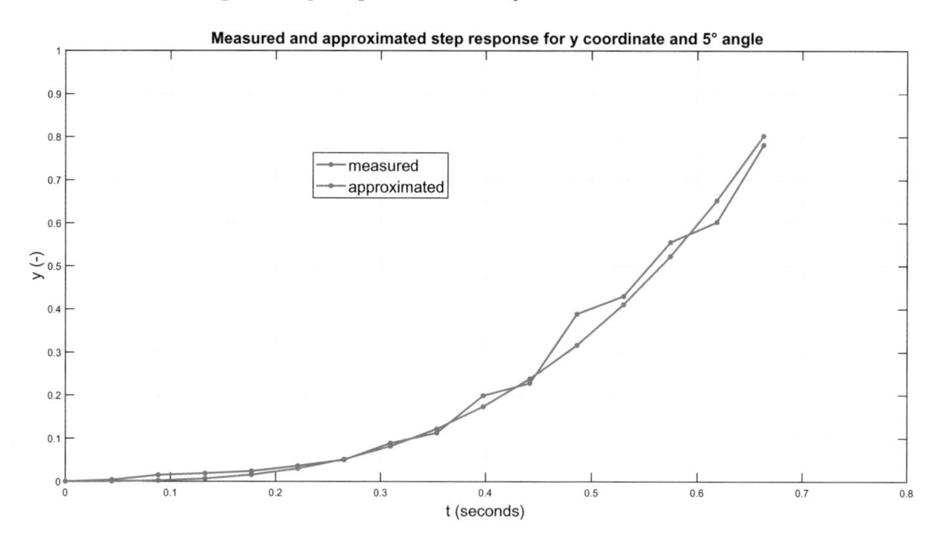

Fig. 5. Step responses of the system for y coordinate

Identified constants C_x and C_y for x and y coordinates respectively are:

$$\begin{bmatrix} C_x \\ C_y \end{bmatrix} = \begin{bmatrix} 2.8548 \\ 3.3091 \end{bmatrix} \tag{3}$$

2.2 Controller Design

Controller designed in this paper is linear quadratic (LQ) 2DoF controller closely described in [5] and [6]. It consists of feed-forward C_f and feed-back C_b parts as presented in Fig. 6, Eqs. 4 and 5. The summation portion $1/K(z^{-2}) = 1/(1-z^{-1})$ of these parts is taken out to keep the outputs of sub-controllers to rise endlessly. Figure 6 also shows reference value $w(k)$, output value $y(k)$, controller output $u(k)$, linearized plant $G(s)$ and disturbances $n(k)$ and $v(k)$.

This controller was chosen because it is easy to implement, design, very reliable and robust. It is able to seamlessly control the unstable system such as Ball & Plate almost "out-of-the-box" and needs very few parameters to be modified during the design. It is possible to move towards more advanced methods after the model will be operating at its maximum potential (most likely by incorporating EGM into the robot - see Sect. 1.2).

$$C_f(z^{-1}) = \frac{R}{P} = \frac{r_0}{1 + p_1 z^{-1} + p_2 z^{-2}}, \tag{4}$$

$$C_b(z^{-1}) = \frac{Q}{P} = \frac{q_0 + q_1 z^{-1} + q_2 z^{-2} + q_3 z^{-3}}{1 + p_1 z^{-1} + p_2 z^{-2}}, \tag{5}$$

where C_f, C_b are feed-forward and feed-back parts of the controller respectively and p_i, q_i, r_i are coefficients of the controller that need to be calculated.

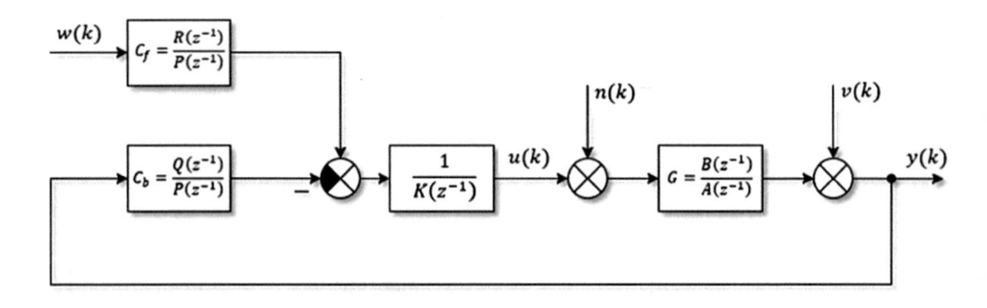

Fig. 6. Structure of 2DoF controller

3 Results

The results are shown in Figs. 7 and 8. The robot was stabilizing the ball in the center of the plate, thus reference value was equal to 0 for both coordinates. Coordinates x and y are normalized and angles of the plate u_x and u_y are measured in degrees. Note that the angle u_x changes the x coordinate and the u_y changes the y coordinate. Close-ups of specific regions are enlarged to better show plots for the angle of the plate, mainly in the nearly-stabilized region.

It can be seen that the stabilization process is not so fast or precise. Note that the noise around the center is not that significant compared to the size of the ball. The controller certainly works, but the process could be faster and more optimal. This is not the fault of the controller, but restrictions of the

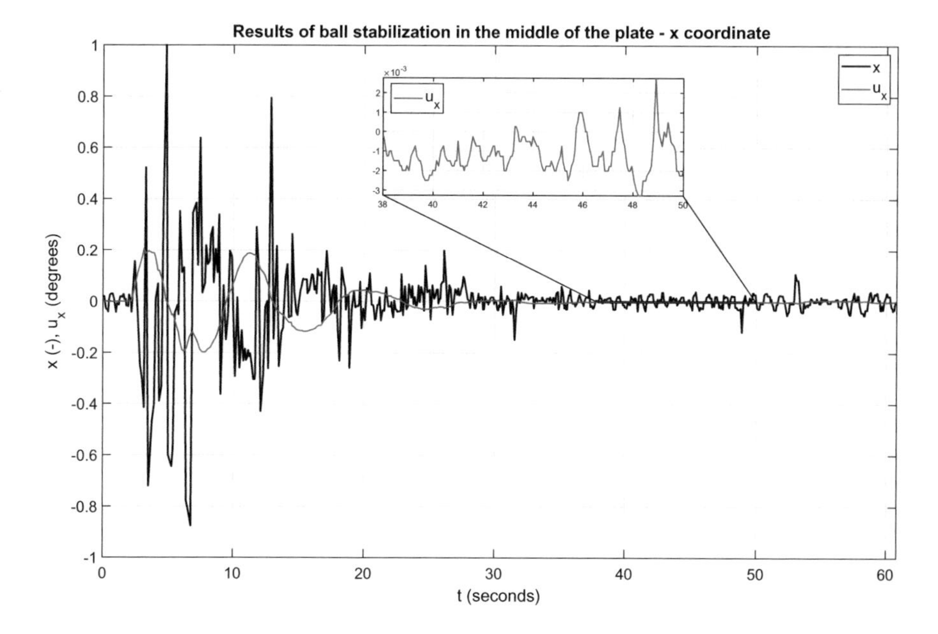

Fig. 7. Results for x coordinate

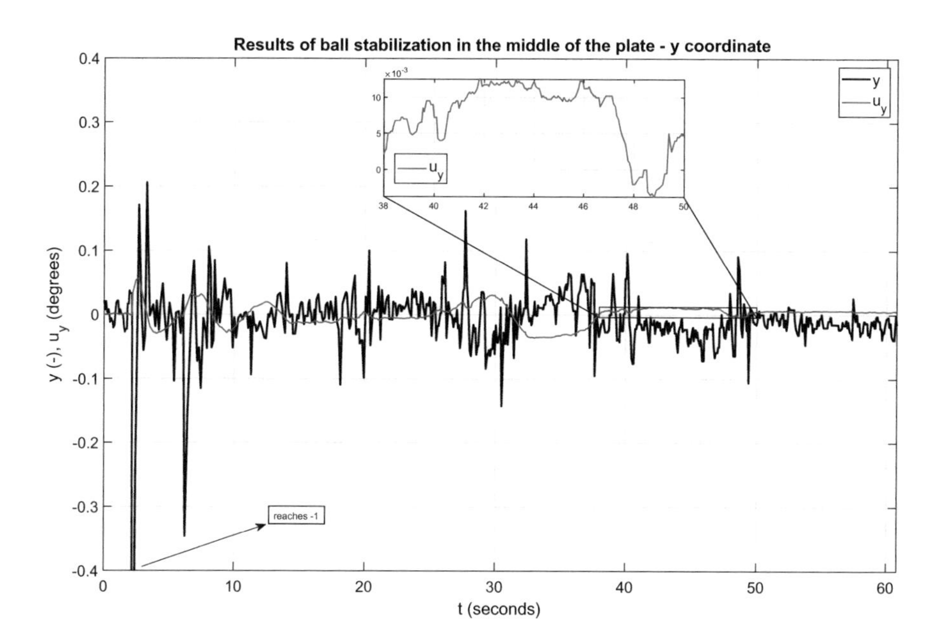

Fig. 8. Results for y coordinate

robot mentioned in Sect. 1.2. It has been found that the robot executes only every sixth command from the controller. This is caused by the point-to-point design of the robot. It has to go to the given target and only after it reached the correct position can continue with another. This process takes more time than is required, but the robustness and quality of the designed controller make up for this drawback. The solution is to use EGM described in Sect. 1.2, which can be used to directly control the robot overcoming the point-to-point strategy.

The output of the 2DoF controller is calculated using Eq. 6 for both the x coordinate and y coordinate.

$$\begin{aligned} u_k &= (1-p_1)u_{k-1} + (p_1-p_2)u_{k-2} + p_2 u_{k-3} + \\ &\quad + r_0 w_k - q_0 y_k - q_1 y_{k-1} - q_2 y_{k-2} - q_3 y_{k-3} , \end{aligned} \tag{6}$$

where u_{k-i} and y_{k-i} are outputs of the controller and plant respectively, w_k is reference value and individual calculated parameters are shown in Eq. 7 for x coordinate controller and in Eq. 8 for y coordinate controller. It is implemented in Raspberry Pi 3, which then sends the desired angle to the robot using the UDP protocol.

$$\begin{bmatrix} r_0 \\ p1 \\ p2 \end{bmatrix} = \begin{bmatrix} 0.0017 \\ -1.5525 \\ 0.6308 \end{bmatrix}, \quad \begin{bmatrix} q0 \\ q1 \\ q2 \\ q3 \end{bmatrix} = \begin{bmatrix} 40.7130 \\ -119.4071 \\ 116.7777 \\ -38.0819 \end{bmatrix}. \tag{7}$$

$$\begin{bmatrix} r_0 \\ p1 \\ p2 \end{bmatrix} = \begin{bmatrix} 0.0015 \\ -1.5510 \\ 0.6298 \end{bmatrix}, \quad \begin{bmatrix} q0 \\ q1 \\ q2 \\ q3 \end{bmatrix} = \begin{bmatrix} 30.5838 \\ -89.6308 \\ 87.5942 \\ -28.5457 \end{bmatrix}. \tag{8}$$

4 Discussions

Industrial robots are not designed for the operation described in this paper, however, presented results show it is not impossible. These are first results for the real system, thus they are satisfactory enough to continue with the project. Authors anticipated it will be interesting to use an industrial robot instead of own hardware solution, but it turned out it is quite challenging to solve problems with black-box devices when they appear. The first major problem that emerged is the robot is not fast enough to execute all instructions. This is solvable by using special software extension for ABB robots called EGM or external robot control software called ROS. More advanced controllers can be designed afterward, compared and results evaluated. Results can be used for fast processes where the movements of the robot are controlled by external process and the position of the robot has to be dynamically synchronized (e.g. during scanning or laser cutting using robot).

Acknowledgments. This article was created with support of the Ministry of Education of the Czech Republic under grant IGA reg. n. IGA/CebiaTech/2019/004.

References

1. Jadlovska, A., Jajcisin, S., Lonscak, R.: Modelling and PID control design of nonlinear educational model Ball & Plate. In: 17th International Conference on Process Control, Strbske Pleso, pp. 475–483 (2009)
2. Knuplez, A., Chowdhury, A., Svecko, R.: Modeling and control design for the ball and plate system. In: IEEE International Conference on Industrial Technology, Maribor, Slovenia, pp. 1064–1067 (2003)
3. ABB Robotics - IRB 14000 YuMi. http://new.abb.com/products/robotics/industrial-robots/yumi
4. Spacek, L., Vojtesek, J., Zatopek, J.: Collaborative robot YuMi in Ball & Plate control application: pilot study. In: Silhavy, R., Senkerik, R., Kominkova Oplatkova, Z., Prokopova, Z., Silhavy, P. (eds.) Cybernetics and Algorithms in Intelligent Systems. CSOC 2018. Advances in Intelligent Systems and Computing, vol. 765, pp. 167–175. Springer, Heidelberg (2018)
5. Bobal, V., Bohm, J., Fessl, J., Machacek, J.: Digital Self-tuning Controllers. Springer, London (2005)
6. Spacek, L., Bobal, V., Vojtesek, J.: LQ digital control of Ball & Plate system. In: 31st European Conference on Modelling and Simulation ECMS 2017, pp. 427–432. European Council for Modelling and Simulation (2017)
7. Externally Guided Motion, Application manual - Controller software IRC5 - 3HAC050798-001 Revision C, pp.326–375
8. Robot Operating System. Springer, New York (2018). ISBN 978-3-319-91589-0

Automatic Parametric Fault Detection in Complex Microwave Filter Using SVM and PCA

Adrian Bilski(✉)

Warsaw University of Life Sciences, Nowoursynowska 159, Warsaw, Poland
blindman26@o2.pl

Abstract. The aim of this paper is to present the diagnostics of complex linear analog systems with parametric faults, using Support Vector Machine (SVM) as a tool for fault location. The diagnostic results of a microwave filter with the help of SVM network are presented. A strategy for finding the optimal kernels and their parameters for the particular system under test is proposed. A method for characteristic points reduction based on the statistical PCA method is also presented.

Keywords: Complex analog systems · Microwave filter · Support Vector Machine · Parametric fault detection · Principal Component Analysis

1 Introduction

Complex analog systems process continuous signals based on large number of parameters. Decisions regarding such a system are difficult and depend on the designer's intuition and experience. Even the system with moderate number of parameters but ill-conditioned, can pose a challenge because of ambiguity groups [1]. The analog system is considered complex for at least twenty parameters. The information about the systems behavior is collected from the measured signals, so the lesser number of parameters that shape these outputs, the easier diagnostics is. In literature mainly simple systems are considered (like induction machines [2], dc motors [3] or pumps [4]).

The aim of the diagnostic process is the assessment of the System Under Test (SUT) functionality in reference to the design specifications, based on the analysis of the observable functions recorded at the accessible or partially accessible nodes. The analysis is focused on determining the differences between the actual features (symptoms) and their nominal counterparts.

Two diagnostic approaches exist [5]: the Specification-Driven Test (SDT) and the Fault-Driven Test (FDT). The first one is used to assess whether the analyzed system meets the specification, while the second one is used to locate the faulty element, responsible for the system's faulty behavior. Both are applied regarding parametric simulation, which allows to acquire information about the systems behavior by changing its parameters values and recording responses. This information can then be

R. Silhavy (Ed.): CSOC 2019, AISC 986, pp. 292–305, 2019.
https://doi.org/10.1007/978-3-030-19813-8_30

applied to the expert system exploiting AI based fault detectors in order to assess the state of the analyzed object.

Although universal methods to effectively diagnose complex analog systems do not exist, some approaches in that field have been made [6]. The decomposition approach being is the most interesting [7]. Test equations are derived for a partitioned network from Kirchhoff current law equations at the partition points. Voltages at these points are used to identify network parameters and reduce the influence of a faulty element on its vicinity, facilitating the procedure. The high computation requirements of such a method limit the size of testable circuits. Modern heuristic approaches made this approach obsolete. The faulty element of the selected subsystem has an impact on the remaining subsystems, leading to the two-staged identification procedure. In the first step the faulty subsystem is located, in the second – the element itself.

Even though such approaches provide useful information on the system's behavior, its diagnostics still requires optimization of set of test nodes, excitation signals and analysis domains.

In the case of complex microwave circuits the probability of a proper fault identification decreases with the increase of the system size. This justifies searching for new, more efficient methods.

This paper presents the diagnostic system based on the Support Vector Machine (SVM) binary classifier as fault detector with PCA method utilized as the means of symptoms reducer.

The SVM was used in analog systems diagnostics many times [2, 8–10]. Although proven useful, the problem of optimal kernel selection and its parameters remains [11]. This paper demonstrates how diagnostics of complex systems can be successfully conducted using SVM networks. The additional problem considered in this research is efficiency of the classifier in noisy conditions. Selection of a kernel parameters proper for the task of diagnostic process based on parametric simulation is also presented. It is a novel approach to diagnosis of complex microwave systems.

Principal Component Analysis (PCA) has already been applied to engineering [12, 13]. Here it is used to establish a relationship between observable symptoms of output signals and assess their relevance in the fault classification process. PCA is only used as a tool for optimal selection of the characteristic points and there is no direct dependency between the Principal Components (PCs) and the SVM kernel parameters.

The paper has the following structure. Section 2 presents the diagnostic principles. The description of data processing methods and its implementation is presented in Sect. 3. Section 4 contains the description of the diagnostic methodology. It provides illustrative example, i.e. diagnostics of a 37-element microwave filter and diagnostic results. In Sect. 5 conclusions and future prospects are included.

2 Diagnostic Principles

This paper is devoted to Linear Time-Invariant Systems (LTI) diagnostics. The output signal of such systems is a superposition of components, which are systems responses to input signal components (Fig. 1).

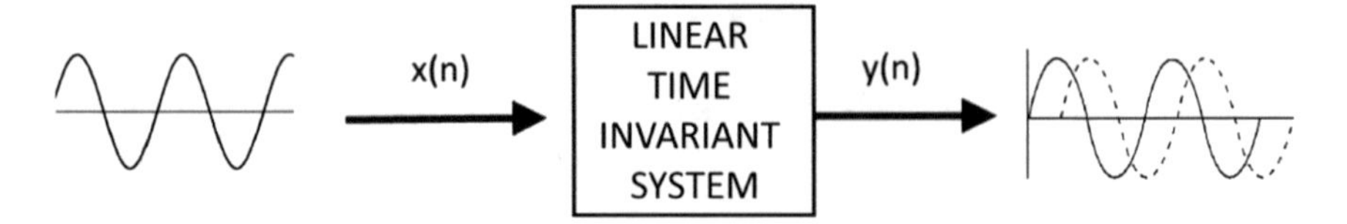

Fig. 1. The input-output relationship in linear time-invariant systems.

For input signals $x_1(t)$, $x_2(t)$ and output signals $y_1(t)$, $y_2(t)$, the additivity and homogeneity conditions are realized:

$$a \cdot x_1(t) + b \cdot x_2(t) \xrightarrow{system} a \cdot y_1(t) + b \cdot y_2(t). \tag{1}$$

The most prominent is the input-output analysis. This is due to the tendency of integrating modern analog circuits in the single chip, limiting the number of accessible nodes. The circuit is treated as a black box, measurable only at external nodes. This often means the single input and single output node. Given the voltage power source and an output voltage signal, the information measured during the circuit analysis is acquired from only two sources: the input current and output voltage. This analysis is limited to the output voltage only, since the input current and the output voltage values for are correlated and their inclusion in the analysis does not improve fault detection.

The proper preparation of datasets for the AI-based diagnostic module requires selection of analysis domains, excitation signals, testing frequencies and the set of the relevant symptoms. Because at least part of this information is unknown a priori, it is reasonable to initially utilize all information acquired from the SUT during the parametric simulation.

2.1 Data Acquisition

The LTI systems can be analyzed in time, frequency and mixed domains. In the time domain the SUT parameters are acquired from the periodic or quasi-periodic waveforms. In the frequency domain, the spectral analysis is performed, through the means of Fast Fourier Transform (FFT) or the Total Harmonic Distortion (THD) [14]. In the mixed domain both time and frequency components are considered. The methods commonly used here include Short Time Fourier Transform (STFT) or Wavelet Transform (WT) [15].

Each parametric simulation produces a measurable output signal, from which a set of characteristic points can be registered. These points provide the data on the SUT deviation from the nominal state. The procedure is repeated for every element separately (the remaining elements are at nominal values).

Numerous feature parameters of the signal analyzed in the time domain can be acquired from the parametric simulation. These include the dimensional parameters (mean, RMS) and non-dimensional ones, like waveform index, pulse index, etc. The frequency spectrum can take into account the features acquired from amplitude spectrum as well as power spectrum.

2.2 Data Creation

The diagnostic process is limited to single faults detection. All single faults are equally likely. During each parametric simulation, the learning (L), testing (T) and validation (V) datasets are created [16]. The classification method based on the AI algorithm assigns the vector of measured symptoms (simulation results) $\boldsymbol{e_i}$ to a particular fault identifier $\boldsymbol{c_l}$, making the supervised learning possible (2). This knowledge is gathered in the learning set $\boldsymbol{L}$, which contains patterns describing the systems behavior dependent on different parameter configurations (those, that simulate parametric faults). Patterns supplied to the set should cover the maximal number of states of the diagnosed object, including its nominal state (and tolerances) and particular faults. It increases the chance to generalize the diagnostic knowledge (that is the proper reaction to patterns, which have not been supplied to the classifier before). Each pattern was assigned multiple values (parametric sweep). The number of patterns in the set should be possibly low, which allows for the simulation process to be short and to avoid creating too complex knowledge during the process of learning.

$$\boldsymbol{V} = \boldsymbol{T} = \boldsymbol{L} = \begin{bmatrix} \boldsymbol{e_1} \\ \vdots \\ \boldsymbol{e_n} \end{bmatrix} = \begin{bmatrix} s_{11} & \cdots & s_{1m} & c_1 \\ \vdots & \ddots & \vdots & \vdots \\ s_{n1} & \cdots & s_{nm} & c_k \end{bmatrix} \tag{2}$$

The knowledge on how to classify patterns is then verified with the utilization of $\boldsymbol{T}$ set of analogous structure, that symbolizes patterns acquired from the real world, utilized to verify the classifiers precision after the process of learning has been conducted. Both $\boldsymbol{L}$ and $\boldsymbol{T}$ sets are created during systems simulation in off-line mode, during which the calculation time is of secondary importance. Often the validation set $\boldsymbol{V}$ is also utilized, with which the optimization of the classifier parameters can be performed. The aim of optimization is to find the best vector of attributes in a particular solution (for example in the configuration of classifiers parameters, that maximize its accuracy).

Element tolerances ($\pm 5\%$) are also introduced in the diagnostic problem. By definition, tolerance inclusion blur the circuits frequency and time characteristics, which makes it difficult to detect faulty element and to determine the systems nominal state. The problem of tolerance increases at the end of the frequency range, due to similarity between the characteristics of the faulty and nominal SUT (5% deviation of the parameter – nominal state, 5.1% deviation – faulty state). Uncertainty conditions both related to the operation of the SUT and its environment influence the measurement accuracy. This in turn may degrade fault detection. Fault identification may be further negatively affected by the existence of Ambiguity Groups (AG). These are subsets of SUTs parameters indistinguishable in the objects response [17].

While the first two phenomena may be dealt with the proper utilization of PCA method, the elimination of the latter one is done with the inclusion of accessible nodes [9]. However, the parametric simulation will not produce satisfying results when the SUTs behavior has low sensibility to the particular parameter.

An open problem is the division of patterns between particular sets, which can have significant influence over the efficiency of classification method. Here, the 4-fold cross-validation method has been utilized.

3 Data Processing Tools

This section provides information about AI algorithms utilized for the purpose of data processing and decision making. The proper mechanics of SVM utilization in parametric fault detection has been introduced here, for both single and multiple fault detectors. The aforementioned classifier operates on data, which were reduced using the basic PCA method.

3.1 Fault Classifier (SVM)

The Support Vector Machine is a binary classifier, whose behavior is similar to that of a single perceptron. It differentiates linearly nonseparable data into distinguishable categories assuming a parametric kernel function. It can be thought of as a quadratic programming problem concerning the inequality constraints. The learning process of a single linear SVM classifier in the context of the problem solved here can be described as follows [18].

Given a training set for the object x (represented by n observable characteristic features: $(\mathbf{s}_1, \ldots, \mathbf{s}_n)$) and corresponding binary class labels $y_i \in \{-1, +1\}$, $i = 1, 2, \ldots, n$, vector s is identified as belonging to one of two categories. The classification function has the form:

$$\begin{cases} if \quad \mathbf{w}^T s_i + b \geq 0 \quad then \quad y_i = +1 \\ if \quad \mathbf{w}^T s_i + b < 0 \quad then \quad y_i = -1 \end{cases} \tag{3}$$

$\mathbf{w}^T \cdot s + b = 0$ is a linear equation of a hyperplane, separating two categories, $\mathbf{w}^T = [w_1, w_2, \ldots, w_n]$ is the transpose weight vector, displacement b is a scalar used to define the position of the hyperplane (offset value), while $\mathbf{s}_i = [s_1, s_2, \ldots, s_n]$ is the symptoms vector (consisting of real numbers) of the analyzed pattern. A two-dimensional hyperplane is of the following form: $w_1 s_1 + s_2 w_2 + b = 0$. The resultative value y_i represents the binary value (-1 or 1) constituting a part of fault category representation assigned to training data. Its index relates to the binary position of the number in the category code for this pattern. There is an infinite number of solutions $(\mathbf{w}, b)$ of Eq. (3). In order for the solution to become unique, a normalization is necessary, executed by calibration of the $(\mathbf{w}, b)$ pair in such a way, so that the closest patterns to the hyperplane would fulfil the condition $w^T \mathbf{s}_i + b = 1$. Figure 2 shows such normalization effect (the separating function is denoted by $w^T \mathbf{s}_i + b = 0$).

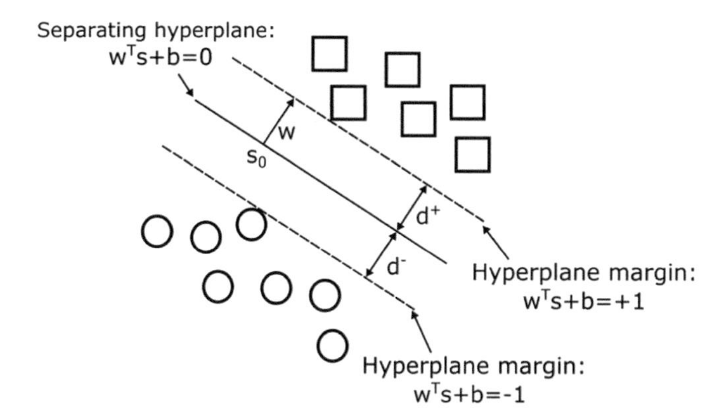

Fig. 2. The normalization effect of patterns belonging to two categories, gathered on a 2D plane.

The margin width *d* is normalized depending on the value of the weight vector:

$$\mathrm{d} = \frac{2}{\|\boldsymbol{w}\|} \tag{4}$$

where $\|\boldsymbol{w}\| = \sqrt{w_1^2 + w_2^2 + \ldots + w_h^2}$ denotes the length of the vector perpendicular to the separation function. To acquire the best pattern separation, the margin *d* should be maximized.

The number of SVMs during the training stage is usually equal to the number of training data. Depending on the value restrains, the complexity of the neural network is reduced and only a part of training vectors become the support vectors. Their number at the end of the training stage therefore denotes the complexity of acquired hyperplane.

As a result of such transformation, a non-linear (in this case elliptic) mapping function $\phi(s)$ is created. Kernel $K(s, s)$ is defined as the square of mapping function:

$$K(s, s) = \phi^T(s)\phi(s) \tag{5}$$

Because $K(s_i, s_j) = K(s_i, s_j)$, the kernel function is symmetrical.

Unlike typical neural networks, in which the output of the network depends on the activation function $\varphi(s)$, the quality of the SVM classification is primarily influenced by the kernel function.

In order to optimize the SVM classifier, proper kernel selection is required. Most of them are parametrized with non-negative real values (such as the width of rbfs or the degree of the polynomial). Among the available functions, nonparametric ones can also be found (eg. splines or linear function). Their classification efficiency is usually low, due to the fact them being dependent only on the quality and the size of the training dataset. The proper description of the most popular SVM kernels together with their parameters can be found in [2].

Identification of multiple faults in the analyzed analog systems requires classifier adjustment. Single SVM realizes binary classification exclusively, allowing only for two-class separation. In multiclass classification case, the combination of multiple such classifiers should be utilized, which in turn is related to the output coding method. Among the most popular ones are Minimum Output Coding (MOC) [19] or Error Output Coding [20]. However simpler strategies exist, like One-versus-All (OVA) or One-versus-One (OVO), which allow for simplification of the diagnostic module. The main flaw of the OVO strategy is the increase of the binary units necessary for separation of equal amount of categories. For that reason, the OVA strategy has been utilized in this paper.

3.2 Feature Reduction Using PCA

The direct use of the measured values of the circuit's output signals to detect damage would be ineffective due to the significant dispersion of the measured signals. During the learning phase, the automatic selection of the most important features should be conducted. The impact of the particular measured data on the analyzed system dynamics is often unknown. In such case a greater amount of data than it is necessary is acquired. The PCA method is necessary to separate those features of the measured system, that well represent the measurement points, while being resistant to the impact of changes during circuits analysis (eg. the value of the power signal or the order in which particular measurements were conducted). These features are chosen so that their variance corresponding to signals belonging in the same group (eg. fault of a particular element) was minimal, and at the same time that the relative differences between groups of data representing different faults were as large as possible. The most important stage of this operation was the elimination of the constant measurement component, which allowed to emphasize the greater relative variability between the signals. Subsequently the set of symptoms should be minimalized in order to make the classifier simpler, which in turn lowers the costs of the diagnostic procedure. PCA has already been used in electrical machine diagnostics [21]. Here it is used as means of characteristic points reduction from the set of all characteristic points acquired from the SUTs response.

Based on the covariance or correlation matrix, constructed based on the input information, a data reduction takes place, to those with the greatest impact. In the first case variables with the greatest variance have a greater impact on the given result, which is desired when the observation range is similar. Alternatively, the utilization of correlation matrix is associated with the preliminary normalization of the input data in such a way, that each variable has identical variance by default, which can be desired when the variable values have different range. For that reason, the correlation matrix was used here.

Given a set of n dimension feature vectors x_t $(t = 1, 2, \ldots, m)$, where $n < m$, let

$$\mu = \frac{1}{m}\sum\nolimits_{t=1}^{m} x_t \tag{6}$$

The covariance matrix of feature vectors is calculated based on:

$$C = \frac{1}{m}\sum\nolimits_{t=1}^{m} (x_t - \mu)(x_t - \mu)^T \tag{7}$$

Principal Components (PCs) are calculated by solving the eigen value problem of covariance matrix C,

$$Cv_i = \lambda_i v_i \tag{8}$$

where λ_i denote eigenvalues sorted in a descending order, while $v_i (i = 1, 2, \ldots, n)$ denote the corresponding eigenvectors.

$$\frac{\sum_{i=1}^{k} \lambda_i}{\sum_{i=1}^{n} \lambda_i} \geq \theta \tag{9}$$

where θ denotes the precision parameter.

Let $V = [v_1, v_2, \ldots, v_k]$, $\Lambda = diag[\lambda_1, \lambda_2, \ldots, \lambda_k]$. The low-dimensional feature vectors (PCs) are determined by

$$P = V^T x_t \tag{10}$$

The number of PCs equals the number of original variables, however usually only the first couple of them have the variance value big enough, to be essential in the classification process. Each variance is normalized, so that its mean value equals zero, while the standard deviation equals one.

4 Diagnostic Methodology

The diagnostic method is divided into two parts. The first one, identical for any simulated system, is responsible for preparing the data set and performing the SVM training. In the presented example, the set is generated from the simulations in the Micorwave Office software. The generated set provides the learning data for the second step of the method. This is used to evaluate the generalization ability of the fault classifier.

The learning-testing procedure is repeated multiple times to verify the optimal configuration of the classification method. The Steve Gunn's SVM toolbox was used in this research. The kernel's parameter must be tuned to the problem, as they significantly influence the SVM behavior. The applied parameter optimization was the exhaustive search on the discretized range of values. For rbf kernels, this range was $1 \cdot 10^{-7} \div 1 \cdot 10^{0}$, while for polynomial kernel it was $1 \div 10^{3}$. The selected interval was divided into 100 parts, with single value randomly taken from each one. The SVM with the particular parameter value is then trained on the learning set and its performance evaluated on the testing one. The optimization gives the parameter providing the best diagnostic result.

The diagnostic quality is measured as the percentage of the correctly classified patterns p_O to the dataset cardinality n

$$c = \frac{p_o}{n} \cdot 100[\%] \tag{11}$$

Two applications of the fault classification were used: fault detection (a go/no go test) and identification (determining the discrete measure of difference between the actual SUT parameters and their nominal counterparts). The fault categories in the second case are determined by parameters' values deviations from their tolerance regions (Fig. 3). Each parameter p_i is divided into (at least three) intervals, each representing the degree of its deviation from the nominal value p_{inom} (in both positive and negative direction). The thresholds between the categories are selected to describe the SUT behavior. The change in the parameter making it fault source was set to be above ±5% of its p_{inom} (the predefined tolerance). This way three categories are determined for p_i: "nominal", "smaller than nominal" and "larger than nominal". Additional intervals can be selected to increase the identification resolution (with the arbitrary thresholds). In the presented case the following intervals were used: ±5% (categories 1 and 2), ±10% (categories 3 and 4) and ±20% (categories 5 and 6). Adding more categories increases the difficulty of the diagnostic task. During the simulations it was assumed that the SUT contains gradual changes of parameters with the uniform distribution up to the 100% of deviation.

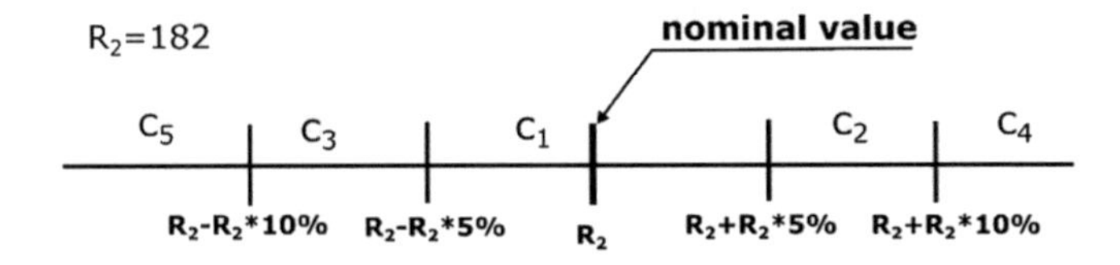

Fig. 3. Assignment of new fault codes based on the deviation from the nominal value of the R2 resistor.

Each element was first assigned a range of actual values fitting the intervals defined above. For instance:

- Rs1 was assigned values from 1 to 10 Ω
- C5 was assigned values from 10 to 100 pF
- Rp2 was assigned values from 10 to 100 Ω
- L4 was assigned values from 50 to 500 nH.

The simulation was performed after inserting the single fault into the SUT structure, remaining parameters were within their nominal ranges. Some fault may not be visible in responses because of the parameter's low sensitivity.

4.1 System Under Test Description

The diagnosed linear circuit consists of 37 elements forming the cascade – 11 capacitors, 10 inductances and 14 resistors (Fig. 4). The information about the proper functioning of the device is acquired from the transmittance characteristic measured on the output node (PORT 2–50 Ω load). In this case only input-output analysis was considered. Because of the large number of elements, such a diagnostic task is difficult, but introducing additional nodes to the model would require making changes in the circuit casing, so it was omitted. This system was not only modeled in the PSpice software, but also evaluated in the laboratory. This allowed for verifying the accuracy of the model during the real-world object analysis.

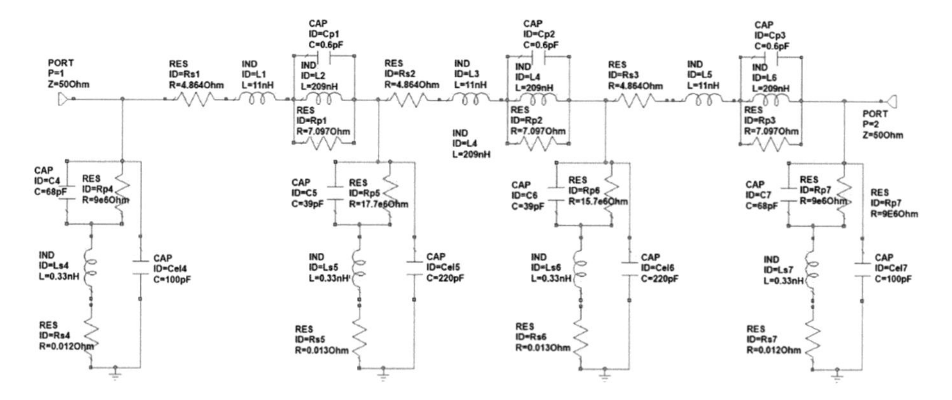

Fig. 4. 37-element microwave filter, modeled in microwave office.

For this purpose, simulated characteristics (denoted with blue color in Fig. 5) are compared with the characteristics measured in the real system (violet color). Transmittance S is thus acquired, understood as the ration of the incident to the reflected wave. The reflection coefficient is below 10 decibels. There are slight differences between the two characteristics in the nominal state, which is why a tolerance of ±10 dB was assumed, indicating that the system is treated as the nominal one.

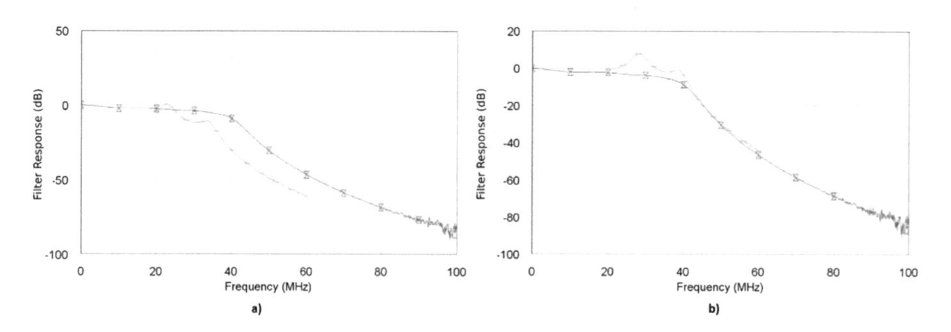

Fig. 5. Exemplary responses of the microwave system in the input-output analysis for the parameter change of (a) Rs_3, (b) Rp_7.

The parameter values of the individual system components are as follows: $Rs_1 = 4.864\ \Omega$, $Rs_2 = 4.864\ \Omega$, $Rs_3 = 4.864\ \Omega$, $Rs_4 = 0.012\ \Omega$, $Rs_5 = 0.013\ \Omega$, $Rs_6 = 0.013\ \Omega$, $Rs_7 = 0.012\ \Omega$, $Rp_1 = 7.097\ \Omega$, $Rp_2 = 7.097\ \Omega$, $Rp_3 = 7.097\ \Omega$, $Rp_4 = 9e6\ \Omega$, $Rp_5 = 17.7e6\ \Omega$, $Rp_6 = 15.7e6\ \Omega$, $Rp_7 = 9e6\ \Omega$, $L_1 = 11nH$, $L_2 = 209nH$, $L_3 = 11nH$, $L_4 = 209nH$, $L_5 = 11nH$, $L_6 = 209nH$, $Ls_4 = 0.33nH$, $Ls_5 = 0.33nH$, $Ls_6 = 0.33nH$, $Ls_7 = 0.33nH$, $Cp_1 = 0.6pF$, $Cp_2 = 0.6pF$, $Cp_3 = 0.6pF$, $C_4 = 68pF$, $C_5 = 39pF$, $C_6 = 39pF$, $C_7 = 68pF$, $Cel_4 = 100pF$, $Cel_5 = 220pF$, $Cel_6 = 220pF$, $Cel_7 = 100pF$.

4.2 Feature Extraction

The following set of characteristic points used to conclude about the actual state of the SUT was prepared:

- time domain: the values of maximum and minimum in the output signal with time instants required to reach them and zero-crossing coordinates.
- frequency domain: amplitude and phase values for the *m* calculated frequencies (this way *2m* values were acquired) and the 3db-frequency.

The sine input signal was used as an excitation. The selected characteristic points allowed for acquiring the most important information from the output signal. Behavior of each diagnosed SUT parameter was represented by eight examples. This is a compromise between the minimum number of patterns required to perform the diagnostics and the complexity of the data acquisition.

Table 1 presents the fragment of the training set, i.e. fault scenarios for 37 elements of the microwave filter. Here (P_1, T_1) and (P_2, T_2) are the coordinates of the first two extremes of the response signal. The element code is the number of the circuit element whose value is currently changed. The fault code is the number of the faulty parameter of "0" in the nominal state. For instance, the resistor R_{s1} has number 1, R_{s2} has number 2.

The size of the training set was 296 patterns, while the size of validating set consisted of 222 patterns. Each element was described by 6 patterns, selected randomly in two from the intervals, respectfully ±5%, ±10% and ±20%.

Table 1. Fragment of the learning dataset prepared for the purpose of the diagnostics.

U1 [V]	T1 [s]	U2 [V]	T2 [s]	Element code	Parameter value	Fault code
3.92e−2	4.22e−6	−3.90e−2	4.72e−6	1	5.00e+2	1
2.22e−2	3.23e−6	−2.24e−2	3.73.e−2	1	7.00e+2	1
1.46e−1	4.19e−6	−1.46e−1	4.70e−6	3	2.50e+1	0
5.44e−2	4.22e−6	−5.19e−1	4.70e−6	4	1.00e+2	1

4.3 Feature Reduction

To reduce the features dimension and to determine dependence between particular symptoms, a covariance matrix with the size of 384 × 18 was constructed: $X^{18\times384}$. The corresponding eigenvectors matrix takes the form of $V^{18\times18}$.

To keep most of the original information, the precision parameter (which denotes the approximation precision of the largest eigenvalues) should be $\theta = 0.9$. After applying (8), the number of symptoms can be reduced to K = 4. This information is then utilized to construct the new matrix V_4. The new sample data can then be calculated based on the formula (9) with four equivalent features $P^{4\times384}$.

Table 2 presents the Principal Component (PC) matrix for analyzed circuit, with columns representing individual PCs, grouped based on the decreasing variance.

Table 2. Exemplar table of the PCA results.

PC1	PC2	PC3	PC4
−0.06560	0.5062	−0.0678	0.5062
−0.0200	0.4933	−0.6785	0.4933
0.7553	0.5156	0.0290	0.5156
−0.1085	0.4844	0.7309	0.4844

4.4 Diagnostic Results

The general structure of the radial type SVM network designed for the purpose fault detection is n-m-1, where n determines the number of input units, while m denotes the number of support vectors (hidden neurons acquired in the process of network training). Because the classifier is meant for fault detection only, it is equipped with a single output. The structure of the neural network obtained as a result of learning and testing procedure on the gathered diagnostic data with zero element tolerance was 10-43-1. In case of ±5% assumed element tolerance, this structure was 10-24-1. The results of the two-class categorization (0% and ±5%) and 6-class categorization are presented in Tables 3 and 4.

Table 3. Fault classification results for input-output analysis

Type of kernel	Algorithm efficiency (with PCA)	Algorithm efficiency (without PCA)	Function parameter	Parameter value
Rbf	76%	76%	Width of rbf	0.000039–0.000096
Erbf	76%	76%	Width of rbf	0.00108–0.00207
Poly	63%	63%	Poly degree	33–103

Table 4. Fault classification results for 6-class classification

Type of kernel	Algorithm efficiency (with PCA)	Algorithm efficiency (without PCA)	Function parameter	Parameter value
Rbf	62%	62%	Width of rbf	0.000039–0.000096
Erbf	62%	62%	Width of rbf	0.00108–0.00207
Poly	55%	55%	Poly degree	33–103

In the process of classification results optimization, obtained for various kernel functions (in addition to a properly constructed learning dataset), the value of the optimization parameter has a significant influence. Since the goal of pattern separation in SVM networks is to keep the margin between the classified patterns and the hyperplane as minimal as possible, the experiments presented here show that the value of the mentioned parameter should be as small as possible. The value for which the best results were obtained are small (0.000039 in case of rbl kernel width). The same can be said about the number of generated support vectors. This suggests a simple form of a hyperplane that separated the training patterns and thus good SVM generalization properties.

5 Conclusions

An automatic diagnostic technique of parametric faults based on SVM and PCA was presented in relation to complex microwave filter analysis. The best fault classification results can be acquired for the rbf kernel. Though the algorithm requires the training data to be put in memory, it has good scalability to learn large amount of sample data. Because the SVM classifier is often combined with the algorithm of initial feature selection, i.e. Principal Component Analysis, its usefulness in the diagnostic process has also been tested.

The PCA analysis conducted for various parameter values with inclusion of unfaulty elements tolerances showed, that the analysis of measurement dataset produces overlapping or partially-overlapping characteristics, which correspond to different fault states. The distances between individual data groups were large enough to separate them into different classes. Hence the transformation of the dataset into the set of input patterns **x** for the neural network allows for obtaining well-separated fault classes, understood as particular faults of a single element or denoting the SUT as nominal. After PCA processing, the redundant features can be removed efficiently. In this work 12 features decrease to 4 efficient features. The reduction ration is about 33.3%. Although most of the features are reduced, the average diagnosis accuracy does not decrease.

The input-output analysis proved to be insufficient for successful diagnosis of multi-class categorization. It can be clearly seen in diagnostic efficiency values presented in Tables 3 and 4, acquired with the utilization of gaussian kernel functions (62% for 6-class categorization). This suggests the need to broaden the future analysis by including a greater number of accessible nodes.

References

1. Stenbakken, G.N., Souders, T.M., Stewart, G.W.: Ambiguity groups and testability. IEEE Trans. Instrum. Meas. **38**(5), 941–947 (1989)
2. Osowski, S., Kurek, J.: Support vector machine for fault diagnosis of the broken rotor bars of squirrel-cage induction motor. Neural Comput. Appl. **19**(4), 557–564 (2010)
3. Bilski, P.: Automated diagnostic system using graph clustering algorithm and fuzzy logic method. In: 18th European Conference on Circuit Theory and Design 2007, ECCTD 2007, pp. 779–782 (2007)
4. Tax, D.M.J., Ypma, A., Duin, R.P.W.: Pump failure detection using support vector data description. Lecture Notes in Computer Science, vol. 1642, pp. 415–425 (1999)
5. Milor, L.S.: A tutorial introduction to research on analog and mixed-signal circuit testing. IEEE Trans. Circ. Syst. II **41**(10), 1389–1407 (1998)

6. Tadeusiewicz, M., Korzybski, M.: A method for fault diagnosis in linear electronic circuits. Int. J. Circ. Theory Appl. **28**(3), 245–262 (2000)
7. Starzyk, J.A., Dai, H.: A decomposition approach for testing large analog networks. J. Electron. Test.: Theory Appl. **3**, 181–195 (1992)
8. Aravindh, K.B., Saranya, G., Selvakumar, R., Swetha, S.R., Saranya, M., Sumesh, E.P.: Fault detection in induction motor using WPT and multiple SVM. Int. J. Control Autom. **2** (2), 9–20 (2010)
9. Bilski, A., Wojciechowski, J.: Automatic parametric fault detection in complex analog systems based on a method of minimum node selection. Int. J. Appl. Math. Comput. Sci. **26** (3), 655–668 (2016)
10. Bilski, A., Bilski, P., Wojciechowski, J.: Overview of optimization methods in diagnostics of analog systems. In: Telecommunication Review + Telecommunication News, vol. 6, nr LXXXIV, pp. 611–617 (2015)
11. Bilski, P.: Automated selection of kernel parameters in diagnostics of analog systems. Przegląd Elektrotechniczny (Electrical review) **5**, 9–13 (2011)
12. Tripathy, M.: Neural network principal component analysis based power transformer differential protection. In: Power Systems 2009, ICPS 2009, pp. 1–6 (2009)
13. Eyoh, J.E., Eyoh, I.J., Umoh, U.A., Udoh, E.N.: Health monitoring of gas turbine engine using principal component analysis approach. J. Emerg. Trends Eng. Appl. Sci. (JETEAS) **2** (4), 717–723 (2011)
14. Seera, M., Lim, C.P., Ishak, D., Singh, H.: Application of the fuzzy min-max neural network to fault detection and diagnosis of induction motors. In: Neural Computing & Application. Springer (2012)
15. Muralidharan, V., Sugumaran, V.: A comparative study of Nad've Bayes classifier fusion methods for chemical processes. Comput. Chem. Eng. 34 (2012)
16. Bilski, P., Wojciechowski, J.: Artificial intelligence methods in diagnostics of analog systems. Int. J. Appl. Math. Comput. Sci. **24**(2), 271–282 (2014)
17. Stenbakken, G.N., Souders, T.M., Stewart, G.W.: Ambiguity groups and testability. IEEE Trans. Instr. Meas. **38**(5), 941–947 (1989)
18. Osowski, S.: Sieci neuronowe do przetwarzania informacji. Oficyna Wydawnicza Politechniki Warszawskiej, Warsaw (2006). (in Polish)
19. Ghani, R.: Using error-correcting codes for text classification. In: Proceedings of the Seventeenth International Conference on Machine Learning, pp. 303–310 (2000)
20. Dietterich, T.G., Bakiri, G.: Solving multiclass learning problems via error-correcting output codes. J. Artif. Intell. Res. **2**, 263–286 (1995)
21. Padilla, M., Perera, A., Montoliu, I., Chaudry, A., Persaud, K., Marco, S.: Fault detection, identification and reconstruction of faulty chemical gas sensors under drift conditions, using Principal Component Analysis and Multiscale-PCA. In: The International Joint Conference on Neural Networks (IJCNN), Barcelona, Spain, 18–23 July (2010)

Mathematical Models Incorporated in a Digital Workflow for Designing an Anthropomorphous Robot

Aleksandar Ivanov(✉), Mihaela Ivanova, and Anton Anchev

University of Ruse, 8 Studentska Street, POB 7017, Ruse, Bulgaria
akivanov@uni-ruse.bg

Abstract. The paper presents a digital workflow for the design and a case study of an anthropomorphous robot in which that workflow was incorporated. A streamlined workflow for the design of robots is presented. The workflow follows a rule-based algorithm, resolving the complex task structure-to-function, with minimum human interaction. The workflow offers a fast, controllable way to generate semiautomatically the different models needed on every step of the design. The geometrical model was verified by a graphical model which itself was verified by FEM static analysis. The kinematic chain is generated semi-automatically by one initial model. The kinematics of the anthropomorphous robot was simulated and visualized. The workflow is such arranged that allows fast semiautomatically development of a final product. It also has the potential to easily incorporate changes into the final product.

Keywords: Computational workflow · Anthropomorphous robot · Topological model · Graphical model · Simulated kinematics · Mathematical model · Denavit-Hartenberg notation · Automated design

1 Introduction

Conventional design methods nowadays begin with defining the task that the robot needs to fulfill, which defines strict design and strict programming, that doesn't allow any changes. File-to-product strict design methods, especially when we talk for robotics, are putting limitations on the workflow.

Through geometrical Computer-aided Design (CAD) models, Computer-aided Engineering (CAE) and Computer-aided Manufacturing (CAM), kinetic models, dynamic models, analytical models of any kind, programming the process to get to the end product primary looks continual. The process involves transfer of data, commonly speaking, generations of mathematical models and in most cases all those models are independently created by a designer. That discrete process, needing models to be generated by a human on every stage, looks like is not using the capacity of the technology available nowadays. The human should only be observant of how those models are created and interfere only if needed, rather than generating them.

The development of mathematical description of very complex 3D models is limited. Driven by the motivation to overcome that limitations, and the complexity of

R. Silhavy (Ed.): CSOC 2019, AISC 986, pp. 306–314, 2019.
https://doi.org/10.1007/978-3-030-19813-8_31

the desired final product, we present that workflow were the mathematical models are generated semiautomatically.

In reality, people need open systems that might be redesigned according to the daily changed tasks for the different applications needed.

In this paper we present a generative design open-system approach, driven by the client's custom needs. As a specific case, in need of an anthropomorphous robot [1], this generative method that flows the custom client's meta-data from-software-into-software, with the corresponding models, was created beginning with the design that incorporates the client's needs. The method uses many functionally and structurally different mathematical models [2, 3] of one initial 3D CAD model. The type of the model depends on the properties of the product that is being designed and of the stage of the development. The workflow is using variety of mathematical models. In this workflow they are generated semiautomatically. This is resource-saving, considering that in other cases they should be developed manually by the designer and in many cases the designer should fill in one and the same information but in different format for the various models.

Firstly, the method starts with verification of the already designed 3D model. The 3D model was verified with stress analysis [4]. Unconventionally, the design was verified by graphical model. In this workflow the FEM analysis was carried out only once to verify a graphical analytical mathematical model, which is actually used to verify the design. It takes considerably less time of the graphical model to verify the design compared to a FEM analysis [4, 5]. The workflow is resource-saving, considering FEM analysis are taking that much of the resource of a computer. By two means the adequacy of the graphical model was proven.

The Denavit-Hartenberg notation [1, 6] was used to describe the anthropomorphous robot, which represents actually a serial-link mechanism, to find the kinematic solutions [7]. The significant difference from the common design process is that the 3D model is directly exported from the CAD as topological model. That model is in fact STL approximation of the 3D model. To that topological model of the 3D model was automatically assigned coordinate frames to the links. The four Denavit-Hartenberg (DH) parameters for the links and for the joint were predefined by the 3D model and were incorporated in that topological model. By contrast in other workflows the 3D CAD model is not taken into account directly when determining the four parameters and the designer would assign the parameters once again, even though they are already included in the 3D CAD model. This is not only making the method resource-saving, but also gives the construction potential to be easily changed.

The process allows the designer to make changes of the design that is transpassed semiautomatically into the workflow, allowing the robot to be as redesigned as needed, yet fully functional, with open programming. That makes this approach in line with the needs of high customization level.

Furthermore, the open system allows newly designed components to be integrated onto the final product easily, in this case in the anthropomorphous robot, and to be redesigned. That gives potential to dynamically apply improvements in the product.

Finally, open systems for mass-customization are becoming widely popular as the 3D printing technology is becoming more and more accessible. And as the 3D printing is fast method, the generation of the mathematical models needed to verify the design and to get solutions for the kinematics, is neither fast, nor resource-saving. There is a need mathematical models to be generated faster and easier, with less resource involved.

2 Method

The method is based on the semiautomatic generation of mathematical models. In this section we discuss the mathematical models generated by the workflow (see Fig. 1). Next, we introduce our case study, presenting the algorithm that the workflow is following, the simulations that were done in order to verify the models.

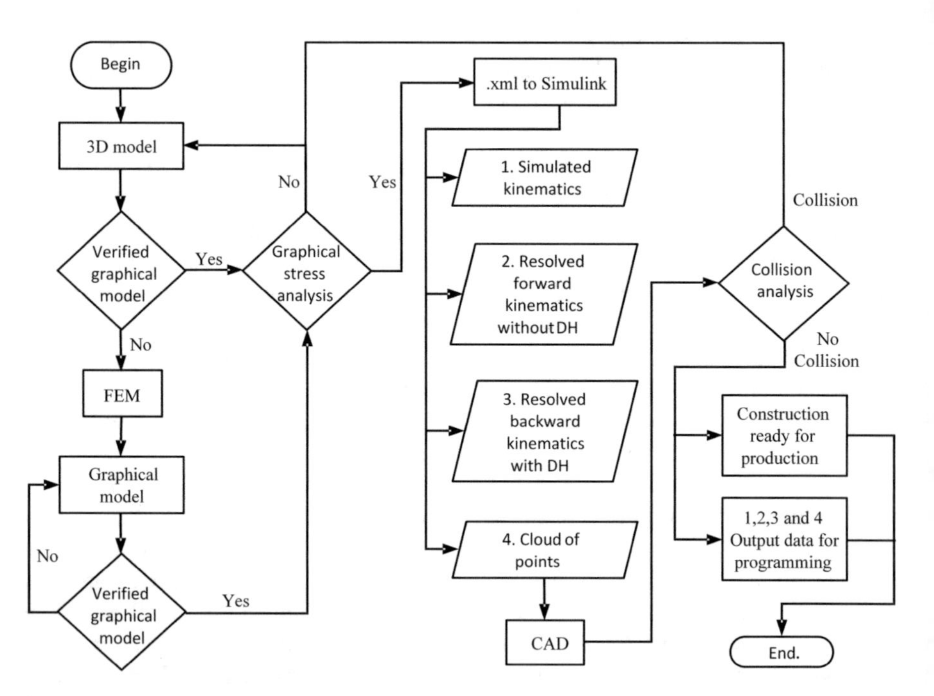

Fig. 1. Overview of the computational workflow for automated design. Constant continuous export from the initial 3D CAD model is demonstrated. The workflow demonstrates how the design is continuous and how the models relate to each other, in contrast with other design methods that are discrete and the models have no relation between each other and are independently generated.

The workflow is working with different models that describe the final product using different data. Those models are flowed from one software into the other, yet on every step they could be changed by the engineer. The mathematical models that the workflow is using are structural and functional.

The structural mathematical models describe the structure and the form of the designed product. They are divided into geometrical and topological ones. The geometrical models are the 3D models. Based on the geometrical model the topological model was generated. It describes the structure, e.g. the complex of components and the connections between them. It was topological model that was used to solve the kinematics.

The functional mathematical models describe the functional and informational processes. Motion study, stress analysis and deformation analysis were carried out. They are divided into analytical and algorithmical.

The workflow (see Fig. 1) is based on the models that are generated from one software to serve other. The initial 3D model is converted and supplemented multiple times, incorporating different type of data.

As a result from that continual workflow the designer's task is completed and final design is ready for production. Information regarding the programming of the robot is also part of the output. So once the designer is finished with designing, programmer could take the final output and continue with programming. The workflow is also fault-protected in case any of the actuators block – in this scenario the kinematics is recalculated with that blocked actuator's parameter.

2.1 Geometrical Model

The workflow starts with the 3D model (see Fig. 2). The workflow is such organized that allows the components to be manufactured with any tolerance class and any technology needed.

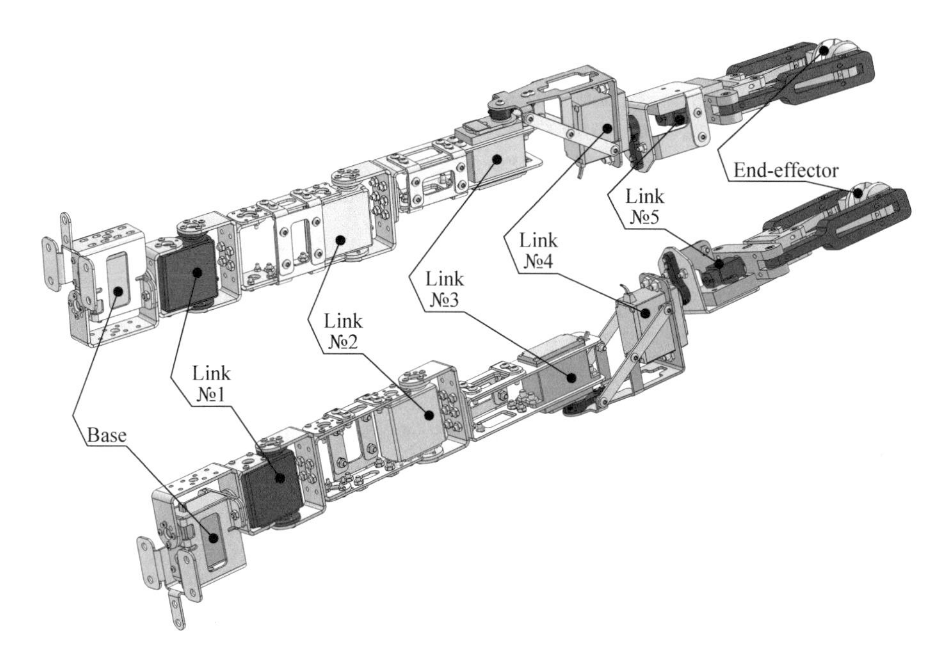

Fig. 2. The zero-angle configuration of the anthropomorphous robot

At this point the designer needs to define the parameters [8] of the links – link length a_i, and the link twist α_i, and the parameters of the joints – link offset d_i, and the joint angle θ_i. In this specific case study, since the designed anthropomorphous robot has only revolute joints, the joint variable is θ_i and d_i is a constant. At this point the links, their parameters, the base and the end-effector are defined. In order to do that the joint and the links are determined. The anthropomorphous robot consists of 5 joints and 6 links. The zero angle configuration (see Fig. 2) is defined. In this case the kinematic zero-angle configuration of the anthropomorphous robot is different to the joint's controller's zero-angle configuration due to the motors specification. The design is such that the top point of the end-effector is positioned absolutely to the starting position.

This 3D model will firstly be graphically represented to verify the strength of the design. The graphical model itself will be verified by FEM analysis.

Then, the 3D model is exported to Matlab to get two other models. One of the model conversions – from geometrical model to topological one. In the topological will be represented the structure, as well as the control. The other output of the export from the CAD to Matlab would be the STL approximation. With the control incorporated in the topological model, the STL approximation will be dynamically controlled and the kinematics will be simulated.

2.2 Analytical Model

Conventionally, to check if the model will endure the stress applied on the structure, analysis should be carried out. The potential of the CAD to verify the structure using static stress FEM analysis seems to be reasonable tool to do that. But as the structure is supposed to be changed and reconfigured FEM should be carried out after any of those actions. Instead, in our case with the Anthropomorphous robot, following the work-flow, the FEM analysis was only used to verify a graphical model. That graphical model would afterwards be used many times during the design and redesign of the anthropomorphous robot. It is used to determine the stress, instead using the FEM.

The Design of the Griper. On Fig. 3a the final design of the 2-jaw griper is represented, which consists of few components ending with jaws, which hold the manipulated object. The construction is symmetrical and to ease the task of the design the calculations were carried out only for one half of it.

One of the objectives of the design of the griper was to secure the force F_D (see Fig. 3) needed to hold the object [9] and to make possible the jaws to be that much opened that they will release the held object, respectively so much closed that they will hold the object, which depends on the dimensions l_1 and l_2 (see Fig. 4).

The design of the griper consists of three main tasks:

1. To secure the minimum force to hold the object;
2. To secure enough space between the two jaws when opened to release the object;
3. To carry out stress analysis to secure the components will endure.

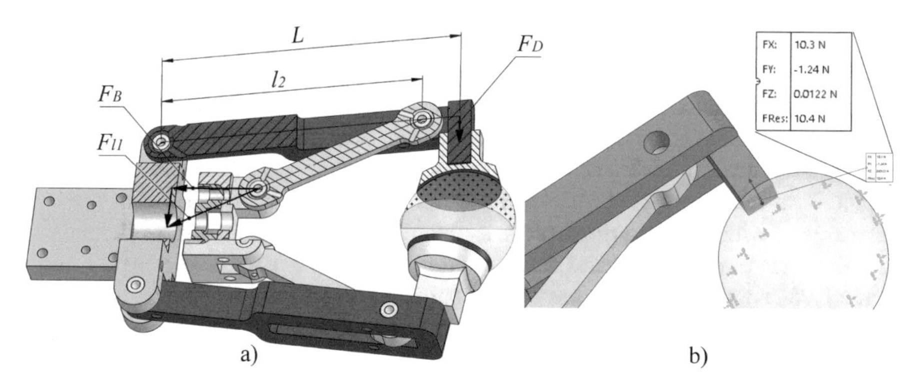

Fig. 3. On the figure the construction of the griper is presented. (a) represents the forces on which the hold force F_D depends. (b) represents the result from the FEM analysis that would be used to check the graphical model.

Following the workflow, a FEM analysis (see Fig. 3b) was carried out once. The results would be later on used to check with the results that the graphical method would calculate.

For our convenience a table (see Table 1) had been drafted before the graphical method was used. With it we did a prestudy to look roughly what dimensions should we check with the graphical model and in what limits the dimensions l_1 and l_2 vary. With it we could easily check that the needed force to hold the manipulated object which in our case is F_D = 10,3 N. At this point the Table 1 and the carried FEM analysis are both calculating the same result.

Table 1. Mathematical model to check limits to which the dimensions of l_1 and l_2 could vary.

№	Parameter	Var.1	**Var.2**	Var.3
1	l_1, mm	**70.0**	**60.0**	**40.0**
2	l_2, mm	**90.0**	**90.0**	**70.0**
3	l_3, mm	21.5	**21.5**	21.5
4	l_4, mm	19.5	**19.5**	19.5
5	l_5, mm	106.6	**106.6**	106.6
6	l_6, mm	30.0	**30.0**	30.0
7	L, mm	103	**103.0**	103.0
8	$x_2=\sqrt{l_5^2+l_6^2}, mm$	110.8	**110.8**	110.8
9	$\gamma=arctg(l_5/l_6),°$	74	**74**	74
10	$\beta=f(x_2,L,l_3,l_4),°$	22	**22**	22
11	$\delta=180°-\beta-\gamma,°$	84	**84**	84
12	$\theta=90°-\delta,°$	6	**6**	6
13	$a_1=a+l_2.cos(\delta), mm$	26.9	**26.9**	24.8
14	$\omega=arccos(a_1/l_1),°$	67	**63**	32
15	$\alpha=90°-\omega-\theta,°$	17	**24**	0.6
16	a, mm	17.5	**17.5**	17.5
17	$b=f(\alpha,\theta,l_1,l_2), mm$	24.9	**35.9**	38.3
18	$x=\sqrt{a^2+b^2}, mm$	30.4	**39.9**	42.1
19	Fb, N	30.0	**30.0**	30.0
20	$\psi=\delta,°$	84	**84**	1.5
21	$F_{l1}=F_b/sin(\alpha)$, N	32.5	**33.6**	38.3
22	$F_C=F_{l1}.sin(\alpha)$, N	9.3	**11.8**	20.5
23	$F_D=F_C.l_2/L$, N	**8.1**	**10.35**	**13.9**

The advantage of Table 1 is that it is directing to the optimal solution. At this case it was the middle column that best fitted. The accuracy of that table depends on to what level the parameters are discrete. In our case we did a complete combination between the parameters and only three of them are shown here in the table.

The graphical model (see Fig. 4) is a rule-based parameterized model. The graphical model for the first check with the results given by the FEM analysis, was dimensioned with the same dimensions as the 3D model. If the results for the resulting F_D is the same then the graphical model is verified. After the graphical model is verified by the FEM analysis the designer could continue with making decisions for the final design. The advantage of the graphical model is that the designer, having the set limits by Table 1, checks only limited amount of possibilities. Other advantage of the graphical model is that it gives results instantly. To carry out a FEM analysis to that construction takes hours, but the graphical model gives results instantly.

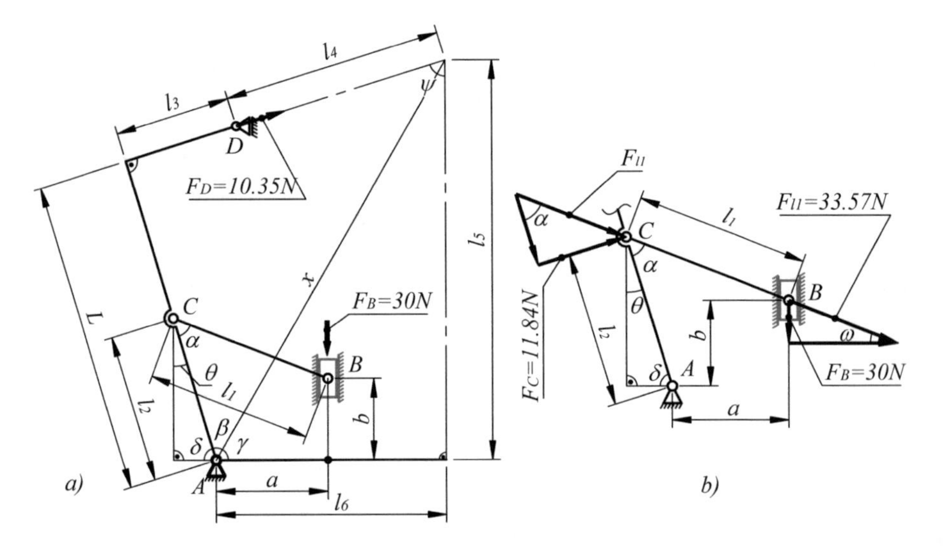

Fig. 4. The graphical model which is rule-based parametrized model, that determines the length of l_1 and l_2. (a) The force F_D that will secure the hold force is represented. (b) Model is set to represent the F_B which depends on the motor's power.

The adequacy of the graphical model represented on Fig. 4 is proven once by the FEM analysis, results from which could be seen on Fig. 3b, that is F_{Res} = 10.3 N, and by the table results shown in Table 1, where F_D = 10.35 N. The proof for the adequacy is that on all three models we get the same results for the resulting F_D.

As the dimension l_4 is the dimension between the two jaws when the griper is closed, l_{41} is the dimension when the jaws are opened and should be as opened as to release the object. That dimension l_{41} was similarly calculated.

Potentially, this graphical model might be incorporated in the 3D model. The graphical model could parametrically dimension the construction based on functions.

2.3 Topological Model

It is only after the construction has proven its strength that the workflow moves to the other stage. The 3D CAD model is turned into a solid body in Simscape Multibody. The 3D CAD system's availability to save the 3D model as .XML and its part files as . STL simplifies the engineer's task to simulate kinematics and is one of the outputs. Mechanical components and electrical ones are represented in a block diagram since it is inserted in Matlab using Simulink and this makes up the topological model (see Fig. 5). Information about the link length a_i, and the link twist α_i is inserted with the solid body. In Simulink the joints are identified. As part of the transition from the CAD system to Matlab automatically a coordinate frame was given to each link. This model represents the initial 3D model topologically, but stays open – controls could be inserted, other limitations, that were not represented in the geometrical model, could also be added, e.g. the possible angles of rotations are defined, dependently on the type of the actuators. As a result, the forward kinematics for the all the points that lay on the axis of the motors, has a solution. If solution for the forward kinematics for other points is needed, Denavit-Hartenberg matrices should be inserted in Matlab. In order to have fully defined solution for the inverse kinematics Denavit-Hartenebrg matrices should be inserted in Matlab, also.

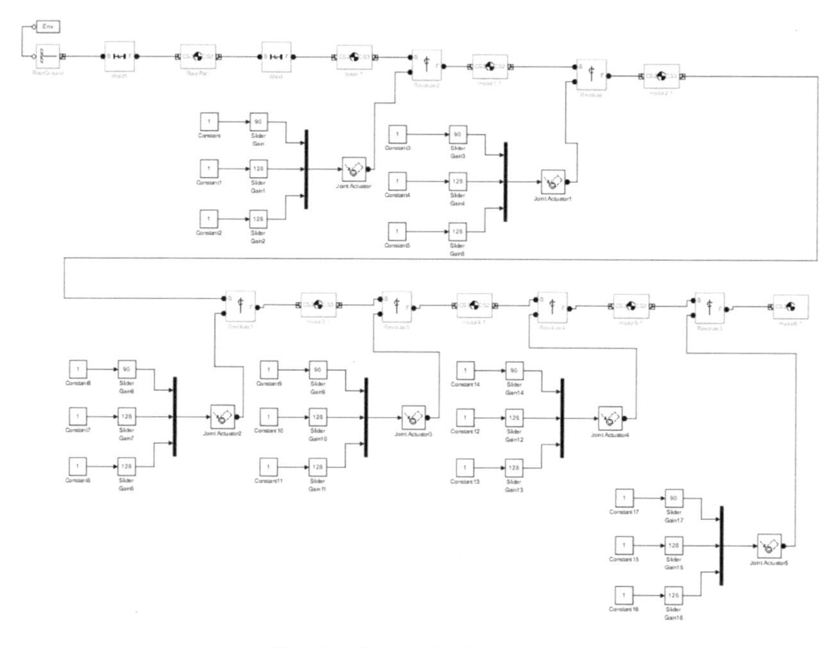

Fig. 5. The topological model

As a result of having the kinematic chains we resolved the task of planning the trajectory [10, 11]. The resulting trajectory is time-ordered set of consistent intermediate configurations of the model between starting and end position. As further advantage of the method, Matlab collects the information for the points in its trajectory, thus making cloud of points. That cloud of points could be graphically represented and could be imported back to CAD to carry out collision analysis.

Not only that Matlab collects the information about the points laying on the axes, but also generates the plan for the velocity and accelerations and the rotational angles.

3 Conclusion

In this paper, we have presented a computational workflow used to design an anthropomorphous robot. The method presented continuous automated transformations of data from one mathematical model that generates other. The ability to relate each of the following models to an initial complex 3D model show completeness that ease designer's job. The method shown introduces design opportunities that supplement the 3D printing technologies as the mathematical models are generated automatically and faster. While our workflow generates mathematical models that describe the design structurally and functionally, future work may explore the combination of mathematical models and programming that is not discrete, but is based on that workflow's output.

References

1. Siciliano, B., Sciavicco, L., Villani, L., Oriolo, G.: Robotics: Modelling, Planning and Control. Springer (2008). https://doi.org/10.1007/978-1-84628-642-1
2. Maliuh, V.: Vvedenie v sovremennce SAPR. (in Russian) DMK Press, Moskva (2012). ISBN 978-5-94074-860-1
3. Ushakov, D.: Vvedenie v matematicheskie osnovы SAPR. (in Russian) DMK Press, Moskva (2013). ISBN 978-5-94074-829-8
4. Sivertsen, O., Haugen, B.: Automation in the Virtual Testing of Mechanical Systems: Theories and Implementation Techniques. CRC Press, Boca Raton (2018). ISBN 9780429879074
5. Erez, T., Tassa, Y., Todorov, E.: Simulation tools for model-based robotics: Comparison of Bullet, Havok, MuJoCo, ODE and PhysX. In: 2015 IEEE International Conference on Robotics and Automation (ICRA), pp. 4397–4404. IEEE (2015)
6. Spong, M., Hutchinson, S., Vidyasagar, M.: Robot Modeling and Control. Wiley, New York (2006). ISBN-13: 978-0471649908
7. Corke, P.: Fundamental Algorithms In MATLAB® Second, Completely Revised, Extended And Updated Edition, vol. 118. Springer (2017). https://doi.org/10.1007/978-3-319-54413-7
8. Corke, P.I.: A simple and systematic approach to assigning Denavit-Hartenberg parameters. IEEE Trans. Rob. **23**(3), 590–594 (2007)
9. Datta, R., Deb, K.: Multi-objective design and analysis of robot gripper configurations using an evolutionary-classical approach. In: Proceedings of the 13th Annual Conference on Genetic and Evolutionary Computation (GECCO 2011), pp. 1843–1850. ACM, New York (2011). https://doi.org/10.1145/2001576.2001823
10. Ivanova, T., Georgieva, Ts.: Software applications for simulation of wireless sensor networks. In: Proceedings of University of Ruse, pp. 25–29 (2018)
11. Karakaya, S., Kucukyildiz, G., Ocak, H.: A new mobile robot toolbox for matlab. J. Intell. Rob. Syst. **87**(1), 125–140 (2017). https://doi.org/10.1007/s10846-017-0480-2

Multi-channel FXLMS Filter as an Array of Processing Blocks and a Method to Maintain Its Performance when Increasing the Number of Channels

Inna V. Ushenina(✉)

Penza State Technological University, Penza, Russia
ivl23@yandex.ru

Abstract. The advantage of field programmable gates arrays (FPGAs) over microprocessors is the possibility of parallel computing based on logical resources organized as a required number of parallel devices. The disadvantage of FPGAs is the risk of performance degradation of implemented devices the performance of which is based mainly on programmable interconnections between resources. The performance can also be decreased because of high signal fanouts and non-optimal positioning of separate units of the device on a chip. In the paper below, the author analyzes the architecture of FPGA-based multi-channel adaptive FXLMS filter implemented as an array of processing blocks. FXLMS filters are widely used in active noise control systems. The paper also analyzes the main reasons for reduction in filter performance if the number of channels increases and the ways of maintaining the filter performance at the level of logical FPGA resources. It is shown that the most efficient way to keep the filter performance high is to use a register tree for signal pipelining.

Keywords: Multi-channel FXLMS filter · FPGA · Register tree

1 Introduction

Controllers of the majority of active noise control systems (ANC systems) are based on adaptive filters (AF) with filtered-reference least mean square algorithm (FXLMS) [1–3]. For multi-channel ANC systems used for such tasks as noise suppression in enclosed space or creation of local zones of silence, multi-channel AFs shall be used [1–4].

In most cases, signal processors are used to implement adaptive filters [1–3]. However, the performance of signal processors may not be enough to implement controllers of multi-channel ANC systems, which contain high order filters in each channel. The paper [5] shows that FPGAs can be considered as an alternative to signal processors and proposes the architecture of a multi-channel AF as an array of processing blocks. The main advantage of this architecture is a balanced combination of parallel and sequential operations which is selected taking into account the features of FPGA resources.

R. Silhavy (Ed.): CSOC 2019, AISC 986, pp. 315–324, 2019.
https://doi.org/10.1007/978-3-030-19813-8_32

The performance of a FPGA-based AF can significantly decrease as the number of channels and (or) filter orders in the channels (*N*) increases. Such decrease is caused by the method of implementing programmable interconnections between logical FPGA resources and by the growth of signal fan-outs.

The ways to maintain the performance of FPGA-based devices are signal pipelining and signal fan-out reduction. Increase in performance can also be ensured by compact placement of device fragments on the chip.

In the paper below, the author evaluates the efficiency of all above mentioned ways to maintain the performance of a multi-channel AF with an array architecture while the number of channels and filter orders in the channels increases. The achieved performance level is compared with the maximum operating frequency of the FPGA logic sources used to implement an AF.

2 Related Work

The description of FPGA implementation techniques for the FXLMS algorithm and other modifications of the adaptive LMS algorithm can be found, for instance, in papers [5–10].

The papers [6–9] describe modifications of the systolic filter architecture implemented on the base of FPGA DSP slices connected in series. The main advantage of this approach is high throughput of the implemented filter. Such architecture ensures high filter performance by using hardware DSP slices connected via special routing FPGA resources. However, in case of implementation of a multi-channel filter with time division of channels, this structure shall be completed with shift chains and circuits of access control to the data of each channel implemented on the programmable FPGA logic. This measure will inevitably reduce filter performance.

The papers [5, 10] analyze approaches to implementing a filter as an array of processing blocks (modules). The advantage of this approach is better suitability to the implemented algorithm. For example, when implementing an array of processing blocks, the author of the paper [5] takes into account the fact that the number of error microphones (*M*) in the ANC systems is usually higher than that of reference microphones (*J*) and loudspeakers (*K*), and that the size of an array depends on *K* and *J*. Each of the processing blocks of an array operates independently and it is possible to arrange processing blocks in a compact manner in order to ensure their high performance. This article analyzes the filter architecture proposed in the paper [5].

3 Architecture of the Multi-channel Filter

The architecture of multi-channel adaptive filters featured as an array of processing blocks is proposed and thoroughly described in the paper [5]. The proposed architecture is shown in Fig. 1 (in simplified form). The following designations are used in this figure [5]: *J* is a number of reference microphones of the ANC system; *K* is a number of loudspeakers emitting compensating signals; X_j stands for memory blocks used for storing samples of reference signals; S_k stands for memory blocks used for storing coefficients of estimations of cancellation path transfer functions (FEs) in the channels; B_{kj} stands for processing blocks, each containing two DSP slices and two memory blocks; Y_k stands for noise-compensating signals. Each row of processing blocks receives one of J signals from the reference microphones. Each column of processing blocks generates one of K noise-compensating signals. Samples of error signals from M reference microphones (e_m) are stored in a memory block used for all processing blocks (not shown in Fig. 1).

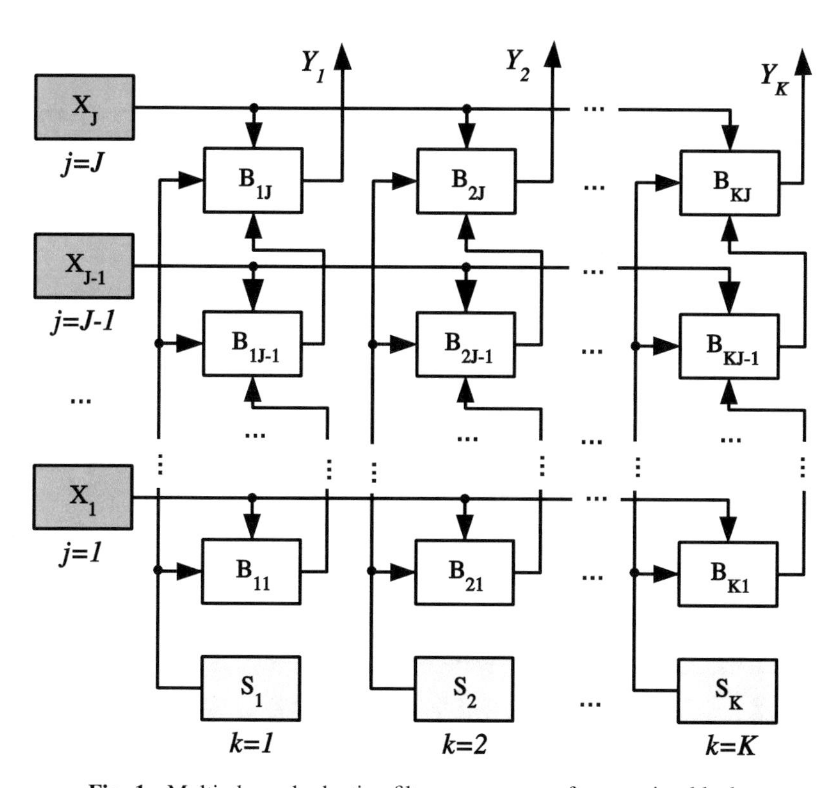

Fig. 1. Multi-channel adaptive filter as an array of processing blocks

3.1 Performance Loss of Separate Processing Blocks

The structure of a single processing block is shown in Fig. 2. FPGA-based implementation of a processing block requires two DSP slices, block RAM (for implementation of memory blocks) and a certain amount of programmable logic for auxiliary operations. Memory block W_{kj}_RAM is used for storing the AF coefficients (N coefficients). Memory block X'_{mkj}_RAM is used for storing the $N + 1$ sets of M samples of the j-th reference signal filtered by M FEs.

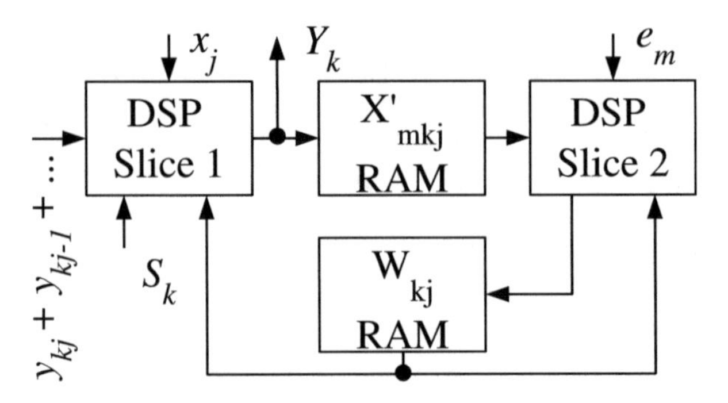

Fig. 2. Processing block structure

The block RAMs and DSP slices are grouped in columns (see Fig. 3). The amount of programmable logic included between the columns significantly exceeds the needs of the processing block.

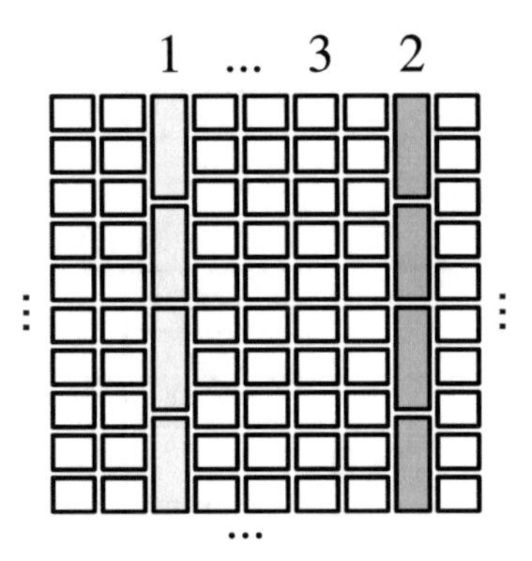

Fig. 3. Topology of an FPGA chip fragment. 1: DSP slices, 2: Block RAM; 3: Programmable logic

The increase of N and (or) M can lead to the need for several block RAMs for implementation of memory blocks. An FPGA block RAM storage capacity is 36 kbit; the blocks may be connected in pairs without loss of performance. Further memory block size enlargement leads to a performance decrease caused by additional delays in signal propagation between the blocks. However, FPGAs have special routing resources to unite memory blocks and minimize such delays.

Each included RAM block significantly increases the installation space required for the processing block (see Fig. 3). The performance loss in this case may be caused by additional signals' propagation delays between memory blocks and DSP slices.

3.2 Performance Loss of Separate Columns of Processing Blocks

The situation with excessive logical resources between the columns of memory blocks gets worse when we add a column of processing blocks. Despite the fact that the calculations in each block are performed independently, the k-th compensating signal is generated by adding its components produced by the J processing blocks in the k-th column (see Fig. 1). The adding is carried out by DSP slices of the processing blocks combined in a chain. Each next slice adds one more component to the accumulated amount. In order to maintain high performance of DSP slices, the chains must be located at adjacent positions in a column. The memory blocks of the processing blocks are located around the DSP slices. Thus, the FPGA chip area is shared not between separate processing blocks but between columns of blocks (see Fig. 4).

3.3 Performance Loss of the Entire Filter Structure

The architecture of the multi-channel filter shown in Fig. 1 suggests that the main reason for performance losses when the number of channels increases is the fact that each block needs to receive the samples of error signals, the samples of reference signals and Fes coefficients of cancellation paths in the channels.

4 Methods

The maximum performance of a processing block is determined by the operating speed of DSP slices and block RAMs. The block RAMs of Artix 7 XC7A200T-3 FPGA used as the basis of a multi-channel AF project, have a lower operating speed, therefore, the results of different methods of maintaining the filter performance were compared with the maximum operating frequency of these devices which is 447 MHz [11].

All studies were carried out using the ISE Design Suite tools produced by Xilinx.

For research purposes, VHDL descriptions of a separate processing block and the entire structure designed for implementation on Xilinx 7 series FPGAs were prepared. The VHDL description of the processing block is based on a template proposed in the ISE Design Suite environment for devices implemented using DSP slices.

The maximum operating frequency calculated based on the results of the project by the timing analyzer of the ISE Design Suite environment was used to assess the performance of the processing block and the filter.

The synthesis and implementation of the projects are carried out with a constraint for a period of clock signal which corresponds to the operating frequency of the block RAM (447 MHz).

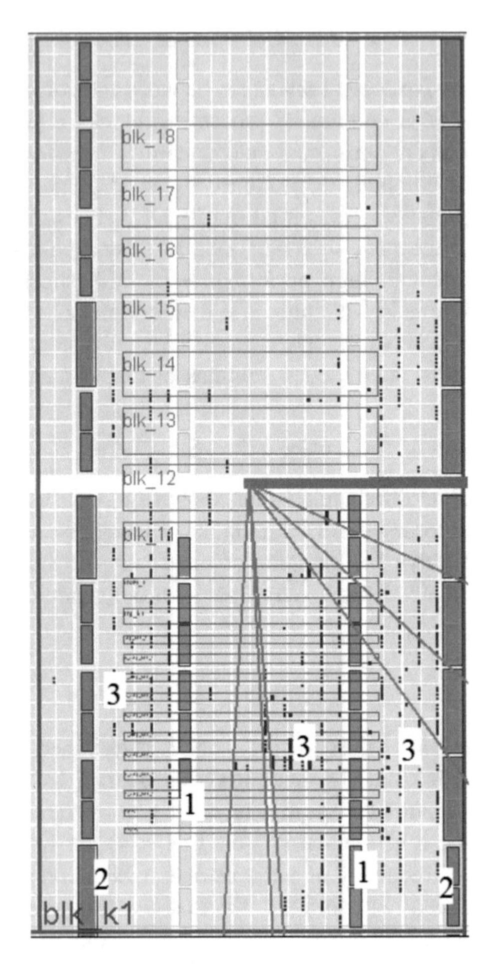

Fig. 4. Location of column elements of the processing blocks when $J = K = 8$; $N = 1024$. 1: DSP slices in use, 2: Memory blocks in use, 3: Elements of configurable blocks in use

The PlanAhead package is used to control placement of the processing blocks on the FPGA chip. It is used to select a chip area limited by the required number of DSP slices and RAM blocks for placement of a separate processing block and columns of processing blocks (see Figs. 3 and 4).

The MAX_FANOUT synthesis option available in the ISE Design Suite was used to assess the effectiveness of reducing signal fan-out as a way to maintain the performance of the filter.

According to the architecture of the multi-channel AF, a register tree was used as a structure of signal routing for pipelining of each signal common for the entire array, row or column of the processing blocks (see Fig. 5). The source sends signals to the central register, the central register sends them to the pair of another registers, etc. As a result, each processing block receives signals from its own register simultaneously with other blocks. Figure 5 shows a tree which corresponds to the eight-column AF. The proposed routing structure allows reducing propagation delays not only due to pipelining, but also due to the possibility of placing the "branches" of the tree closer to "own" processing blocks (signal receivers).

When dealing with a single processing block, *N* parameter was assumed to be 1024 which is close to the maximum number of the AF coefficients. *M* parameter varied from 2 to 64. When studying the dependency between the performance of the filter and its size, $K \times J$ ratio varied from 2×2 to 10×10. *N* parameter was assumed to be 1024; *M* parameter was assumed to be 12.

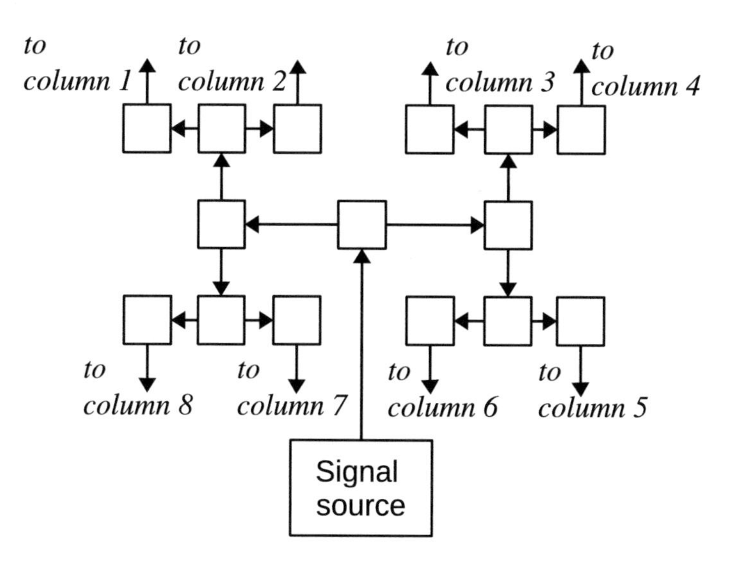

Fig. 5. Tree-like structure of routing of signals to the processing blocks with $K = 8$

5 Results

Table 1 shows the maximum operating frequencies of a processing block determined by the ISE Design Suite timing analyzer after its placing and routing with varying value of *M* parameter.

Table 1. Maximum operating frequencies of the processing block.

M	2	6	12	24	32	64
f_{max}, MHz	447	447	447	446	400	337

Table 2 shows the maximum operating frequencies of a multi-channel filter determined by the ISE Design Suite timing analyzer after its placing and routing for different $K \times J$ ratios, $N = 1024$ and $M = 12$, without limiting the signal fan-out and pipelining of signals.

Table 2. Dependency of the maximum performance (the maximum operating frequency) of the processing blocks array on its dimension.

$K \times J$	2×2	4×4	6×6	8×8	10×10
f_{max}, MHz	409	324	268	241	220

Table 3 shows the maximum operating frequencies of a multi-channel filter determined by the ISE Design Suite timing analyzer after its placing and routing for different $K \times J$ ratios, $N = 1024$ and $M = 12$, with the limited signal fan-out. The signal fan-out was selected for each size of the array individually and did not exceed 10.

Table 3. Dependency of the maximum performance (the maximum operating frequency) of the processing blocks array on its dimension with the limited signal fan-out.

$K \times J$	2×2	4×4	6×6	8×8	10×10
f_{max}, MHz	421	339	293	256	227

Table 4 shows the maximum operating frequencies of a multi-channel filter determined by the ISE Design Suite timing analyzer after its placing and routing for different $K \times J$ ratios, $N = 1024$ and $M = 12$, using the register tree for pipelining of signals common for the entire array, row or column of the processing blocks.

Table 4. Dependency of the maximum performance (the maximum operating frequency) of the processing blocks array on its dimension when pipelining signals common for the entire array, row or column of the processing blocks.

$K \times J$	2×2	4×4	6×6	8×8	10×10
f_{max}, MHz	447	447	438	398	370

6 Discussion

Table 1 confirms that the maximum performance of the processing blocks corresponds to the cases when memory blocks are fully mapped into single or paired 36-kbit block RAMs.

It is obvious from Table 2 that without taking special measures the performance of the processing blocks array is rapidly decreasing with the increase of its size. The reports generated by the timing analyzer confirm that this is mainly caused by the necessity to deliver the same set of signals to each block: the same input signal to each row of the processing blocks, the same set of the FEs coefficients to each column of the processing blocks, and the same set of error signals to all the processing blocks.

According to Table 3, a limited signal fan-out has minor effect on the performance. This result proves that the main reason for performance loss is a significant distance between signal receivers and their sources, rather than a large number of receivers. When duplicating signal propagation circuits are located closer to the receivers, thereby reducing propagation delays, this causes a slight increase in filter performance.

According to Table 4, the use of the register tree for pipelining of signals allows maintaining the AF performance at a level close to the maximum one when the array dimension is up to 8 × 8. Further increase of K and J leads to a decrease in filter performance. Nevertheless, the AF performance is much higher when pipelining of signals is used rather than when there is no pipelining. Including the additional registers in the "branches" of the register tree can allow maintaining the filter performance in a more effective manner when the values of K and J are high. The number of registers used for pipelining must be selected individually for each combination of K and J.

7 Conclusion

The above mentioned studies show that the main reason for the reduction in multi-channel FXLMS filter performance when the number of channels increases is an increase of propagation delays of signals received by the processing blocks from the common sources. These signals include error signals, reference signals, and FEs coefficients of cancellation paths.

The use of the register tree for pipelining of these signals when placing the processing block columns in chip areas limited by DSP slices and block RAMs allows to significantly improve the filter performance.

References

1. Kuo, S.M., Morgan, D.R.: Active Noise Control Systems: Algorithms and DSP Implementations. Wiley (1995)
2. Elliott, S.: Signal Processing for Active Control. Academic Press (2000)
3. Hansen, C.H., et al.: Active Control of Noise and Vibration, 2nd edn. CRC Press (2012)
4. Morgan, D.R., Quinlan, D.A.: Local silencing of room acoustic noise using broadband active noise control. In: IEEE Workshop on Applications of Signal Processing to Audio and Acoustics. Final Program and Paper Summaries. IEEE (1993)
5. Ushenina, I.V.: FPGA-based multi-channel adaptive FXLMS filter implemented as an array of processing blocks. Electron. J. "Technical Acoustics" (2016). http://www.ejta.org
6. Yi, I., Woods, R.: FPGA-based implementations of Delayed-LMS Filters. J. VLSI Sig. Proc. **39**, 113–131 (2005)
7. Thomas, J.: Pipelined Systolic architecture for DLMS adaptive filtering. J. VLSI Sig. Proc. **12**, 223–246 (1996)
8. Dai, J., Wang, Y.: NLMS adaptive algorithm implement based on FPGA. In: 3rd International Conference on Intelligent Networks and Intelligent Systems (ICINIS), pp. 422–425. IEEE (2010)

9. Boroujeny, S.G., Eshghi, M.: FPGA implementation of a modular active noise control system. In: 18th Iranian Conference on Electrical Engineering, pp. 658–661. IEEE (2010)
10. Bahoura, M., Ezzaidi, H.: FPGA-implementation of parallel and sequential architectures for adaptive noise cancelation. Circ. Syst. Sig. Process. **30**(6), 1521–1548 (2011)
11. Artix-7 FPGAs Data Sheet: DC and AC Switching Characteristics. http://www.xilinx.com/support/documentation/data_sheets/ds181_Artix_7_Data_Sheet.pdf

Stochastic Discrete Nonlinear Control System for Minimum Dispersion of the Output Variable

Svetlana Kolesnikova(✉)

Institute of Computational Systems and Programming,
St. Petersburg State University of Aerospace Instrumentation,
St. Petersburg, Russia
skolesnikova@yandex.ru

Abstract. This paper deals with the technique application of adaptation on manifolds for stochastic nonlinear models. The analysis and control approach are used to implement the principle of directed self-organization and decomposition of non-linear dynamic systems is used. A new problem statement for the synthesis of a robust stochastic regulator based on the principles of control on manifolds is discussed. The paper presents a new algorithm to design a stochastic discrete control system and provides illustrative example of solving nonlinear stochastic control problem based on the designed regulator. The properties of the control algorithm for a second-order object are formulated. It is shown that the designed control gives a minimum of variance of the response variable and a minimal dispersion of the Lagrange-Euler function, which is used for the structure of the stochastic control system.

Keywords: Nonlinear multidimensional stochastic object · Robust regulator · Minimal variance of response variable · Target manifold

1 Introduction

The nonlinear control approaches implementing the principle of directed self-organization and decomposition of non-linear dynamic systems are currently becoming popular in the theory of nonlinear systems [1–10], since they correspond to physical control theory [1].

This paper deals with the issue of extending the classical method of analytical design of aggregated regulators (ADAR) [3], which was developed earlier for the deterministic case, to nonlinear objects with stochastic uncertainties to achieve a robust control system. This work is a generalization of the result obtained earlier in [11], where a solution was proposed for a one-dimensional object.

The first part describes the new algorithm to synthesize a robust regulator based on the method of analytical design of aggregated regulators for a nonlinear stochastic object; the second part presents sample solution of applied problem on the above given algorithm: control over the economic model of a market (a second-order object).

R. Silhavy (Ed.): CSOC 2019, AISC 986, pp. 325–331, 2019.
https://doi.org/10.1007/978-3-030-19813-8_33

2 Formulation of the Problem

Let a probability space $(\Omega, \mathscr{F}, (\mathscr{F}_k)_{k\geq 0}, \mathbf{P})$ be given, where $\mathscr{F}_k = \sigma\{\xi[t] = (\xi_1[t], \ldots, \xi_m[t]),\ t \leq k\}$ and $\{\xi_i[t]\}_{t\geq 0},\ i = 1, \ldots, m$ are the sequences of equally independent distributed quantities with the average zero and variances $\sigma_i^2,\ i = 1, \ldots, m$.

Let us discuss a stochastic vector object with a description

$$Y_i[t+1] = F_i[t] + u_{i_j}[t] + \xi_i[t+1] + c_i\xi_i[t], \tag{1}$$

where $0 < c_i < 1,\ i \geq 1,\ i \in \mathrm{N};\ \ F_i[t] = F_i(Y[t]),\ u_{i_j}, i_j \in I_u, I_u = \{n_1, \ldots, n_k\}$ – are nonlinear functions and controls, respectively; the set I_u contains the numbers of the equations in the original description with controlled variables; $Y[t] = (Y_1[t], \ldots, Y_n[t])$.

Remark 1. Coefficients $0 < c_i < 1,\ i = 1, 2$ characterize the velocity of prehistory "forgetting".

Remark 2. In [11] the same problem was considered only for a one-dimensional object. Note that in an one-dimensional object described by

$$Y[k+1] = F[k] + u[k] + \xi[k+1] + c\xi[k],\ k \in \{0, 1, 2, \ldots\}$$

where $\{\xi[k]\}_{k\geq 0}$ are the independent identically distributed stochastic variables characterized by $\mathbf{E}\{\xi[k]\} = 0,\ \mathbf{D}\{\xi[k]\} = (\sigma)^2,\ 0 < c < 1,\ k \geq 0,\ F[k] = F(Y[k]),\ u[k]$ are the nonlinear function and control, respectively; the response signal variance, $D(Y[k+1]) = D(F[k] + u[k] + c\xi[k]) + \sigma^2 \geq \sigma^2$ cannot be smaller that of the noise presented in any description of an object under any type of control. In this case, it is reasonably that there is a control, where the variance of the response variable is minimal [7, 8].

A problem is formulated to synthesize a robust control, which brings an object (1) into a target manifold of states described as a mean value of the target manifold

$$\mathbf{E}\{\psi[k]\} = 0,\ \psi[k] = \psi(Y[k]) = 0,\ Y[k] = (Y_1[k], \ldots, Y_n[k]), k \in \{0, 1, \ldots\},$$

where $\psi(\cdot)$ is a known function of states, which is referred to as a macro variable [3].

It is required that $\mathbf{E}\{\psi[k+1] + \omega\psi[k]\} = 0$, and that the variance $\mathbf{D}\{\psi[k+1] + \omega\psi[k]\}$ of the stochastic values $\psi[k+1] + \omega\psi[k]$ and the response variable $\psi[k]$ be equal to the lowest possible values $\forall k \geq 0$ at the given initial conditions $Y[0]$.

3 Problem Solution of Synthesizing a Robust Regulator for a Discrete Model

Let us list the main assumptions of an algorithm based on the combination of ADAR and conditional mathematical expectations techniques [8]:

(1) search for the structure of ADAR controls $\tilde{u}^A[k],\ k \in \{0, 1, \ldots\}$ with fixed noises; here we obtain an expression for basic ADAR-control $\tilde{u}[k]$, given that the quantities $\{\xi_i[k]\}_{k \geq 0},\ i = 1, 2$ are known;

(2) determination of the value of conditional expectation $u[k] = \mathbf{E}\{\tilde{u}^A[k]\,|\xi^k\}$ taking into account $\mathbf{E}\{\xi_i[k]\} = 0,\ i = 1, 2$ and pairwise independence of random variables $\xi_i[k], \xi_j[k], X_i[k], X_j[k],\ i \neq j$, where

$$\xi^k = (\xi[k], \xi[k-1]\ldots\xi[0]), \xi[k] = (\xi_1[k], \ldots, \xi_n[k]),\ k \in \{0, 1, \ldots\};$$

(3) decomposition of the initial description by substituting the relations obtained for $u[k] = \mathbf{E}\{\tilde{u}^A[k]\,|\xi^k\}$;

(4) transformation of the basic functional equation $\psi[k+1] + \lambda\psi[k] = 0$ by substituting the relations obtained for $u[k] = \mathbf{E}\{\tilde{u}^A[k]\,|\xi^k\}$;

(5) formation of an algebraic system of equations to determine the explicit form of expressions containing random variables, based on the ratios obtained at stages 3, 4 and its solution;

(6) replacing random expressions in the formula for the controller taking into account step 5.

The synthesis of a robust regulator for a discrete model is completed.

4 Technique Applications of Stochastic Control Synthesis for a Second-Order Nonlinear Object

4.1 Synthesizing a Stochastic Discrete Regulator

Let us discuss a stochastic vector object with a description (see (1), $n = 2$, $m = 1$)

$$\begin{aligned} Y_1[k+1] &= F_1[k] + \xi_1[k+1] + c_1\xi_1[k], \\ Y_2[k+1] &= F_2[k] + u[k] + \xi_2[k+1] + c_2\xi_2[k]. \end{aligned} \tag{2}$$

Let us consider the following problem of control with the descriptions of the control targets

$$\psi[k] = Y_1[k] - \rho\, Y_2[k] = 0,\ k \to \infty,\ \rho = const \neq 0, \tag{3}$$

According to the algorithm above (Sect. 3) in the *first step* we obtain the expression for basic ADAR-control $\tilde{u}[k]$, given that the quantities $\{\xi_i[k]\}_{k \geq 0},\ i = 1, 2$ are known. From the technique of the classical ADAR-method it follows that

$$\tilde{u}[k] = -F_2[k] - \xi_2[k+1] - c_2\xi_2[k] \\ + \rho^{-1}(F_1[k] + \xi_1[k+1] + c_1\xi_1[k] + \omega\psi[k]). \tag{4}$$

Let us give a more detailed explanation of the form (4).

It is known [3, 4] that ADAR-control will be referred to as the (vector) variable $u^A(x(t)) = \left(u_1^A, \ldots, u_m^A\right)$, delivering a solution to the variation problem

$$\Phi_D = \sum_{k=0}^{\infty}\sum_{j=1}^{m}\left(\alpha_j^2\psi_j^2[k] + (\Delta\psi_j[k])^2\right) \to \min$$

with the restrictions $\psi(Y[k]) = 0$, $\psi(Y[k]) = (\psi_1(Y[k]), \ldots, \psi_m(Y[k]))$ in the discrete cases. Note that equation

$$\psi[k+1] + \omega\psi[k] = 0, \ |\omega| < 1, \ k \geq 0 \tag{5}$$

is the Euler-Lagrange equation for Φ_D, and the scalar quantities α and ω are related by the following formula: $\omega = 0,5\left(1 + \alpha^2 - \sqrt{(1+\alpha^2) - 4}\right)$, $|\omega| < 1$.

We obtain the expression (4) from the Eq. (5) taking into account the condition (3) and description (2).

In the *second step* we obtain an expression for conditional expectation

$$u[k] = \mathbf{E}\left\{\tilde{u}[k]\,|\xi^k\right\} = \rho^{-1}F_1[k] - F_2[k] + \rho^{-1}\lambda\psi[k] - c_2\xi_2[k] + \rho^{-1}c_1\xi_1[k].$$

Within the conditions of the law of control $u[k]$ for the response variable $Y_2[k]$ and macro variable (3) we have the following relations in the *3–4 steps*:

$$\psi[k+1] + \omega\psi[k] = \xi_1[k+1] - \rho\xi_2[k+1], \\ Y_2[k+1] = \rho^{-1}F_1[k] + \rho^{-1}\omega\psi[k] + \rho^{-1}c_1\xi_1[k] + \xi_2[k+1]. \tag{6}$$

We take $c_1 = c_2 = c$ to simplify the problem and consider (6), and we obtain

$$u[k] = \rho^{-1}F_1[k] - F_2[k] + \left(\rho^{-1}\omega + c\rho\right)\psi[k] + c\rho\lambda\psi[k-1]. \tag{7}$$

The synthesis of a robust control system for the problem (2), (3) is completed.

4.2 Properties of a Stochastic Discrete Controller

The structure of the application of a stochastic regulator for problem (1) can be schematically presented as follows (see Fig. 1).

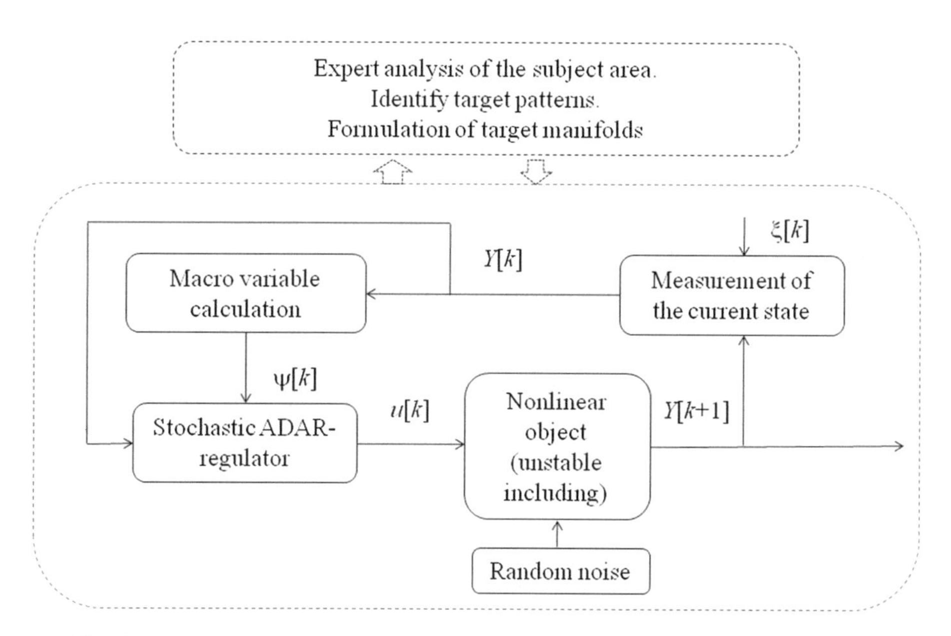

Fig. 1. Block diagram to implement stochastic control of a nonlinear discrete object

The dotted line in Fig. 1 shows two main blocks of the control synthesis process based on the control design algorithm discussed here. The upper block means that the subject knowledge in target manifold is "the price of this issue" for a successful design of a regulator in the class of ADAR-controls. The lower block is a graphic illustration in detail above of the presented control synthesis technique for a nonlinear stochastic object.

Statement. The control (7) for the problem (2), (3) ensures an asymptotically stable, on average, behavior of the system of Eqs. (2), (3), (7) in the neighborhood of the manifold $\psi[k] = Y_1[k] - \rho Y_2[k] = 0$ and possesses the following properties:

(a) $\mathbf{E}\{\psi[k+1] + \omega\psi[k]\} = 0$;
(b) dispersions of the stochastic quantities $\psi[k+1] + \omega\psi[k]$ and the response quantity $\psi[k] = Y_1[k] - \rho Y_2[k]$ under the action of control (7) are equal to the lowest possible values $\forall k \geq 0$ and obey the relation

$$\mathbf{D}\{\psi[k+1] + \omega\psi[k]\} = \sigma_1^2 + \rho^2\sigma_2^2,\ |\omega| < 1,\ k \geq 0.$$

Example. Let us take a nonlinear object (2) relied on the basis of the Feigenbaum's model (having chaotic states among the limiting states)

$$\begin{aligned} F_1[k] &= Y_1[k](\alpha C_0 - \mu\beta_1 Y_1[k]Y_2[k]), \\ F_2[k] &= A(\alpha C_0 - \mu\beta_2 Y_1[k]Y_2[k]),\ k \in \{0, 1, \ldots\}, \end{aligned} \tag{8}$$

where constant parameters $\alpha, C_0, \mu, \beta_1, \beta_2$ are the proportionality coefficients with a physical conceptual meaning.

Remark 3. In particular, here variables $Y_1[k], Y_2[k]$ can be interpreted as the quantities proportional to the production volumes supplied to the market by two entities, respectively; parameters α, C_0, μ are the proportionality coefficients with a conceptual economical meaning; parameters β_1, β_2 denote the prices set by the producers of the same type of products $Y_1[k]$, $Y_2[k]$, respectively [3].

According to the algorithm of synthesizing a stochastic discrete regulator we obtain a system of stochastic control given by (2), (3), (7), (8), including concrete descriptions, whose behavior is illustrated by the trajectories given in Fig. 2, and which suggests an explicit preference to use a stochastic regulator over the deterministic one under the conditions where noise cannot be calculated (or under the unmodeled dynamics conditions).

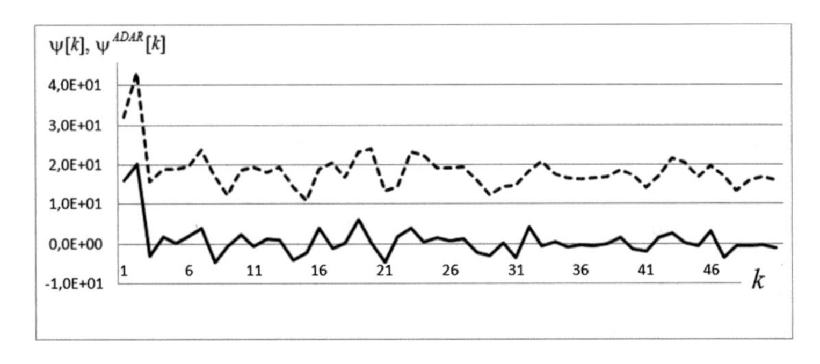

Fig. 2. Trajectories of the macro variables $\psi[k]$, $\psi^{ADAR}[k]$, solid and broken lines, respectively; $\xi_1[k]$, $\xi_2[k] \sim N(0,2)$

5 Conclusion

A new algorithm to design a robust stochastic regulator has been proposed, which possesses the following advantages:

(1) the control synthesis technique ensures the robust properties of the regulators;
(2) provision of a minimal variance of the response variable;
(3) minimal dispersion of the Lagrange-Euler function, which is used for the structure of the stochastic control system;
(4) clear priority of our method over the deterministic method of analytical design of aggregated regulators in the case of unmodelled stochastic dynamics [12], which is inevitably presented in unstable control objects;
(5) the quality functional of the synthesized control system meets the required properties of the target states, which meets the of the physical control theory [1, 3, 4].

The method of synthesis of a reliable stochastic regulator, discussed in this article, will be useful in the systems of decision-making support as it concerns the problems of control over a wide range of poorly formalized objects from different applied areas.

Acknowledgment. The reported study was funded by RFBR according to the research project №. 17-08-00920.

The author expresses her acknowledgment to Professor Kolesnikov A.A., Honored worker of science and engineering of the Russian Federation, for useful discussions.

References

1. Krasovskiy, A.A.: Mathematical and Applied Theory. Selected Works. Nauka, Moscow (2002)
2. Khalil, H.K.: Nonlinear Systems. Prentice Hall, Upper Saddle River (1996)
3. Kolesnikov, A.A.: Synergetics and problems of control theory: collected articles, ed. by Kolesnikov, A.A. FISMATLIT, Moscow (2004)
4. Tyukin, I.Yu., Terekhov, V.A.: Adaptation in nonlinear dynamical systems. LKI, SPb (2008)
5. Ott, E., Grebodi, C., Yorke, J.A.: Controlling chaos. Phys. Rev. Lett. **64**, 1196–1199 (1990)
6. Chen, G., Yu, X.: Chaos Control: Theory and Applications. Springer, Berlin (2003)
7. Astroem, K.J., Wittenmark, B.: Adaptive Control. Dover Publications, New York (2008)
8. Astroem, K.J.: Introduction to the Stochastic Theory of Control. Mir, Moscow (1973)
9. Emelyanov, S.V., Korovin, S.K.: Control of Complex and Uncertain Systems: New Types of Feedback. Springer, Heidelberg (2012)
10. Fradkov, A.L., Miroshnik, I.V., Nikiforov, V.O.: Nonlinear and adaptive control of complex systems. Kluwer Academic Publishers, Dordrecht (1999)
11. Kolesnikova, S.I.: A multiple-control system for nonlinear discrete object under uncertainty. Optim. Methods Softw. (2018). https://doi.org/10.1080/10556788.2018.1472258
12. Kolesnikova, S.I.: Synthesis of the control system for a second order non-linear object with an incomplete description. Autom. Remote Control **9**(79), 1556–1566 (2018)

An Approach to Developing Adaptive Electronic Educational Course

R. Yu. Tsarev[1(✉)], T. N. Yamskikh[1], I. V. Evdokimov[1], A. V. Prokopenko[1], K. A. Rutskaya[1], V. N. Everstova[2], and K. Yu. Zhigalov[3,4]

[1] Siberian Federal University, Krasnoyarsk, Russia
tsarev.sfu@mail.ru
[2] North-Eastern Federal University, Yakutsk, Russia
[3] V.A. Trapeznikov Institute of Control Sciences of Russian Academy of Sciences, Moscow, Russia
[4] Moscow Technological Institute, Moscow, Russia

Abstract. Modern information and telecommunication technologies allow not only to provide information required for training students in electronic form, they promote changes in educational process. The electronic information educational environment provides the necessary mechanism for implementing student's individual trajectories of training. Depending on his or her learning progress the process can vary from basic to intensive, with changes in volumes of information provided. One of the approaches to developing an adaptive electronic educational course is described in this paper. The tree of concepts mastered by students is considered as a basis for knowledge component organization. The tree of concepts will allow not only to organize theoretical material of the course in the form of the separate blocks of educational information intended for studying but also to determine a necessary and sufficient set of questions for testing students.

Keywords: Electronic educational course · Individual trajectory · Tree of concepts · Thesaurus

1 Introduction

Application of information technologies in education allows to expand the range of educational technologies, to use new methods and ways to organize educational process. It is necessary to understand information technologies more widely, than direct representation of information required for training students in electronic form. Today a number of electronic information educational environments and platforms have been created. They allow to place educational and methodical materials within an electronic course, and transform presentation of the material and assessment results.

The purpose of electronic information educational environments and platforms application consists in providing interactive access to information educational resources of the course: theoretical materials, practical and independent tasks, tests, methodical instructions for working with the course and its separate elements. As a

R. Silhavy (Ed.): CSOC 2019, AISC 986, pp. 332–341, 2019.
https://doi.org/10.1007/978-3-030-19813-8_34

rule, electronic educational courses are located on the higher education institution website. However, educational and methodical materials representing intellectual property are in the access closed for external users, and this access can be provided only to the authorized users of the system [1, 2].

In the course of work in the electronic information educational environment of higher education institution students have an opportunity to use various training courses, information resources, including multimedia components of training materials, they can independently master training material in convenient time and in necessary amount, take tasks and tests in a desirable order [3]. Regular and up-to-date information provision on assessment of tasks and tests taken by students is the advantage of electronic educational courses [4]. These courses provide a student with an opportunity to interact with the teacher and the other students out of classroom by means of interactive forms of an electronic educational course [5]. Implementation of interactive forms of education promotes intensification of training process, allow to activate and increase the efficiency of student independent work. Moreover, application of electronic tutorials creates students responsibility, self-motivation and independence [6].

Nowadays one of the actual directions in education development is organization of such training process which considers individual capabilities and needs of the trainee as the subject of educational process. Information and telecommunication technologies and electronic management systems in education make forming of individual trajectories of the course studying possible and reasonable.

The principles of one of the approaches to developing an adaptive electronic educational course where adaptation is realized by means of student's trajectory in the course depending on his or her current assessments are described in this article.

2 Form and Content of the Adaptive Electronic Educational Course Section

2.1 General Scheme of an Adaptive Electronic Educational Course Functioning

To consider the procedures allowing to create the adaptive electronic educational course (AEEC), we will start with the essence of adaptive information and telecommunication technologies in education.

To understand what AEEC should do, it is necessary to provide a situation which arises in the course of studying the discipline, and that lack of necessary knowledge and practical experience faced by the student in the course of learning. Having provided a task mentally, it is necessary to choose decision strategy and to determine necessary resources.

Most often at this stage a student asks himself the following questions: What should I do? How should I do it? How do I know whether I do it correctly?

To answer these questions, it is necessary to contact with students (teacher and/or AEEC): to show, to instruct, to give advice, etc.

Table 1 contains standard questions arising in the course of performing new or difficult tasks in the block of training material, and answers to them.

Table 1. Questions arising in the course of working with AEEC.

Questions or requirements of the student	AEEC answers
What should I do it for?	Explanations, example, consequences
What is it?	Definitions, illustrations, descriptions
What refers to it?	Available connections
How I do it?	Procedure, interactive reference books, ready methods (flowcharts, algorithms, consultations)
How and why did it happen?	Explanation, example or demonstration
Give me an example…	Examples
Teach me…	Interactive training, practical activities with feedback
Help me…	Interactive consultations
Advise me…	Ready methods, flowcharts, algorithms, consultations
Let me try…	Practical work
Supervise me	Tracking systems
Estimate me	Assessments or tests
Understand me	Feedback with the report, reasoning, interpretation; the tracking systems monitoring actions of the user or communication
How does it work?	Explanations and examples
Compare this and that for me	Comparative explanations and descriptions
Predict for me	Descriptions and demonstration of consequences
Where am I?	Management systems, tracking systems, communication types (You are here)
What's next?	Directions, hints, trainings, list of options or prompts

The table is not limited with the listed points, but allows to create a base for creative thinking in AEEC planning. It is necessary to remember that the number of questions is limited, and all related information can be determined precisely. Each trainee can use more convenient and understandable questions-answers connections, realizing thereby individual training. It is clear, that individual classes with a student (teacher/AEEC) is the most perspective educational technique.

The sequence of material presentation, depth and knowledge of information will individually vary for the specific student within an educational trajectory. When one of educational methods does not work, the student will try to choose another or concretizes his request to AEEC. The teacher can choose various help strategies. Besides, he can adjust the student's work, approving the strategy chosen or offering another one, more convenient for the achievement of an effective objective. It also helps to implement one of the promising pedagogical methods – model of mastering levels regarding cultural action pattern [7].

So, the offered approach to developing AEEC is based not only on educational concepts but also on educational resources and objectives. Resource and objective approach to developing information systems in education is not new. However there are not enough effective, simple and easily implemented in practice methods of their development. Besides, the existing IT decisions do not contain the means of choosing student's individual trajectories of training, depending on his or her progress, i.e. means of AEEC. The approach offered is especially intended for the solution of these problems.

Procedures of this technology are guided by a tree of concepts that allows not only to organize a theoretical course material in the form of the separate blocks of educational information intended for learning but also to determine a necessary and sufficient set of questions for testing students in the course of studying, or a set of competences (competence-based approach in education). Thus, all procedures can be divided into three levels:

1. The top level, based on a tree of objectives, provides preliminary determination of an educational trajectory option.
2. The middle level is guided either by a tree of concepts, or by a set of competences, or by both simultaneously. The task of this level consists in determination of a final decision.
3. The lower level is intended for ensuring execution of educational methods: intellectual information support of students.

The interrelation between procedures of the offered approach to developing AEEC is provided in Fig. 1.

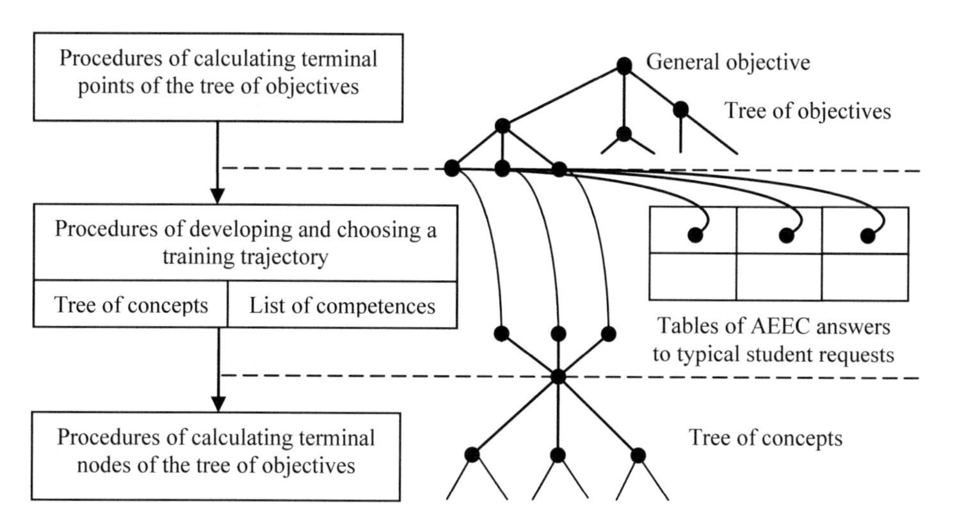

Fig. 1. Interrelation of levels in procedures of the offered approach to developing AEEC.

The general scheme of AEEC functioning allowing to trace the sequence of system blocks inclusion in operation is shown in Fig. 2.

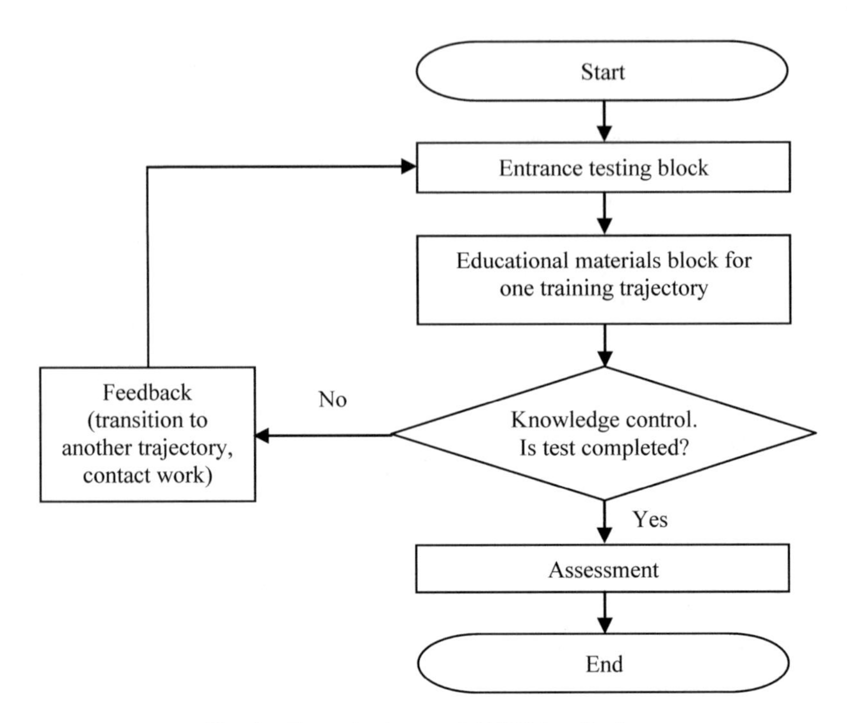

Fig. 2. General scheme of AEEC functioning.

In Fig. 2 the block of entrance testing is based on expanded psychological testing of students with control of residual knowledge in the competences of the disciplines preceding studying this specified curricular course. The results of such testing are subsequently used to form conclusions about a trajectory of training in the block of training material.

2.2 Tree of Concepts as Formal Representation of Theoretical Material of an Electronic Course

Developing an adaptive electronic educational course assumes first of all the organization of theoretical material representing knowledge component. "Knowledge is in turn determined as a form of existence and systematization of the person cognitive activity results or as a form of social and individual memory, the result of an object structuring and reasoning in the course of learning" [8].

When studying theoretical material, connectivity of the provided information, integrity of perception is reached only by means of concepts [9]. Thus, to describe any knowledge we need a necessary condition - the developed conceptual framework.

This idea is proved by Kondakov, claiming that the concept is a complete set of judgments, that is thoughts in which something is affirmed about distinctive signs of the object researched, which kernel are judgments about the most general and at the same time essential signs of this object, while knowledge – complete and systematized set of scientific concepts about regularities of the nature, society and thinking, stored by mankind in the course of vigorous reformative productive activity and directed to further knowledge and changes of the objective world [10].

The organization of theoretical material within an adaptive electronic educational course is carried out on the basis of a tree of concepts. A root of a tree is the most general concept, during mastering the course this concept reveals and disaggregated that is formally presented in the form of tree branches.

The set of concepts of one level having the general ancestor represent structural model of this concept ancestor and completely determine it in total. Each concept provided in a tree is also described by the phenomenological model allocating it from a set of similar concepts, the definition of the concept itself is provided in a separate component of the course – the thesaurus reflecting interrelation between concepts.

Studying of the course can be formally presented in the form of a bypass of a tree of concepts in depth. The advantage of the approach to organizing theoretical course material on the basis of a tree of concepts is the possibility to allocate separate logically connected blocks of educational information including one or several nodes of a tree of concepts depending on the concept complexity. The concept structural model allows to create the bank of the test questions necessary and sufficient to control knowledge acquired by students.

So, we have come to a conclusion that theoretical material within AEEC is carried out on the basis of a tree of concepts which, in turn is nothing else than a kind of thesauruses (dictionaries). To organize information search and to maintain such thesaurus the descriptor classification system of information (still called by descriptive) often is effectively used as its language approaches a natural language of the description of information objects.

The majority of thesauruses standards versions specify connection of terms with concepts of subject domain. The ISO standard (ISO 5963:1985) underlines that the indexing term – is representation of concept preferably in the form of a noun or a noun phrase [11]. At the same time the concept is considered as a unit of thought which is created mentally to reflect all or some properties of specific or abstract, real-life or mental object. Concepts exist as abstract entities, irrespective of terms which express them.

Thus, thesauruses developers assume that the concept of subject domain usually has several possible options of lexical representation in the text which are considered as synonyms. The descriptor is chosen among such synonyms – the term which is considered as the main method of reference to concept within the thesaurus.

The descriptor method of classification in our case consists in the following.

First, a set of the keywords or phrases describing subject domain of AEEC or a set of uniform objects is selected; at the same time there can be synonyms among keywords. To form such set of keywords or phrases three main sources of keywords or phrases selection for a tree of concepts are used.

The first source – professional and common cultural competences defined by Federal state educational standard of higher education bound to a lecture subject provided by the author of the course in the working program of the discipline (in the curriculum of a profile of preparation).

The second source – a subject of specific lecture from the working program of the discipline and other materials relating to this subject.

The third source – phenomenological model as a set of knowledge determining interrelation between various observations of the phenomena (phenomena) according to the fundamental theory, but which do not follow from this theory directly. The phenomenological model transform request like "formulation of a problem" by means of modeling the content of a request situation with thesaurus concepts into Boolean expression "conjunction of disjunctions" over the concepts of the thesaurus:

$$\bigcap_i \bigcup_j c_{ij},$$

where c_{ij} – thesaurus concepts.

Elements of disjunction can be concepts of the thesaurus which are considered as similar in their meaning - they are connected among themselves by thesaurus ways of a certain type.

Secondly, the chosen keywords and phrases are exposed to normalization, i.e. one or several most common are taken from a set of synonyms – descriptors. Other terms from a synonymic row included in the thesaurus are called askriptor or nondescriptors. They are used as the auxiliary elements, text entrances helping to find suitable descriptors. The descriptors are available in both brief and verbose formats.

The set of descriptors should meet the following requirements:

- the opportunity to describe the subject of the majority of texts in subject domain should be provided by means of the descriptors allocated;
- to reduce subjectivity of indexing the set of descriptors should not include sets of close descriptors; for this purpose classes of relative equivalence, when sets of close, but different concepts are reduced to one descriptor are created;
- the descriptor should be formulated unambiguously, its value implied within the thesaurus should be clear to the user. If it is not possible to find the unambiguous and clear descriptor, the term taken as a descriptor is supplied with the comment.

Such substantial types of relations between descriptors are most often not reflected in the detailed list of the thesaurus relations, and are registered by means of a small set of relations which are usually divided into two types: hierarchical and associative.

Many guidebooks and standards emphasize that the hierarchical relations in the thesaurus should be established when the relations are true irrespective of a context – only in such cases descriptors of the thesaurus can be organized in hierarchy. This recommendation is connected with the fact that in information search it is usually very difficult to determine accurately a context of the term usage and to understand whether this or that relation is applicable in this context.

So, for *the software* it is possible to specify that it is a *software product* as it is the internal characteristic of the software as the result of human activities exposed in the market of the mass buyer as goods. At the same time it is wrong to specify that the *software* is *application program packages* as there are, for example, *operating systems* which are not application-oriented.

The basic purpose of establishing associative relations between thesaurus descriptors – specifying on additional descriptors, useful for indexing or search. The relation of association is associative and not hierarchical. It is difficult to define the associative relation. All types of the relations, except a synonymy and the relation "sort - type" are allowed to be included in the associative relation.

Thirdly, the dictionary of descriptors, i.e. the dictionary of the keywords and phrases selected as a result of normalization procedure is created.

2.3 Tree of Concepts as Formal Representation of Theoretical Material of an Electronic Course

Developing individual training trajectory represents the multidimensional process aimed at providing the trainee independence and initiative, possibility to use his or her personal and cognitive potential and professional growth within educational process. For example designing different educational situations for bachelor students need specific educational information units to initiate students' reflection on content assimilation [12]. This kind of reflection being implemented in an activity mode could help students to develop themselves as professionals consciously and purposefully.

The process of individual training trajectory formation assumes building a sequence of educational information units within a certain course which mastering will provide a trainee with a chance to achieve educational goal and to obtain a set of the competences assumed by the curriculum and the working program of the discipline.

Separation of logically complete information blocks to study theoretical material allows to create different trajectories taking into account personal abilities and progress of the student. It was offered to select three trajectories relatively corresponding to studying the material with estimation marks "satisfactory", "good", "excellent".

The content of each educational information block is fixed, however, the form of material representation for each individual trajectory of training differs. For the students studying the discipline following a trajectory "excellent", information is presented in a more squeezed form, assuming that this category of students possesses ability to acquire theoretical material without its detailed explanation. On the contrary, for the students following a trajectory "good" the material is presented in details, with examples and illustrations.

The volume of each block of educational information, and, therefore, the corresponding block of test questions should not be large. It is offered to use minimum possible set of interconnected concepts determined by a tree of concepts for this discipline.

Separate blocks of educational information are formed on the basis of a set of the interconnected concepts, terms, keywords which define precisely the elementary block of theoretical material provided to students during their work with an electronic educational course. Within a tree of concepts, blocks of educational information represent one or several interconnected tree nodes. The concepts presented in a tree are given in the glossary of an electronic educational course.

The most important element of the electronic information educational environment is to control student knowledge throughout the trajectory of his learning [13]. Proceeding from this situation, the student takes the test defining his further training trajectory after studying each block of educational information. Besides assessing assimilation of theoretical knowledge blocks of test questions have to be directed to support student motivation and stimulate interest in further studying of the discipline.

In the beginning of studying the discipline with electronic educational course the student takes entrance test which defines his initial trajectory within a course. As a rule, entrance testing estimates residual knowledge of the disciplines, previous in structure of educational program.

If the test shows that a student hasn't satisfied the condition sufficient for continuing the course there is a kickback, and a student is forced to study repeatedly the same block of theoretical information and to take the test once again (see Fig. 3).

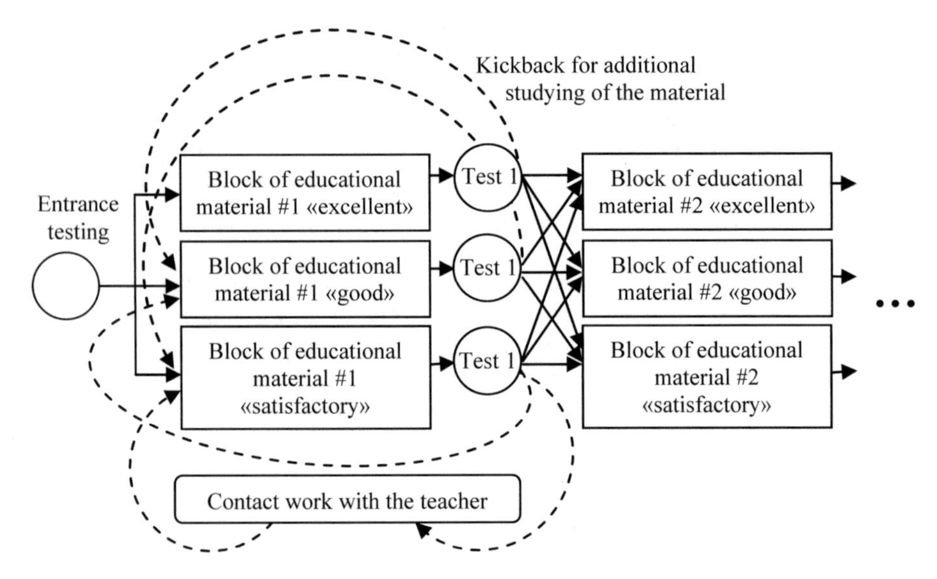

Fig. 3. Developing individual training trajectory with categories of estimation marks satisfactory-good-excellent with kickbacks.

In case of additional studying the block of educational information represented in different form, as a rule, from the lower trajectory is offered to the student. If a kickback has happened after taking the trajectory "good", it is possible to execute return on higher trajectory, believing that the new representation of the same information taking into account the material studied will be more available to the student.

Contact work with the teacher for studying this block of theoretical information is required in case of repeated negative test result.

3 Conclusion

The result of applying the above described approach to developing an adaptive electronic educational course is formalization of the basic principles, methods and technology of implementing adaptive approach to studying the disciplines with individual educational trajectories taking into account the current level of mastering educational material by a certain student.

The created electronic educational course considers the student individual capabilities and on the basis of assessment automatically transfers it to higher or low trajectory. Thus we see the adaptive nature of taking the course and studying theoretical material.

Application of this approach is directed to decrease the classroom loading connected with transferring lectures to student independent work when all theoretical material is provided in appropriate units of an adaptive electronic educational course. It will encourage activity and independence of students that is important not only for studying a separate course, but also for training students as future specialists capable to solve problems independently.

References

1. Elliott, D., Hopfgartner, F., Leelanupab, T., Moshfeghi, Y., Jose, J.M.: An architecture for life-long user modelling. In: Lifelong User Modelling Workshop Held in Conjunction With User Modeling, Adaptation and Personalisation, Torento, Italy, pp. 1–8 (2009)
2. Nguyen, L., Do, P.: Learner model in adaptive learning, world academy of science. Eng. Technol. **45**, 395–400 (2008)
3. Li, Q., Zhong, S., Wang, P., Guo, X., Quan, X.: Learner model in adaptive learning system. J. Inf. Comput. Sci. **7**(5), 1137–1145 (2010)
4. You, D., Shen, L., Peng, S., Liu, J.: Flexible collaborative learning model in e-learning with personalized teaching materials. Adva. Intell. Soft Comput. **105**, 127–131 (2011)
5. Jia, B., Yang, Y., Zhang, J.: Study on learner modeling in adaptive learning system. J. Comput. (Finland) **7**, 2585–2592 (2012)
6. Jia, B., Yang, Y., Zhang, J.: Representation and acquisition of feature value of learner model in adaptive learning system. Adv. Intell. Soft Comput. **122**, 371–377 (2011)
7. Bazhenova, K., Znamenskaya, O.: Teacher's beliefs as a factor of individual progress of pupil. In.: European Proceedings of Social and Behavioural Sciences EpSBS, Kazan, Russia, pp. 100–108 (2017)
8. Hsu, W.-C., Li, C.-H.: A competency-based guided-learning algorithm applied on adaptively guiding e-learning. Interact. Learn. Environ. **23**(1), 106–125 (2015)
9. Op Den Akker, R., Hofs, D., Hondorp, H., Op Den Akker, H., Zwiers, J., Nijholt, A.: Supporting engagement and floor control in hybrid meetings. In: Lecture Notes in Computer Science (Including Subseries Lecture Notes in Artificial Intelligence and Lecture Notes in Bioinformatics), LNAI, vol. 5641, pp. 276–290 (2009)
10. Kondakov, N.I.: Logicheskij slovar' [Logical dictionary]. Science, Moscow, Russia (1971)
11. ISO 5963:1985 Documentation: Methods of documents analysis, determinations of their subject and selection of indexing terms
12. Bazhenova, K.A.: Designing educational situations for bachelor students in Pedagogics Moscow University Bulletin, vol. 3, pp. 73–83 (2014)
13. Rus, V., D'Mello, S., Hu, X., Graesser, A.C.: Recent advances in conversational intelligent tutoring systems. AI Mag. **34**(3), 42–54 (2013)

Model-Algorithmic Support for Abilities Calculating of Control System Based on Projection Operators

Boris Sokolov[1] and Vitaly Ushakov[1,2(✉)]

[1] St. Petersburg Institute for Informatics and Automation of the Russian Academy of Science, 14th Line, 39, St. Petersburg 199178, Russia
sokol@iias.spb.su, mr.vitaly.ushakov@yandex.ru
[2] Saint-Petersburg State University of Aerospace Instrumentation, Bolshaya Morskaya Street, 67, St. Petersburg 190000, Russia

Abstract. One of the important problems in moving objects control system is the calculating of goal abilities, i.e., potential of the system to perform its missions in different situations. Thus, the preliminary analysis of information and technological and goal abilities of moving objects control system is very important in practice and can be used to obtain reasonable means of the moving objects exploitation under different conditions. In the paper model-algorithmic support for abilities calculating of control system based on projection operators are proposed.

Keywords: Moving objects control system · Optimal control · Reachability area · Orthogonal projection · Projection operator · Complex dynamic objects · Interdisciplinary approach · Python

1 Introduction

The main objects of this study are moving object control system (MO CS). The notion "Moving object" summarizes the features of moving elements working with various complex technical systems (CTS) types [1–3]. Depending on the type of CTS, MO can move and interact in space, in the air, on the ground, in the water or on the water surface. Analysis of the main trends of modern MO CS shows their features, for example: multiple aspects and uncertainty of behavior, hierarchy, structural similarity and excess for the main elements and MO CS subsystems, interconnections, diversity of control functions related to each MO CS level, territorial distribution of MO CS components.

One of the main features of modern MO CS is the variability of their parameters and structures at different phases of the MO CS life cycle [2–4]. Under existing conditions incrementing (stabilizing) or reducing the opportunity of MO CS potentials makes it necessary to manage MO CS structures (including the control of structure reconfiguration). There are many possible options for controlling the dynamics of the MO CS structure. For example, this change in the means and purposes of MO CS functioning; changing the order of monitoring tasks and solving control tasks; the

R. Silhavy (Ed.): CSOC 2019, AISC 986, pp. 342–348, 2019.
https://doi.org/10.1007/978-3-030-19813-8_35

redistribution of functions, tasks and control algorithms between MO CS levels; control over reserve resources; motion control of elements and subsystems MO CS; reconfiguration MO CS various structures.

2 Research Methodology

The proposed approach is based on the fundamental scientific results of optimal control program theory. For example, a qualitative analysis based on control theory as applied to a dynamic system gives significant results described in [5]. The research methodology is based on the following basic principles.

The first feature of the research methodology is the original dynamic representation of MO CS schedule as the optimal control program vector, representing MO control programs (MO CS operation plans) [1, 6]. MO CS scheduling is interpreted as dynamic process of operations control, which differs from the concept of dynamic scheduling in traditional rescheduling methods (compare with [7]). The benefits of scheduling with the help of an optimal control program were widely discussed in [8–10].

The calculation procedure for the optimal control program is based on the Pontryagin's maximum principle. Therefore, the parameters and their changes in dynamics are explicitly expressed in the scheduling model and can be used to analyze robustness to integrate the sustainability goal as a non-stationary performance indicator in MO CS scheduling.

The second feature is a dynamic view of the implementation of the MO CS schedule with various uncertainties based on reachability areas [2, 11–13].

In principle, the reachability area is a fundamental characteristic of any dynamic system. If the reachability area is known, its main characteristics essentially replace all the necessary information about the system dynamics, the stability of its operation and performance. The reachability area characterizes all possible states of the MO CS schedule, taking into account different variations of MO CS parameters in nodes and channels.

The theorem which expresses characteristics of the attainability set and aggregated variants of the general dynamic model of MO CS functioning for the attainability set construction were shown in [14].

3 Main Phases and Steps of a Program-Construction Procedure for Optimal Structure-Dynamics Control in a Complex Dynamic Objects

It is proposed to use an algorithm for solving complex dynamic objects (CDO) optimal control theory from [2].

At the first phase forming (generation) of allowable CDO multi-structural macro-states is being performed or structural-functional synthesis of a new CDO shape corresponding to the intended (required) environment.

At the second phase a single multi-structural macro-state is being selected, and adaptive plans (programs) of CDO transition to the selected macro-state are constructed. These plans should specify transition programs, as well as programs of stable CDO operation in intermediate multi-structural macro-states.

Let us consider in more detail the main stages of the first phase [2, 10]:

Step 1. Formation, analysis, and interpretation of input data for the synthesis of CDO multi-structural macro-states. Construction or correction of the appropriate models that are used in the structural-functional synthesis of the shape CDO.
Step 2. Planning of a solving process for the problem of the CDO macro-states synthesis. Estimation of time and other resources needed for the problem.
Step 3. Construction and approximation of reachability area for dynamic system. This set contains indirect description of different variants of CDO make-up (variants of CTS multi-structural macro-states).
Step 4. *Orthogonal projection of a set defining CDO macro-state requirements for the new shape to reachability area.*
Step 5. Interpretation of output results and their transformation to a convenient form for future use.

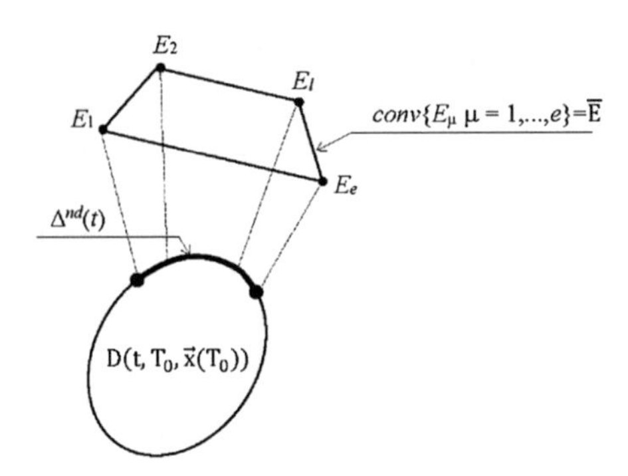

Fig. 1. Option 1 for the relative position of the target set and reachability area

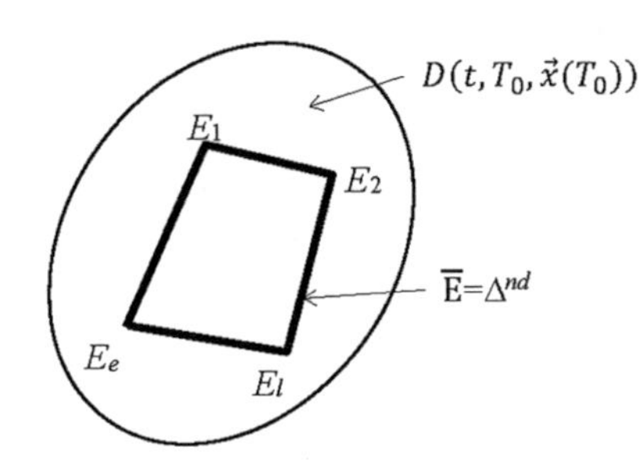

Fig. 2. Option 2 for the relative position of the target set and reachability area

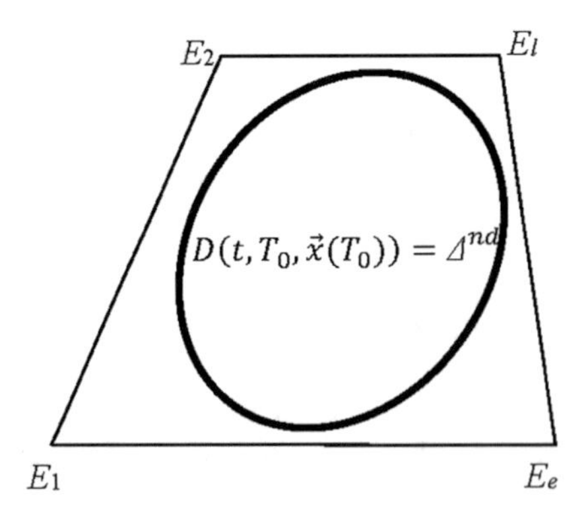

Fig. 3. Option 3 for the relative position of the target set and reachability area

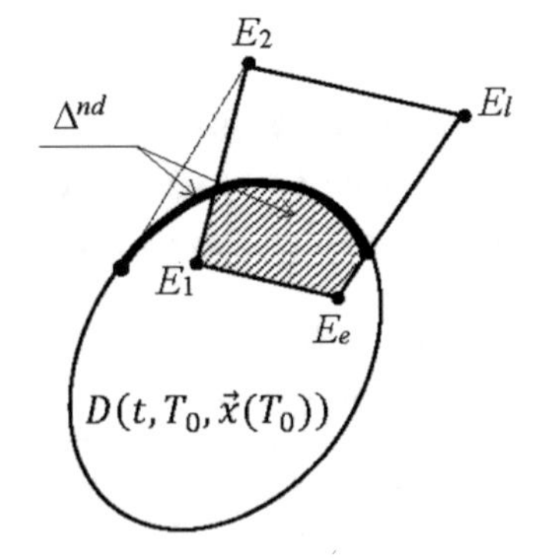

Fig. 4. Option 4 for the relative position of the target set and reachability area

In this article, we dwell in more detail on the orthogonal projection of reachability area. The analysis performed in [2, 11, 15] shows that there are four options for the relative position of the target set (in the form of a convex hull $\bar{\mathrm{E}}$) and reachability area $\mathrm{D}(\mathrm{t}, \mathrm{T}_0, \vec{\mathrm{x}}(\mathrm{T}_0))$.

Option 1: $\bar{\mathrm{E}} \cap \mathrm{D}(\mathrm{t}, \mathrm{T}_0, \vec{\mathrm{x}}(\mathrm{T}_0)) = \emptyset$ (Fig. 1). Convex hull $\bar{\mathrm{E}}$, built on points $\mathrm{E}\mu$ (μ = 1, …, e).
Option 2: $\bar{\mathrm{E}} \subset \mathrm{D}(\mathrm{t}, \mathrm{T}_0, \vec{\mathrm{x}}(\mathrm{T}_0))$ (Fig. 2). Rarest case.
Option 3: $\mathrm{D}(\mathrm{t}, \mathrm{T}_0, \vec{\mathrm{x}}(\mathrm{T}_0)) \subset \bar{\mathrm{E}}$ (Fig. 3).
Option 4: $\widetilde{\mathrm{E}} = \bar{\mathrm{E}} \cap \mathrm{D}(\mathrm{t}, \mathrm{T}_0, \vec{\mathrm{x}}(\mathrm{T}_0)) \neq \emptyset$ (Fig. 4). This is the most common case where the target set and reachability area are located.

We will perform orthogonal projection (Step 4) on the basis of the projection operators theory (Sect. 4) in this paper.

4 Orthogonal Projections in Reachability Area

Now consider the projection operators. The projection operator is the mathematical apparatus that allows us to realize the orthogonal projection of the target set on the reachability area. The projective plane is called the extended Euclidean's plane.

Usually, operator A is defined using a system of linear inequalities:

$$\begin{cases} y_1 \geq a_{11}x_1 + a_{12}x_2 + \cdots + a_{1n}x_{n1} \\ y_2 \geq a_{21}x_1 + a_{22}x_2 + \cdots + a_{2n}x_{n1} \\ \cdots \\ y_n \geq a_{n1}x_1 + a_{n2}x_2 + \cdots + a_{nn}x_{n1} \end{cases}, \tag{1}$$

where $a_{ij} \in \mathbb{R}$ for all i, j = 1, 2, …, n. Then A is the matrix of the operator A in the standard basis, where $A = (a_{ij})$ is the matrix composed of the coefficients in the inequalities system (1).

Equalities of the system of linear Eq. (1) can be written as:

$$\boldsymbol{x} = \begin{pmatrix} x_1 & x_2 & \cdots & x_n \end{pmatrix}^T, \tag{2}$$

$$\boldsymbol{y} = \begin{pmatrix} y_1 & y_2 & \cdots & y_n \end{pmatrix}^T. \tag{3}$$

We note one of the most important properties of projection operators. If A is the matrix of the operator in some basis, and **x** (2) is the coordinates vector column x in the same basis, then the column **y** (3) of the coordinates of the vector A(x) in the same basis is calculated by the formula **y** = A**x**.

As a result, it is possible to talk about the possibility of using projection operators for orthogonal projection of the target set onto the reachability area. In this section, the basic formulas of the projection operators were given, but in general this can be implemented in automatic mode, for example, in Python [16] using the NumPy library [17] and Matplotlib [18]. The first library allows you to get a numerical solution of the problem (coordinates) (Fig. 5), and the second - a graphical (graph). In Fig. 6 shows an example of the orthogonal projection of a curve onto the 371.22x + 9.53y + 5.39z = 1 plane.

```
[371.21600099   9.53083015   5.38549886]
[[-0.18054443  0.48252541 11.77646002]
 [-0.19413951  0.40011517 12.85939621]
 [-0.16092404  0.17808289 10.96282842]
 ...
 [-0.14640947  0.0816043  10.13309699]
 [-0.1368446   0.18147615  9.29705669]
 [-0.13055316  0.30718474  8.64092616]]
```

Fig. 5. Numerical solution

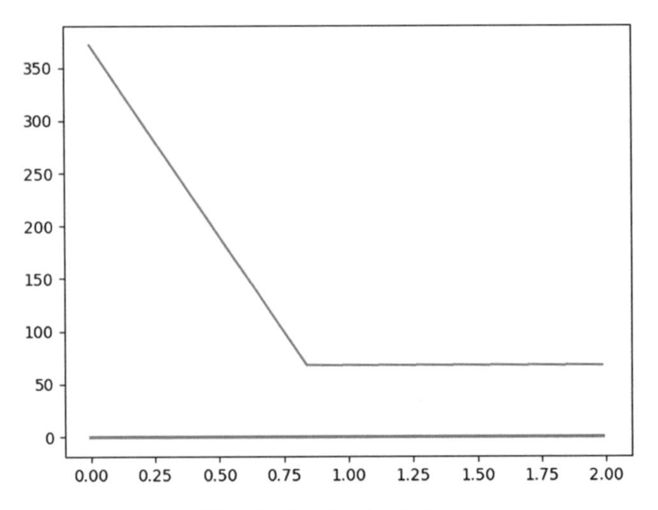

Fig. 6. Graphical solution

NumPy [17]—this is an extension of the Python language that adds support for large multi-dimensional arrays and matrices, along with a large library of high-level math functions for operations with these arrays.

Matplotlib [18]—this is a 2D graphics library for the python programming language with which you can create high-quality drawings of various formats and graphics.

5 Conclusion

Reachability area is a fundamental characteristic of any dynamic system (in our case – MO CS).

Thus, the article provides a description of the main phases and steps of a program-construction procedure for optimal structure-dynamics control in a complex dynamic objects.

In the paper model-algorithmic support for abilities calculating of control system based on projection operators are proposed.

Also, the article provides a way to use projection operators for orthogonal projection of reachability area.

Acknowledgements. The research described in this paper is partially supported by the Russian Foundation for Basic Research (grants 16-29-09482-ofi-m, 17-08-00797, 17-06-00108, 17-01-00139, 17-20-01214, 17-29-07073-ofi-m, 18-07-01272, 18-08-01505, 19–08–00989), state order of the Ministry of Education and Science of the Russian Federation №2.3135.2017/4.6, state research 0073–2019–0004, and International project ERASMUS+, Capacity building in higher education, # 73751-EPP-1-2016-1-DE-EPPKA2-CBHE-JP.

References

1. Kalinin, V.N., Sokolov, B.V.: Optimal planning of the process of interaction of moving operating objects. Int. J. Differ. Eqn. **21**(5), 502–506 (1985)
2. Okhtilev, M.Y., Sokolov, B.V., Yusupov, R.M.: Intellectual Technologies of Monitoring and Controlling the Dynamics of Complex Technical Objects. Nauka, Moskva (2006). (in Russian)
3. Ivanov, D., Sokolov, B.: Adaptive Supply Chain Management. Springer, London (2010). https://doi.org/10.1007/978-1-84882-952-7
4. Klir, G.: Uncertainty and Information: Foundations of Generalized Information Theory. Wiley, Hoboken (2005)
5. Sokolov, B., Dolgui, A., Ivanov, D.: Optimal control algorithms and their analysis for short-term scheduling in manufacturing systems. Algorithms **11**(5), 99–110 (2018). https://doi.org/10.3390/a11050057
6. Kalinin, V.N., Sokolov, B.V.: A dynamic model and an optimal scheduling algorithm for activities with bans of interrupts. Autom. Remote Control **48**(1–2), 88–94 (1987)
7. Vieira, G.E., Herrmann, J.W., Lin, E.: Predicting the performance of rescheduling strategies for parallel machine systems. J. Manuf. Syst. **19**(4), 256–266 (2000)
8. Khmelnitsky, E., Kogan, K., Maimom, O.: Maximum principle-based methods for production scheduling with partially sequence-dependent setups. Int. J. Prod. Res. **35**(10), 2701–2712 (1997)
9. Ivanov, D., Sokolov, B., Kaeschel, J.: A multi-structural framework for adaptive supply chain planning and operations with structure dynamics considerations. Eur. J. Oper. Res. **200** (2), 409–420 (2010). https://doi.org/10.1016/j.ejor.2009.01.002
10. Chauhan, S.S., Gordon, V., Proth, J.-M.: Scheduling in supply chain environment. Eur. J. Oper. Res. **183**(3), 961–970 (2007). https://doi.org/10.1016/j.ejor.2005.06.078
11. Gubanov, V.A., Zakharov, V.V., Kovalenko, A.N.: Introduction to Systems Analysis. LGU, Leningrad (1988). (in Russian)
12. Chernousko, F.L.: State Estimation of Dynamic Systems. CRC Press, Boca Raton (1994)
13. Clarke, F.H., Ledyaev, Yu.S, Stern, R.J., Wolenskii, P.R.: Qualitative properties of trajectories of control systems: a survey. J. Dyn. Control Syst. **1**(1), 1–48 (1995)

14. Sokolov, B., Kalinin, V., Nemykin, S., Ivanov, D.: Models and algorithms for abilities evaluation of active moving objects control system. In: Claus, T., Herrmann, F., Manitz, M., Rose, O. (eds.) European Council for Modeling and Simulation, pp. 467–473. Digitaldruck Pirrot, Dudweiler (2016). https://doi.org/10.7148/2016-0467
15. Petrosjan, L.A., Zenkevich, N.A.: Game Theory. World Scientific Publishing, Singapore (1996)
16. Python 3.7.2 documentation. https://docs.python.org/3/
17. NumPy Reference. https://docs.scipy.org/doc/numpy/reference/
18. MatPlotLib User's Guide. https://matplotlib.org/users/index.html

OFMDC: Optimal Framework for Microarray Data Classification Using Eigenvector Decomposition for Cancer Disease

V. Sudha(✉) and H. A. Girijamma

Department of Information Science and Engineering,
RNS Institute of Technology, Bengaluru, India
sudhavinayakam@gmail.com

Abstract. In the recent era, the research interest has increased among different computer and communication societies towards microarray gene expression detection and profiling. Despite of having a wide range of applications, the more emphasize has been kept towards cancer and its sub-type classifications. It has been seen in the past that the existing data mining approaches impose more cost of computation during pattern discovery and correlation establishment. Thereby, it is needed to address this shortcoming to strengthen the reliable cancer detection and classification process cost-effectively. An efficient machine learning tool has a better scope of optimization towards handling margin and error factors. Addressing this open research issue, the current study has come up with a novel method namely Optimal Framework for Microarray Data Classification (OFDMC) which incorporates Eigenvector decomposition to perform dimension reduction of gene expression data without compromising the complexity and accuracy aspects. The study also validates the performance of the proposed system by introducing a numerical analysis.

Keywords: Machine learning · Gene analysis · Cancer classification · Graph analysis

1 Introduction

The growing impact of cancer worldwide witnesses alarming statistical data. The latest reports of the national cancer institute for the year 2018 reveal that the united states of America alone encounter 1.7 million suspicious subjects out of which approximately 6 million has a fatal condition [1]. The accurate prognosis of the categories of cancer is directly correlated to the method the treatment. The profiling of the Microarray gene express (MGE) is gaining popularity for the classification of the cancer diseases [2–5]. The typical approach of MGE is acting on the tissue to classify as per the profile as either its type as cancerous or non-cancerous as well to distinguish it as the type of cancer. The Fig. 1 shows the typical classification of MGE based cancer classification methods that includes three types namely (1) statistical and data mining, (2) machine learning and (3) ensemble-based classification method [6].

In the initial evolution of the statistical method [7], the emphasis is to establish the correlation in the quantified data, whereas the datamining approach focuses on pattern

R. Silhavy (Ed.): CSOC 2019, AISC 986, pp. 349–356, 2019.
https://doi.org/10.1007/978-3-030-19813-8_36

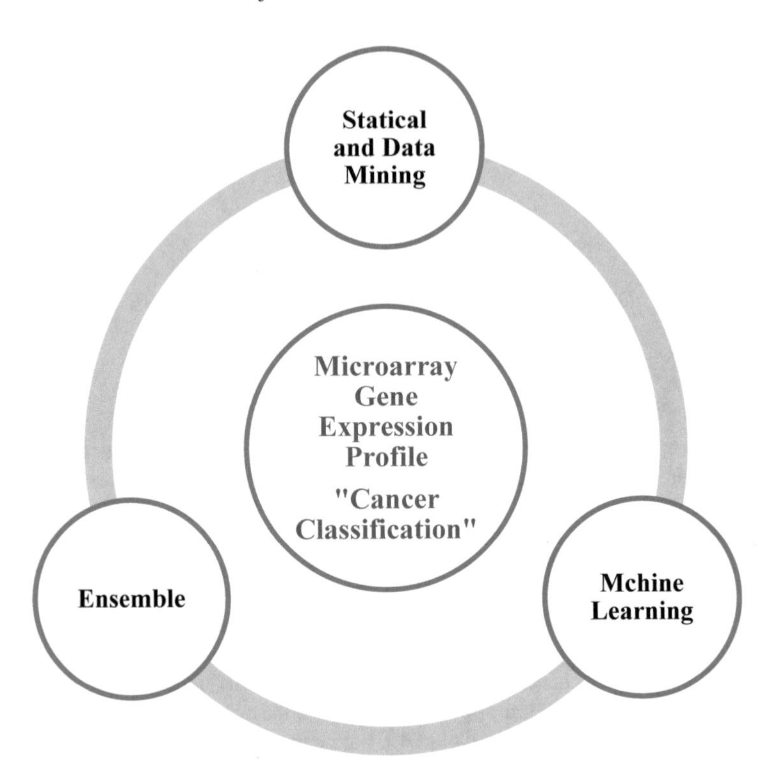

Fig. 1. Types of classification approach for cancer on microarray gene expression profile

discovery and relationship in MGE [8]. In the other hand machine learning approaches to classifying MGE yield high scope for better classification as compared to the histological data. [9]. The machine learning-based classifiers suffer a trade-off for optimizing margin and errors [10]. The ensemble method is hybrid of machine learning for both training and prediction in comparatively more accurate and reliable [11]. There are two typical ensemble methods namely wrapper and filter method. The wrapper method is quite slower due to their approach of search, whereas the selected gene in the filter approach is non-optimal. There is a need of a method which adopts a dimension reduction process and at the same time maximize the accuracy of classification on the reduced data set to ensure minimization of computational complexity [12]. This paper present a method for dimension reduction and classification an a hybrid approach using Eigenvector Decomposition. The rest of the section of the paper is organised as Sect. 2 describes the related work in the Sects. 2, 3 explains about the OFDMC framework and system model, section followed by results and discussion in Sect. 4, finally Sect. 5 conclude the paper in section conclusion and future research direction.

2 Review of Literature

The approach of machine learning for the accurate and real-time classification of a complex disease like cancer adopts unsupervised learning, where one of the most challenging tasks is to have a very robust and efficient clustering or grouping technique, Sudha et al. [13] has reviewed various clustering algorithms pros and cons especially for medical data. In the work of Ainhoa Perez-Diez et al. [14], it is clearly emphasized that the analysis of microdata for the purpose of cancers classification is yet in immature stage to be widely accepted. Some of the predominant methods like comparison, prediction and discovery of the classes are gaining popularity. The authors Sudha et al. [15, 16] have discussed a method for the clustering of the complex dataset of gene profile using fuzzy based method, where simulation results provides promising results. Towards microarray cancer classification, Alshamlan et al. [17] have presented an artificial bee colony (ABC) approach based on particle swarm optimization (PSO) technique where the approach is applied for analyzing the expression profile of microarray gene. The outcomes of [17] have suggested that the proposed hybrid gene selection ABC approach was able to achieve superior improvement in classification. The next work of Alshamlan et al. [18] have addressed the gene selection concern issue in both multi-class and binary classification. In order to overcome these issues, a genetic bee colony algorithm was presented and achieved highest degree of classification accuracy than [17]. Similar kind of work which also found with microarray gene data classification was found in Kar et al. [19]. In this [19], the gene expression classification was done by using K-nearest neighborhood technique and tested on three different microarray datasets like accte lymphoblastic (AL) leukemia and acute myeloid (AM) leukemia and the mixed-lineage (ML) leukemia data. The work [19] have attained better higher accuracy in cancer classification. A review work of Ang et al. [20] have discussed supervised, semi supervised and unsupervised features selection from DNA datasets and it was observed that the unsupervised and semi supervised feature selection is more competitive with supervised feature selection. The use of ABC and artificial neural network (ANN) algorithm for DNA microarrays were found in Garro et al. [21]. The algorithm able to classify the cancer samples at 93.2% of accuracy. A modified analytical hierarchy (AH) for gene selection and cancer classification was presented in Nguyen and Nahavandi [22] by using type-2 fuzzy logic approach. Using this approach [22], the classification performance was improved to a greater extent. Another kind of cancer disease classification with respected to gene expression profiles was found in Salem et al. [23] and achieved feature selection-based cancer classification. Similarly, another work with feature selection-based Hybrid filter was presented in Chuang et al. [24] for microarray classification and obtained better classification accuracy. An interesting work of Xi et al. [25] have also presented feature selection-based cancer classification mechanism by using PSO and support vector machine (SVM). Through this [25], robust, accurate cancer classification results were obtained. An independent component subspace for DNA microarray data classification with respect to feature selection was found in Aziz et al. [26]. Here [26], the independent components of microarray were selected by using fuzzy logic and achieved higher degree of accuracy than SVM and Naive Bayes (NB) classifiers. An artificial intelligence concept for micro array cancer classification was found in Dashtban and Balafar [27]. The use of artificial intelligence in classification has out formed other existing classifiers. Towards classifying the microarray having gene expression-based cancer data, Wang et al. [28] have presented discrete bacterial mechanism and come up with better accuracy in cancer classification. A comparative analysis of Chao et al. [29] have compared surgically respected and paired microarray specimens in large lung cancer. A Memetic algorithm was proposed in Begum et al. [30] for gene selection and cancer diagnosis in microarray data and has obtained the significant results than existing tabu search (TS), simulated annealing (SA) and genetic algorithm (GA).

3 OFMDC Framework and System Model

The OFMDC framework includes methodology of graph and dimension reduction with statistical analysis. The Table 1 describes the complex dataset structure of "M" subject.

Table 1. Dataset description

Sl. No.	Dataset	Description
1	Subject genetic graph signal	Gene of the subject is muted if the flag is '1', non-muted of the flag is 0
2	Histology	For flag '1' type I cancer, for flag '2', type -2
3	Gene Network	Vector of the gene

The phenotype differentiates on the basis of genetic profile as variations occurs even in the same disease. The histology data related the type of cancer as type-I and type-II. The visualization of non-sparse and spare is shown in the Fig. 2.

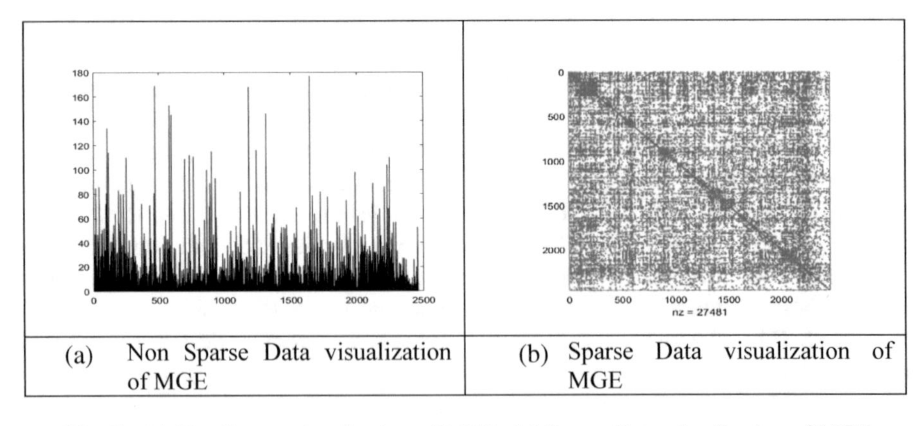

(a) Non Sparse Data visualization of MGE

(b) Sparse Data visualization of MGE

Fig. 2. (a) Non-Sparse visualization of MGE, (b) Sparse Data visualization of MGE

3.1 OFMDC System Model

The OFMDC Model adopts analytical modeling for the graph classification on a genetic network of **‘N’** genes of the database of complex type microarray database, where the connection **‘k’** exists among the genes only if the common genetic operation exists in encoded constitute. The model aims for the complex disease exploration process like cancer.

Algorithm for classification using KNN and filtering

```
Input: G_net
Output: var
Start
    1. L(G_net)←Δf(G_net)
    2. For G_s = L(G_net)
    3. Compute, E_g←f(GFT)
    4. Var←x^T Lx
End
```

The gene network (G_{net}), is connection labeled graph with ‘0’ and ‘1’, the Laplacian of the G_{net} is computed as $L(G_{net})$ in order to illuminate the self-loop. The Laplacian matrix is computed using equation (Eq. 1)

$$x^T LG^x = \sum_{(i,j)\in E} (x_i - x_j)^2 \tag{1}$$

For the graph shift (Gs) equal to the L(Gnet), the eigenvectors (E_g) using graph Fourier theorem using Eq. (2) for the variation.

$$TV_G(x) = \sum_{n=1 m\in N(n)}^{N} (x_n - x_m)^2 \omega_{mn} = x^T Lx \tag{2}$$

The Fig. 3 illustrates the corresponding value of total variation vs eigen value.

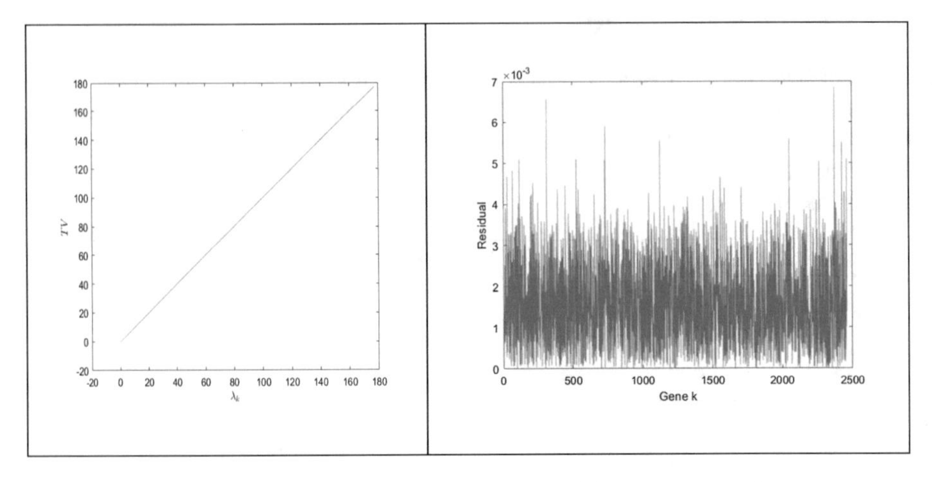

Fig. 3. (a) Eigen value vs Total variation and (b) Gene vs Residual

4 Result and Analysis

The frequent alternation of signal orients towards the bigger eigenvalues of corresponding vectors, a linear correlation is observed.

The plot of the gene vs residual does not clary indicate the criticalities as there exits outlier that categories the type of the cancer the subject has (Fig. 4).

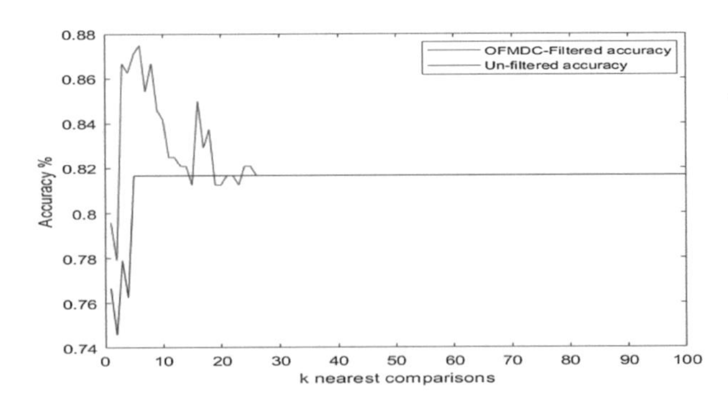

Fig. 4. Comparison between OFMDC- filtered accuracy vs un-filtered accuracy with different k-Nearest neighbor

5 Conclusion

In the recent times the microarray data classification got much attention from the researchers owing to its potential towards detecting various aspects from the medical view point. This study introduces a novel analytical model namely OFDMC got introduced where it has targeted to classify the cancer oriented micro-array data in a cost-effective manner. It also applies a novel Eigenvector Decomposition to enhance the performance of the classifier to a significant extent. The experimental outcome further demonstrated that OFDMC filter accuracy is superior as compared to the un-filtered accuracy. The system also handled the data complexity problem by applying dimension reduction which makes it more applicable to the futuristic cancer research direction.

References

1. National Cancer Institute: Cancer Statistics (2018). https://www.cancer.gov/about-cancer/understanding/statistics. Accessed 16 Jan 2019
2. Hou, J.Y., Wang, Y.G., Ma, S.J., et al.: J. Cancer Res. Clin. Oncol. **143**, 619 (2017)
3. Salem, H., Attiya, G., El-Fishawy, N.: Pattern Anal. Appl. **20**, 567 (2017)
4. Kim, H., Bredel, M.: Predicting survival by cancer pathway gene expression profiles in the TCGA. In: 2012 IEEE International Conference on Bioinformatics and Biomedicine Workshops, Philadelphia, PA, pp. 872–875 (2012)
5. Li, C., Lee, J., Ding, J., et al.: BioData Mining (2018)
6. Alshamlan, H., Badr, G., Alohali, Y.: A comparative study of cancer classification methods using microarray gene expression profile. In: Proceedings of the First International Conference on Advanced Data and Information Engineering, DaEng-2013, pp. 389–398. Springer, Singapore (2014)
7. Shon, H.S., Ryu, K. H.: Predicting cancer from microarray data using statistical method. In: Sixth International Conference on Advanced Language Processing and Web Information Technology, ALPIT 2007, Luoyang, Henan, China, pp. 474–479 (2007)
8. Fiori, A., Grand, A., Bruno, G., Brundu, F.G., Schioppa, D., Bertotti, A.: Information extraction from microarray data: a survey of data mining techniques. In: Business Intelligence: Concepts, Methodologies, Tools, and Applications, pp. 1180–1211. IGI Global (2016)
9. Kourou, K., Exarchos, T.P., Exarchos, K.P., Karamouzis, M.V., Fotiadis, D.I.: Machine learning applications in cancer prognosis and prediction. Comput. Struct. Biotech. J. **13**, 8–17 (2015)
10. Huynh, P.H., Nguyen, V.H., Do, T.N.: A coupling support vector machines with the feature learning of deep convolutional neural networks for classifying microarray gene expression data. In: Modern Approaches for Intelligent Information and Database Systems, pp. 233–243. Springer, Cham (2018)
11. Mohammed, M., Mwambi, H., Omolo, B., Elbashir, M.K.: Using stacking ensemble for microarray-based cancer classification. In: 2018 International Conference on Computer, Control, Electrical, and Electronics Engineering (ICCCEEE), Khartoum, pp. 1–8 (2018)
12. Ghorai, S., Mukherjee, A., Sengupta, S., Dutta, P.K.: Cancer classification from gene expression data by NPPC ensemble. IEEE/ACM Trans. Comput. Biol. Bioinform. **8**(3), 659–671 (2011)
13. Sudha, V., Girijamma, H.A.: Appraising research direction & effectiveness of existing clustering algorithm for medical data. Int. J. Adv. Comput. Sci. Appl. (IJACSA) **8**(3), 343–351 (2017)
14. Perez-Diez, A., Morgun, A., Shulzhenko, N.: Microarrays for cancer diagnosis and classification. In: Madame Curie Bioscience Database [Internet]. Landes Bioscience, Austin (TX) (2000–2013)

15. Sudha, V., Girijamma, H.A.: Novel clustering of bigger and complex medical data by enhanced fuzzy logic structure. In: 2017 International Conference on Circuits, Controls, and Communications (CCUBE), Bangalore, pp. 131–135 (2017). https://doi.org/10.1109/ccube.2017.8394147
16. Sudha, V., Girijamma, H.A.: SCDT: FC-NNC-structured complex decision technique for gene analysis using fuzzy cluster based nearest neighbor classifier. Int. J. Electr. Comput. Eng. (IJECE) **8**(6), 4505–4518 (2018)
17. Alshamlan, H., Badr, G., Alohali, Y.: mRMR-ABC: a hybrid gene selection algorithm for cancer classification using microarray gene expression profiling. BioMed Res. Int. **2015**, 15 (2015)
18. Alshamlan, H.M., Badr, G.H., Alohali, Y.A.: Genetic Bee Colony (GBC) algorithm: a new gene selection method for microarray cancer classification. Comput. Biol. Chem. **56**, 49–60 (2015)
19. Kar, S., Sharma, K.S., Maitra, M.: Gene selection from microarray gene expression data for classification of cancer subgroups employing PSO and adaptive K-nearest neighborhood technique. Expert Syst. Appl. **42**(1), 612–627 (2015)
20. Ang, J.C., Mirzal, A., Haron, H., Hamed, H.N.A.: Supervised, unsupervised, and semi-supervised feature selection: a review on gene selection. IEEE/ACM Trans. Comput. Biol. Bioinform. **13**(5), 971–989 (2016)
21. Garro, B.A., Rodríguez, K., Vázquez, R.A.: Classification of DNA microarrays using artificial neural networks and ABC algorithm. Appl. Soft Comput. **38**, 548–560 (2016)
22. Nguyen, T.T., Nahavandi, S.: Modified AHP for gene selection and cancer classification using type-2 fuzzy logic. IEEE Trans. Fuzzy Syst. **24**(2), 273–287 (2016)
23. Salem, H., Attiya, G., El-Fishawy, N.: Classification of human cancer diseases by gene expression profiles. Appl. Soft Comput. **50**, 124–134 (2017)
24. Chuang, L.-Y., Ke, C.-H., Yang, C.-H.: A hybrid both filter and wrapper feature selection method for microarray classification. arXiv preprint arXiv:1612.08669 (2016)
25. Xi, M., Sun, J., Liu, L., Fan, F., Wu, X.: Cancer feature selection and classification using a binary quantum-behaved particle swarm optimization and support vector machine. Comput. Math. Methods Med. **2016**, 9 (2016)
26. Aziz, R., Verma, C.K., Srivastava, N.: A fuzzy based feature selection from independent component subspace for machine learning classification of microarray data. Genomics Data **8**, 4–15 (2016)
27. Dashtban, M., Balafar, M.: Gene selection for microarray cancer classification using a new evolutionary method employing artificial intelligence concepts. Genomics **109**(2), 91–107 (2017)
28. Wang, H., Jing, X., Niu, B.: A discrete bacterial algorithm for feature selection in classification of microarray gene expression cancer data. Knowl.-Based Syst. **126**, 8–19 (2017)
29. Li, C., Huang, C., Mok, T.S., Zhuang, W., Xu, H., Miao, Q., Fan, X., et al.: Comparison of 22C3 PD-L1 expression between surgically resected specimens and paired tissue microarrays in non–small cell lung cancer. J. Thorac. Oncol. **12**(10), 1536–1543 (2017)
30. Begum, S., Chakraborty, S., Banerjee, A., Das, S., Sarkar, R., Chakraborty, D.: Gene selection for diagnosis of cancer in microarray data using memetic algorithm. In: Intelligent Engineering Informatics, pp. 441–449. Springer, Singapore (2018)

Conversion of Meteorological Input Data Implemented in the Algorithm of Storm Prediction

David Šaur(✉), Jaromír Švejda, and Roman Žák

Faculty of Applied Informatics, Tomas Bata University in Zlin,
Nad Stranemi 4511, Zlin, Czech Republic
{saur, svejda}@utb.cz

Abstract. This article focuses on a new way of converting input meteorological parameters, based on which outputs are computed in the Algorithm of Storm Prediction (Algorithm). This Algorithm has been developed to forecast precipitation and dangerous phenomena associated with convective severe storms that may cause floods in the Czech Republic. Transformation Algorithm of Input Meteorological Parameters is shown in the methodological part for data conversion purposes. The result section evaluates the accuracy of converted outputs using three aggregation methods. The outputs of this article will be used to improve the quality of converted data for computing outputs in the Algorithm of Storm Prediction.

Keywords: Weather forecast · Severe storm · Convective precipitation · Algorithm · Data conversion · Data mining · Aggregation method

1 Introduction

Flash floods are a current phenomenon, which is one of the most frequent natural disasters in the Czech Republic. This type of floods occurs several times a year and causes damage to property at around several hundred million crowns each year [1, 2]. Flash floods and drought are two natural disasters affected by global climate change, especially in the Czech Republic. Its uneven local occurrence is a characteristic feature and in combination with high intensity reaching negative effects [3].

Primarily, flash floods caused by strong convective precipitation with the intensity higher than 20 mm/h [4, 5]. Forecasting of convective storms is very inaccurate due to insufficient input data, especially from ground meteorological stations and technical limitations of forecasting systems [6]. For example, numerical weather prediction models (NWP models) operate with too high horizontal resolution and the hydrostatic core model unsuitable for convection modelling [7–9]; inexactitudes of meteorological measurements and other distance methods of precipitation and the cloud measurement in the atmosphere [10–12]. Now, accurate and quality prediction information with sufficient time in advance is one of prerequisites for the early deployment of preventive measures and subsequent to minimize potential damage. This assumption is solved by the Algorithm of Storm Prediction, the purpose of which is to forecast the occurrence

R. Silhavy (Ed.): CSOC 2019, AISC 986, pp. 357–367, 2019.
https://doi.org/10.1007/978-3-030-19813-8_37

of convective severe storms. Currently, this algorithm has being developed as software and currently tested in experimental mode. On the other hand, the predictive accuracy was higher than standard meteorological forecasting systems based on verified 63 storm events between 2015 and 2017 [6].

Input data are mainly data from NWP models that are processed by the Algorithm stored in standard PNG, GIFF, and JPEG graphic formats [13]. Conversion data is realized by aggregation data methods which were implemented in the Transformation Algorithm of Input Meteorological Parameters.

Data processing methods differ in predictive systems. Raw data is usually distributed in binary form, which is then converted into maps, charts and images. Forecasting meteorological systems such as NWP models and nowcasting systems are implemented in applications targeted for a specific field, such as classical, aeronautical meteorology, agrometeorology, and the like. The demand for information from these applications depends on the trend in the number of natural disasters that affect negatively our society [14–16].

The purpose of this article is to provide information on the input data processing in the Algorithm of Storm Prediction. Article results are focused on the accuracy evaluation of data conversion used in the Transformation Algorithm of Input Meteorological Parameters. In conclusion, the main results of the accuracy evaluation of the Algorithm's predictive outputs are compared with other meteorological forecasting systems for the Czech Republic for the year of 2018.

2 Methods

Forecasting of convective storms is usually performed by NWP models, nowcasting and expert meteorological systems. This chapter is focused on the description of the input data conversion used in the Algorithm of Storm Prediction. Input data conversion is solved by the Transformation Algorithm of Input Meteorological Parameters.

2.1 Algorithm of Storm Prediction

The Algorithm of Storm Prediction (Algorithm) is an application developed as an analytical-assessment tool for forecasting of symptoms of severe storms. These symptoms are convection and torrential rainfall, hailstorm, strong wind gusts and tornadoes on which the risk of flash floods is calculated for individual territorial units like 13 regions and 205 municipalities with extended powers in the Czech Republic. The risk of flash floods is one of the most important outputs because summarizes all other Algorithm's outputs [6, 17].

Table 1 list the predicted phases and outputs of the Algorithm. The first phase is focused on the processing of input data from used meteorological systems. The largest amount of input data is used from NWP models and Czech Hydrometeorological Institute (CHMI) applications.

Table 1. Predicted phases and outputs of Algorithm [6, 17].

Forecast phase	Forecast output
0. Preparatory phase	Definition and conversion input data
1. Time intervals	Time occurrence of precipitation
2. General characteristic	A general characteristic of the situation
3. Air mass of conditions	Instability, Trigger a Support mechanism
4. Local conditions	Temperature, Moisture, Wind and Orographic conditions
5. Storm intensity	1-hour rainfall intensity
6. Dangerous phenomena	Torrential rainfall, Hail, Strong wind gusts, Tornadoes
7. Phases summary	Probability of precipitation, Risk of flash floods
8. Statistical forecast	Historical situation + Storm tracks
9. Forecast report	Visualization of algorithm outputs

NWP models are the main source of input data. These are NWP models of GFS 0.50 and 0.25°; GDPS, EURO4, HIRLAM, ALADIN CHMI, WRF ARW and WRF NMM in 3.4. Input data are mainly data of convection indexes and meteorological variables such as a temperature, humidity and wind for different atmospheric levels [6, 13, 17].

These data from the CHMI portal is publicly available from these applications and alert systems:

- System of Integrated Warning System of Czech Hydrometeorological Institute - predictive warning information on dangerous phenomena,
- METEOALARM – European early warning system,
- ESTOFEX – experimentally early warning system managed by leading European scientists and experts for forecasting of severe storms and phenomena,
- Flash Flood Guidance-CZ (FFG-CZ) - the current estimate of the risk of flash floods for the territory of the Czech Republic,
- Diagnostic system MERGE_CZ - combined radar and station estimation of precipitation [6, 17].

The following stage is the prediction of three-hour intervals calculated from conditions with the probability in the range 0, 57 to 1, when precipitation is forecasted at least four NWP models. If the condition is not met then the cycle is terminated. If the condition is fulfilled, 2 to 8 phases are computed according this formula:

$$P_{h_i} = \frac{\sum_{j=1}^{k} v_j y_{ij}}{3\sum n}, \tag{1}$$

where y_{ij} are values of the critical matrix Y (values of coefficients, converted predictive parameters from NWP models) and v_j is the weight of the j-th criterion which is weighted coefficient values of predictive parameters. Σn represents the sum of predictive parameters in a partial or main output [17–19].

The last output is focused on summarizing and visualizing calculated prediction outputs that always acquire coefficient values ranging from 0 to 3 (Fig. 1):

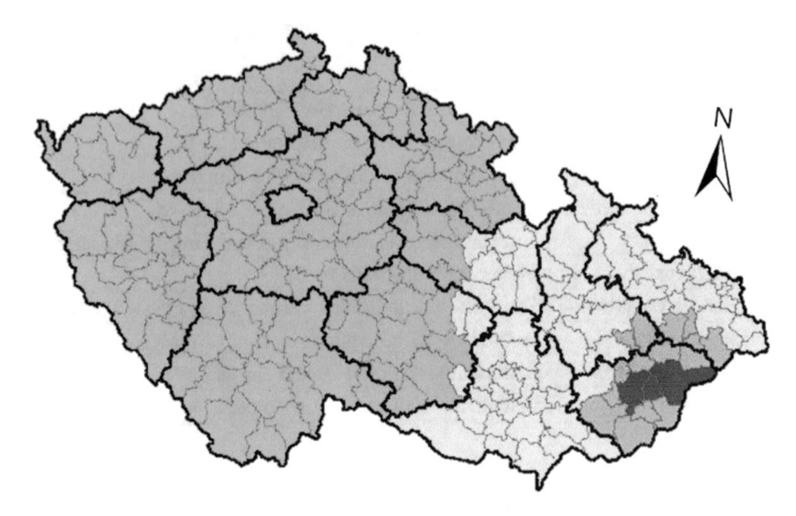

Fig. 1. Map output of the ninth forecast phase of the Algorithm [17].

The significant phase of input data processing is the null phase of the Algorithm. This phase addresses the definition of input conditions and data conversion for Algorithm´s outputs calculations. The entry conditions define the distribution of interest areas to regions and municipalities with extended powers and NWP models specifications, such as its update time, horizontal resolution and other information related to other NWP model characteristics. Input data conversion is implemented in the Transformation Algorithm of Input Meteorological Parameters application as described in the following chapters.

2.2 Transformation Algorithm of Input Meteorological Parameters

The second part of the preparatory phase is data conversion from seven NWP models for determining the time intervals on the basis of which is calculated seven Algorithm´s outputs. Transformation Algorithm of Input Meteorological Parameters application is included with the software of the Algorithm.

Input data of NWP models are processed in raster formats which are converted on data for interval values ranging from coefficient values of 0 to 3 [17]. This application consists of the following major components:

- scale model coefficients,
- model mask,
- colours pallet,
- data aggregation.

As can be seen in Fig. 2, it is first necessary to create the Scale Model Coefficients separately, on the basis of which input coefficients are assigned from input raster formats. Masks model are determined to define the boundaries of forecasted areas. The third component is the colour palette, in which colours are assigned to different areas of regions or municipalities with extended powers. Subsequently, individual pixels are

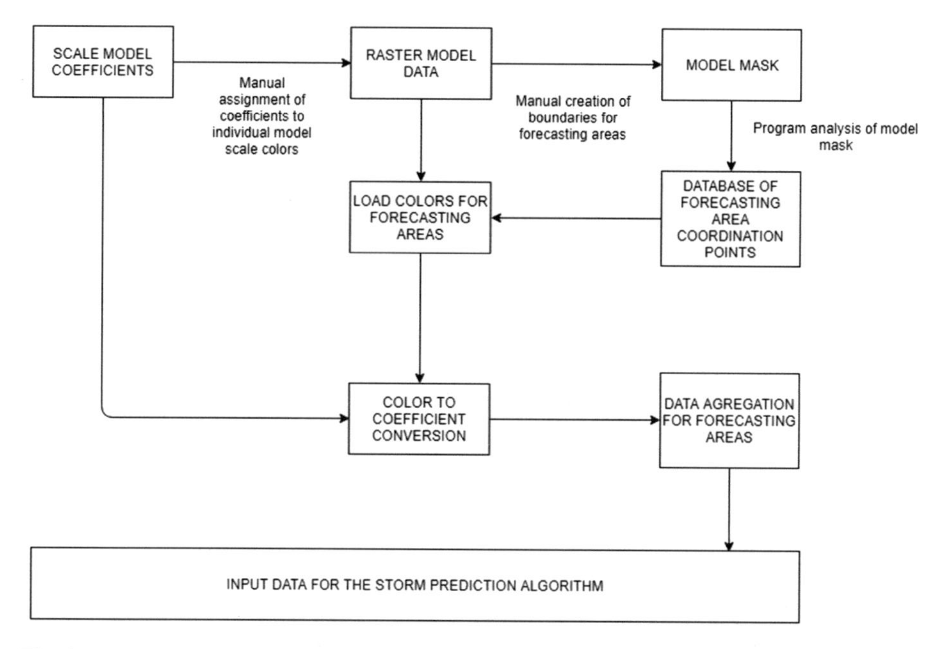

Fig. 2. Flowchart diagram of the Transformation Algorithm of Meteorological Input Parameters.

loaded and converted by the data aggregation from input raster formats. The Algorithm calculates partial and major outputs based on these computed values for various meteorological parameters.

Data Aggregation

Data aggregation is performed for each prediction area separately for a purpose the calculation of the numeric data that is rounded to the value coefficient. Data aggregation is implemented through these methods:

1. Average,
2. Sum,
3. Majority.

Figure 3 show the source code of data aggregation. The setting for each model determines whether data aggregation is to be done by Sum, Average, or Majority. The appropriate aggregation method is retrieved from the database of Forecasting Area Coordination Points and then executed for the given NWP model according to the above source code [17].

Each input data source type has a specified aggregation method:

Table 2 represent aggregation methods, which are computed various meteorological parameters of Algorithm's outputs. Aggregation method "Average" is used in most meteorological parameters focused mainly on the prediction of air mass conditions [17].

```
if (p != null)
            {
                switch (p.Value.ToString()) {
                    default:
                    case "sum":
                        Util.curCountMethod = "suma";
                        value = (float)GetValueFromSpectrumBar(colors, sizeRegion);
                        break;
                    case "average":
                        Util.curCountMethod = "průměr";
                        value = GetValueFromSpectrumBarAverage(colors, sizeRegion);
                        break;
                    case "majority":
                        Util.curCountMethod = "majorita";
                        value = GetValueFromSpectrumBarMajority(colors, sizeRegion);
                        break;
                }
            }
            else
            {
                Util.l($"Chybí specifikace metody, nastavte v
{Util.curModelName}/{Util.pathSource["model_cfg"]}|Chyba modelu");
                return;
            }
```

Fig. 3. Source code of data aggregation methods.

Table 2. Data aggregation methods [17].

Agregation methods	Forecast outputs
Sum	Time intervals and precipitation occurrence
Average	Probability of precipitation occurrence, Rainfall intensity (mm/3 h), Storm intensity (mm/h), Risk of dangerous phenomena, Risk of flash Foods
Majority	Trigger and support mechanism of convection (Moisture Transport Vector and Relative Humidity in levels of 1000–600 hPa)

3 Verification of Aggregation Methods

Forecasts verification is performed by standard methods for which a pivot table with these verification criteria:

Table 3 show the verification criteria for evaluating the accuracy of the converted prediction outputs of the Algorithm and NWP models. Value 1 represents the occurrence of a predicted or measured phenomenon, and the exact opposite is true for value 0. The exact forecast corresponds in particular to the HIT criterion. Even though Correct rejection criterion is assessed as positive, and in particular to evaluate rare phenomenon does not too conclusive result.

Table 3. Pivot table for determination percentage values of the forecasting accuracy.

Criterion	Forecast	Reality	Result
HIT	1	1	1
MISS	0	1	0
FALSE ALARM	1	0	0
CORRECT REJECTION	0	0	1

Subsequently, the resulting percentage of the accuracy of convective precipitation predictions is calculated according to the formula:

$$\overline{X} = \frac{1}{n}\sum\nolimits_{i=1}^{n} x_i, \tag{2}$$

where $\sum x_i$ is the sum of the coefficients (0 or 1) of criteria for evaluating the success rate of convective precipitation forecasts, expressed as a percentage [13].

4 Results

Verification of predictions is focused on the Sum, Average, and Majority aggregation method that are validated at approximately 70 storm situation for year of 2018.

Figure 4 demonstrate the average accuracy of daily and nightly convective precipitation predictions using the SUM aggregation method. This method is used to predict the time and occurrence of convective precipitation to determine the three-hour intervals for Algorithm´s outputs.

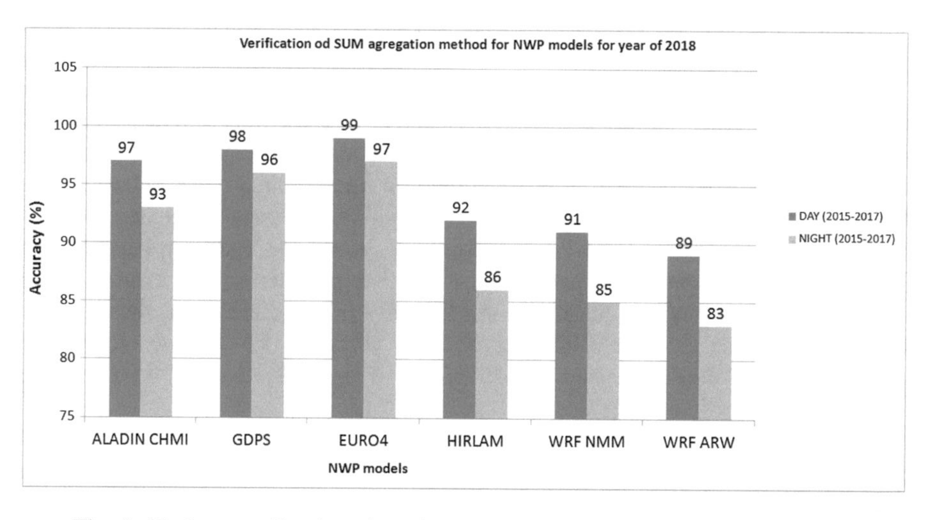

Fig. 4. Evaluate verification of precipitation using Sum data aggregation method.

Higher accuracy of predictions was achieved for the hydrostatic NWP models of ALADIN CHMI, GDPS and EURO4. The predicted occurrence of convective precipitation was one of the main reasons for detecting a minor error in terms of precipitation. On the contrary, non-hydrostatic models of the WRF NMM and ARW achieved lower accuracy due to the local precipitation occurrence, although these models were paradoxically developed for convection modelling. Generally, the lower accuracy was also recorded for nocturnal convection. It was given an uneven surface appearance of convective precipitation.

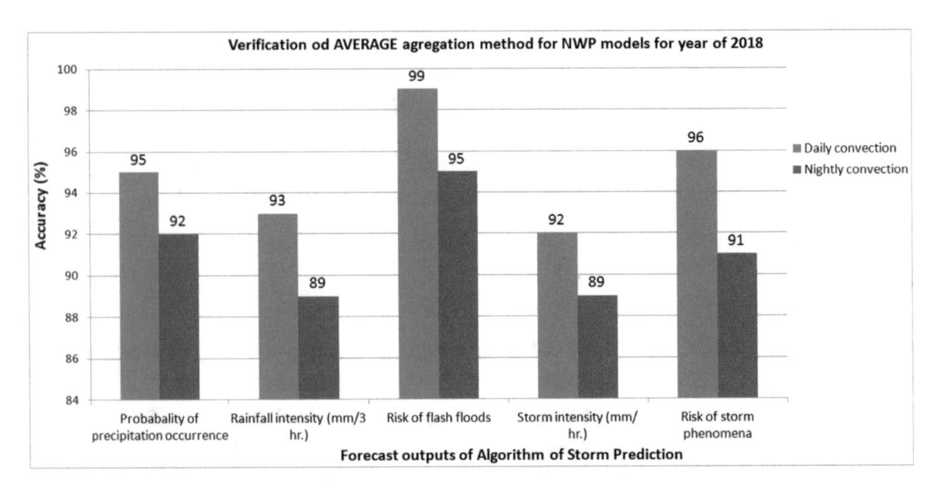

Fig. 5. Evaluate verification of precipitation using Average data aggregation method.

Figure 5 illustrate the verification evaluation of converted outputs of the Algorithm for the daily and nocturnal occurrence of convective precipitation for the territory of the Czech Republic in year of 2018. As in the previous case (Fig. 4), a higher accuracy of predictions was achieved in the daily occurrence of convective precipitation due to a higher number of assessed situations and a more flattening of rainfall than during the night hours. The highest accuracy was recorded for forecasting the risk of flash floods, dangerous phenomena and probability of precipitation occurrence. Higher precision was due to characteristics of the verified Algorithm´s outputs. The main reason is that these outputs are combined plurality of predictive parameters, for example, than Rainfall intensity (mm/3 h) and the Storm intensity (mm/h). However, these outputs are used for rough estimate rainfall in order to compare, for example, forecasting warning information of the System Integrated and Warning System of CHMI and their subsequent verification.

Figure 6 show the evaluation and comparison of the average accuracy of the converted output (Average), precipitation forecast of NWP models (Sum) and the accuracy of the Moisture Transport Vector (MTV) and Relative humidity from levels of 1000 to 600 hPa (Majority). Highest accuracy values were reached for Algorithm´s outputs. Conversely, MTV’s and Relative Humidity of 1000 to 600 hPa output had the lowest accuracy due to the problematic estimate of the degree of orographic effect on windward sides of the mountains and hills in relation to the measured rainfall. Another reason is the evaluation of a single output compared to a larger number of parameters, where the lower accuracy of individual outputs is averaged by other parameters. On the other hand, this output is partial prediction of the Triggers and Convective Supporting Mechanisms, which is calculated at the same time on the basis of seven additional prediction parameters.

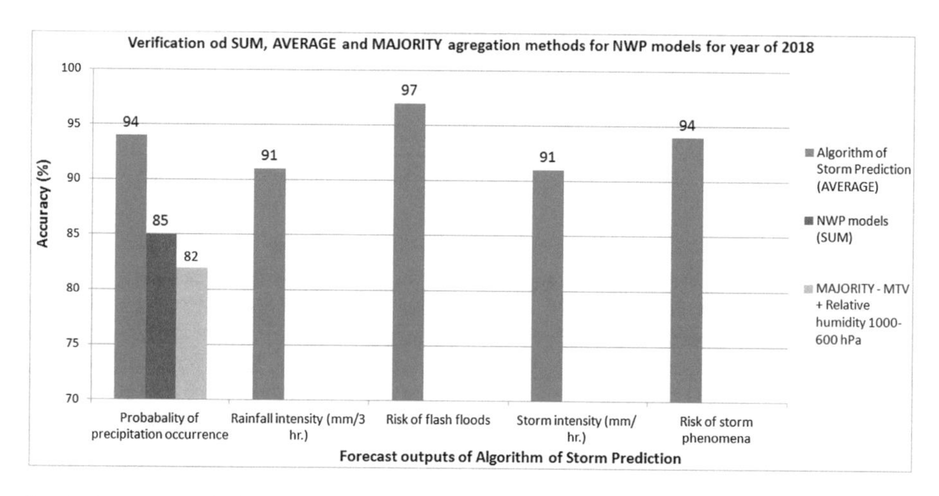

Fig. 6. Rating prediction outputs, including verification of NWP models using all data aggregation methods.

5 Conclusion

The purpose of this article was to provide information about an application intended to convert input meteorological parameters for subsequent calculations of the Algorithm´s outputs. The article (chapter 4) was to evaluate the accuracy of the Algorithm´s outputs based on three aggregation methods which are Sum, Average, and Majority.

The accuracy of converted outputs was calculated for 70 storm situations in the Czech Republic in year of 2018. The average accuracy of the Algorithm's output is computed by the Average method was 93%, precipitation forecast calculated by Sum of NWP models was 85% and the MTV parameter reached 82%. Paradoxically, the predictions accuracy of daily convective precipitation was higher for hydrostatic models that forecast all types of precipitation flatly. In contrast, non-hydrostatic models WRF NMM and ARW are directly designed for convection modelling; however, these models have reached lower accuracy due to local precipitation forecasts. Very high accuracy of converted parameters is determined by a large number of second prediction parameters, suitably selected aggregation methods for each partial forecasts and just use hydrostatic models for modelling precipitation and other parameters. Only MTV parameter achieved the lowest accuracy because of difficult predictions of wind orographic effects that is experimentally deployed for the prediction of the probable precipitation occurrence in the hilly terrain.

Higher accuracy of forecasts within the data conversion, which form the initial processing of input data, is essential to obtain accurate data as coefficient values. For these reasons, future research will focus on reviewing and optimizing meteorological parameters, depending on aggregation methods, to increase the accuracy of converted data for calculating outputs in the Algorithm. This Algorithm will be verified for large number of storm events for the next few years, and on the basis of compelling verification results, it will be offered for implementation to the System Integrated Warning System of Czech Hydrometeorological Institute.

Acknowledgments. This work was supported by the project No. CEBIA-Tech LO1303, A2.4 – ICT for support of crisis management.

References

1. Rapant, P., Inspektor, T., Kolejka, J., Batelková, K., Zapletalová, J., Kirchner, K., Krejci, T.: Early warning of flash floods based on the weather radar. In: Proceedings of the 2015 16th International Carpathian Control Conference (ICCC), pp. 426–430. IEEE (2015). https://doi.org/10.1109/CarpathianCC.2015.7145117. Accessed 3 May 2018. ISBN 978-1-4799-7370-5
2. Safarik, Z., Vicar, D., Strohmandl, J., Masek, I., Musil, M.: Protection from flash floods. In: The International Business Information Management Conference 26th IBIMA, 11–12 November 2015, Madrid, Spain (2015). ISBN 978-0-9860419-5-2
3. Rulfová, Z., Beranová, R., Kyselý, J.: Climate change scenarios of convective and large-scale precipitation in the Czech Republic based on EURO-CORDEX data. Int. J. Climatol. **37**(5), 2451–2465 (2017). https://doi.org/10.1002/joc.4857. ISSN 08998418
4. Šercl, R.: Flash Flood Guidance. Forecast possibilities of Flash floods in the Czech Republic, vol. 60, pp. 10–28 (2015). ISBN 978-80-87577-27-1
5. Meteorological Dictionary Interpretative and Terminological (eMS) (1993). http://slovnik.cmes.cz
6. Šaur, D.: Information Support for Crisis Management of the Region in Terms of Evaluation of Flood Events, Dissertation thesis. Academia Centrum TBU in Zlín, Zlín (2017). 172s. ISBN 978-80-7454-712-6
7. Flora, M.L., Potvin, C.K., Wicker, L.J.: Practical predictability of supercells: exploring ensemble forecast sensitivity to initial condition spread. Mon. Weather Rev. **146**(8), 2361–2379 (2018). https://doi.org/10.1175/MWR-D-17-0374.1. ISSN 0027-0644
8. Manola, I., Van den Hurk, B., De Moel, H., Aerts, J.C.J.H.: Future extreme precipitation intensities based on a historic event. Hydrol. Earth Syst. Sci. **22**(7), 3777–3788 (2018). https://doi.org/10.5194/hess-22-3777-2018. ISSN 1607-7938
9. Wang, Y., Belluš, M., Ehrlich, A., et al.: 27 years of regional cooperation for limited area modelling in Central Europe. Bull. Am. Meteorol. Soc. **99**(7), 1415–1432 (2018). https://doi.org/10.1175/BAMS-D-16-0321.1. ISSN 0003-0007
10. Mejsnar, J., Sokol, Z., Minářová, J.: Limits of precipitation nowcasting by extrapolation of radar reflectivity for warm season in Central Europe. Atmos. Res. **213**, 288–301 (2018). https://doi.org/10.1016/j.atmosres.2018.06.005. ISSN 01698095

11. Apke, J.M., Mecikalski, J.R., Bedka, K., McCaul, E.W., Homeyer, C.R., Jewett, C.P.: Relationships between deep convection updraft characteristics and satellite-based super rapid scan mesoscale atmospheric motion vector–derived flow. Mon. Weather Rev. **146**(10), 3461–3480 (2018). https://doi.org/10.1175/MWR-D-18-0119.1. ISSN 0027-0644
12. James, P.M., Reichert, B.K., Heizenreder, D.: NowCastMIX: automatic integrated warnings for severe convection on nowcasting time scales at the German weather service. Weather Forecast. **33**(5), 1413–1433 (2018). https://doi.org/10.1175/WAF-D-18-0038.1. ISSN 0882-8156
13. Šaur, D.: Forecasting of convective precipitation through NWP models and algorithm of storms prediction. In: Artificial Intelligence Trends in Intelligent Systems: Proceedings of the 6th Computer Science On-line Conference 2017, CSOC 2017, pp. 125–136 (2017). https://doi.org/10.1007/978-3-319-57261-1_13. ISSN 2194-5365
14. Ahijevych, D., Pinto, J.O., Williams, J.K., Steiner, M.: Probabilistic forecasts of mesoscale convective system initiation using the random forest data mining technique. Weather Forecast. **31**(2), 581–599 (2016). https://doi.org/10.1175/WAF-D-15-0113.1. ISSN 0882-8156
15. Bankert, R.L., Hadjimichael, M.: Data mining numerical model output for single-station cloud-ceiling forecast algorithms. Weather Forecast. **22**(5), 1123–1131 (2007). https://doi.org/10.1175/WAF1035.1. ISSN 0882-8156
16. Moghadam, A.N., Ravanmehr, R.: Multi-agent distributed data mining approach for classifying meteorology data: case study on Iran's synoptic weather stations. Int. J. Environ. Sci. Technol. **15**(1), 149 (2018). https://doi.org/10.1007/s13762-017-1351-x. ISSN 1735-1472
17. Šaur, D.: Algoritmus předpovědi bouří: Technická specifikace. Tomas Bata University in Zlin, Faculty of Applied Informatics, CEBIA-Tech (2018)
18. Šaur, D., Víchová, K., Mastorakis, N., Mladenov, V., Bulucea, A. Forecasting of flash floods by algorithm of storm prediction. In: MATEC Web of Conferences, vol. 210 (2018). https://doi.org/10.1051/matecconf/201821004033. ISSN 2261-236X
19. Hovorka, M.: Evaluation of multicriteria methods in security practice, Diploma thesis, Zlín (2013)

Author Index

R. Silhavy (Ed.): CSOC 2019, AISC 986, pp. 369–370, 2019.
https://doi.org/10.1007/978-3-030-19813-8

Printed in the United States
By Bookmasters